Professional Fiber Optic Installation, v.9
-The Essentials For Success

Version 9.0

Eric R. Pearson, CFOS

Pearson Technologies Inc.
Acworth, GA

Disclaimer

All instructions contained herein are believed to produce the proper results when followed exactly with appropriate equipment. However, these instructions are not guaranteed for all situations.

Notice To The Reader

The publisher does not warrant or guarantee any of the products described herein or perform any independent analysis in connection with any of the product information contained herein. Publisher does not assume, and expressly disclaims, any obligation to obtain and include information other that provided to it by the manufacturer.

The reader is specifically warned to consider and adopt any and all safety precautions that might be indicated by the activities herein and to avoid any and all potential hazards. By following the instructions contained herein, the reader knowingly and willingly assumes all risks in connection with such instructions.

The publisher makes no representation or warranties of any kind, included but not limited to, the warranties of fitness for particular purpose or merchantability, nor are any such representations implied with respect to the material set forth herein, and the publisher takes no responsibility with respect to such material. The publisher shall not be liable for any special, consequential, or exemplary damages resulting, in whole or part, from the readers' use of, or reliance upon, this material.

The procedures provided herein are believed to be accurate and to result in low installation cost and in high reliability. However, there is a possibility that these procedures may be unsuitable for specific products or in specific situations. Because of this possibility, the installer should review, and follow, the instructions provided by product manufacturers. The instructions contained herein are not meant to imply that conflicting instructions from manufacturers are in error.

Trademark Notice

All trademarks are the property of the trademark holder. Trademarks used in the installation documents include Kevlar™ (DuPont), Hytrel™ (DuPont), and ST-compatible (Lucent).

Copyright © by Pearson Technologies Inc. All rights reserved. Reproduction or translation of any part of this work beyond that permitted by section 107 or 108 of the 1976 United States Copyright Act without permission of the copyright owner is unlawful. No part of this book may be reproduced or transmitted in any form or by any means, electronic or mechanical, including photocopying, recording or by any information storage and retrieval system without the written permission of the publisher, except where permitted by law.

Published by

Pearson Technologies Inc.
4671 Hickory Bend Drive
Acworth, GA 30102
770-490-9991
www.ptnowire.com
fiberguru@ptnowire.com

Version 9.0/ file: PFOI-v9.08

Printed in the United States of America.

10 9 8 7 6 5 4 3 2 1

978-1500792237

TABLE OF CONTENTS

PART ONE
ESSENTIAL INFORMATION

1 INTRODUCTION TO FIBER NETWORKS 1-1
    1.1   NINE ADVANTAGES 1-1
        1.1.1   Unlimited Bandwidth 1-1
        1.1.2   Transmission Distance 1-1
        1.1.3   EMI And RFI 1-2
        1.1.4   Low Bit Cost 1-2
        1.1.5   Dielectric Construction 1-2
        1.1.6   Small Size 1-2
        1.1.7   Light Weight 1-2
        1.1.8   Installation Ease 1-3
        1.1.9   Security 1-3
    1.2   NETWORKS AND FULL DUPLEX LINKS 1-3
    1.3   FIBER LINK EVOLUTION 1-4
        1.3.1   Stage 1 1-4
        1.3.2   Stage 2 1-4
        1.3.3   Stage 3 1-5
        1.3.4   Stage 4 1-5
        1.3.5   Stage 5 1-5
        1.3.6   Stage 6 1-7
        1.3.7   Stage 7 1-8
        1.3.8   Stage 8 1-8
        1.3.9   Stage 9 1-9
    1.4   TOPOLOGIES 1-9
    1.5   LINK COMPONENTS 1-9
    1.6   SUMMARY 1-10
    1.7   REVIEW QUESTIONS 1-10

2 A LIGHT OVERVIEW 2-1
    2.1   INTRODUCTION 2-1
    2.2   PROPERTIES 2-1
        2.2.1   Wavelength 2-1
        2.2.2   Spectral Width 2-2
        2.2.3   Light Speed 2-2
        2.2.4   Optical Power 2-2
        2.2.5   Volume 2-3
        2.2.6   Pulse width 2-3
        2.2.7   Polarization 2-3
        2.2.8   Phase 2-3
    2.3   BEHAVIOR 2-3
        2.3.1   Reflection 2-4
        2.3.2   Refraction 2-4
        2.3.3   Dispersion 2-5
        2.3.4   Attenuation 2-8
        2.3.5   Skew 2-8

2.4	SUMMARY	2-8
2.5	REVIEW QUESTIONS	2-9

3 FIBERS 3-1

3.1	INTRODUCTION	3-1
3.2	FUNCTION	3-1
3.3	STRUCTURE	3-1
3.3.1	Diameters	3-1
3.3.2	Tolerances	3-2
3.3.3	Offset And Non Circularity	3-2
3.4	TYPES	3-3
3.4.1	Multimode SI	3-3
3.4.2	Multimode GI	3-4
3.4.3	Singlemode	3-5
3.5	PERFORMANCE	3-8
3.5.1	Dispersion	3-9
3.5.2	Attenuation	3-9
3.5.3	Bend Insensitive Fiber	3-10
3.6	SUMMARY	3-10
3.7	REVIEW QUESTIONS	3-10

4 CABLES 4-1

4.1	INTRODUCTION	4-1
4.2	STRUCTURE	4-1
4.2.1	Buffer Tubes	4-1
4.2.2	Water Blocking Materials	4-3
4.2.3	Strength Member Materials	4-3
4.2.4	Binding Tapes	4-3
4.2.5	Jackets	4-3
4.2.6	Armor	4-4
4.3	DESIGNS	4-4
4.3.1	MFPT Design	4-4
4.3.2	Central Loose Tube Design	4-5
4.3.3	Ribbon Design	4-6
4.3.4	Premises Design	4-7
4.3.5	Break Out Design	4-7
4.3.6	Blown Products	4-8
4.4	NEC COMPLIANCE	4-9
4.5	DIELECTRIC DESIGN	4-10
4.6	CHARACTERISTICS	4-10
4.6.1	Loads	4-10
4.6.2	Bend Radii	4-11
4.6.3	Dimensions	4-11
4.6.4	Temperature Ranges	4-11
4.7	STANDARDS	4-12
4.7.1	TIA/EIA-568-C	4-12
4.7.2	Color Coding	4-12
4.8	SUMMARY	4-13
4.9	REVIEW QUESTIONS	4-13

5 CONNECTORS 5-1

5.1	INTRODUCTION	5-1
5.2	FUNCTION	5-1
5.3	STRUCTURES	5-1
5.3.1	*Ferrules*	*5-1*
5.3.2	*Latching Structures*	*5-2*
5.3.3	*Mating Structures*	*5-3*
5.3.4	*Connector Structures*	*5-3*
5.3.5	*Colors*	*5-5*
5.4	PERFORMANCE	5-5
5.4.1	*dB Per Pair*	*5-5*
5.4.2	*Maximum loss*	*5-5*
5.4.3	*Typical Loss*	*5-6*
5.4.4	*Repeatability*	*5-6*
5.4.5	*Reflectance*	*5-6*
5.5	TYPES	5-7
5.5.1	*SFF Connectors*	*5-7*
5.5.2	*Common Types*	*5-10*
5.5.3	*Legacy Connectors*	*5-12*
5.6	INSTALLATION METHODS	5-13
5.6.1	*Epoxy*	*5-13*
5.6.2	*Hot Melt™*	*5-14*
5.6.3	*Quick Cure Adhesive*	*5-15*
5.6.4	*Crimp And Polish*	*5-15*
5.6.5	*Cleave-and-Crimp*	*5-16*
5.6.6	*Fuse On Connectors*	*5-16*
5.7	PIGTAIL SPLICING: THE LOWEST COST METHOD	5-17
5.8	SUMMARY	5-17
5.9	REVIEW QUESTIONS	5-17

6 SPLICES 6-1

6.1	FUNCTIONS & LOCATIONS	6-1
6.2	PROCESS	6-1
6.3	TYPES	6-1
6.3.1	*Introduction*	*6-1*
6.3.2	*Tools*	*6-1*
6.3.3	*Fusion Splicing*	*6-1*
6.3.4	*Mechanical Splicing*	*6-3*
6.4	SPLICE HARDWARE	6-4
6.4.1	*Splice Cover*	*6-5*
6.4.2	*Splice Tray*	*6-5*
6.4.3	*Splice Enclosure*	*6-6*
6.5	PERFORMANCE	6-7
6.6	SUMMARY	6-7
6.7	REVIEW QUESTIONS	6-7

7 PASSIVE DEVICES 7-1

7.1	INTRODUCTION	7-1
7.2	TYPES	7-1
7.2.1	*Couplers*	*7-1*
7.2.2	*Splitters*	*7-1*
7.2.3	*Wavelength Multiplexing*	*7-2*

7.2.4	*Optical Amplifiers*	*7-3*
7.2.5	*Dispersion Compensators*	*7-4*
7.2.6	*Switches*	*7-4*
7.2.7	*Rotary Joints*	*7-4*
7.3	TWO INSTALLATION CONCERNS	7-5
7.4	SUMMARY	7-5
7.5	REVIEW QUESTIONS	7-5

8 OPTOELECTRONICS 8-1
8.1	INSTALLATION CONCERN	8-1
8.2	TYPES	8-1
8.2.1	*Transceivers*	*8-1*
8.2.2	*Transponders*	*8-1*
8.2.3	*Light Emitting Diodes*	*8-1*
8.2.4	*Laser Diodes*	*8-2*
8.2.5	*Vertical Cavity Surface Emitting Lasers*	*8-2*
8.3	RECEIVER TYPES	8-3
8.4	PERFORMANCE	8-3
8.4.1	*Power Budget*	*8-3*
8.4.2	*Optional Optical Power Budgets*	*8-4*
8.5	DIRECT VS. COHERENT DETECTION	8-4
8.6	SUMMARY	8-4
8.7	REVIEW QUESTIONS	8-4

9 HARDWARE 9-1
9.1	FUNCTIONS	9-1
9.2	TYPES	9-1
9.2.1	*Enclosures*	*9-1*
9.2.2	*Patch Panels*	*9-2*
9.2.3	*Inner Duct/Sub Duct*	*9-3*
9.2.4	*Storage Loop Holders*	*9-3*
9.2.5	*Hanging Hardware*	*9-3*
9.2.6	*Routing Hardware*	*9-3*
9.3	SUMMARY	9-4
9.4	REVIEW QUESTIONS	9-4

PART TWO

PRINCIPLES OF INSTALLATION

PRINCIPLE

10 PLANNING AND MANAGEMENT ISSUES 10-1
10.1	INTRODUCTION	10-1
10.2	EQUIPMENT AND SUPPLIES	10-1
10.3	EQUIPMENT LOCATIONS	10-1
10.4	DATA SHEETS	10-1
10.5	INSTALLATION TECHNIQUES	10-1
10.6	RECOMMENDED TECHNIQUES	10-2
10.7	PERSONNEL	10-2
10.7.1	*Requirements*	*10-2*

10.7.2 Selection *10-2*
10.8   TESTING NEEDS 10-2
   10.8.1 As-Received Test *10-2*
   10.8.2 Post Cable installation Test *10-3*
   10.8.3 Post Splicing *10-3*
   10.8.4 Post Installation *10-3*
10.9   CREATE DATA FORMS 10-3
10.10   RECORDS 10-3
10.11   IDENTIFY POTENTIAL PROBLEMS 10-4
10.12   IDENTIFY SAFETY ISSUES 10-4
   10.12.1 Eye Safety *10-4*
   10.12.2 Hand Safety *10-4*
   10.12.3 Clothing Safety *10-5*
   10.12.4 Mouth Safety *10-5*
   10.12.5 Chemical Safety *10-5*
10.13   SUMMARY 10-5
10.14   REVIEW QUESTIONS 10-5

11 CABLE INSTALLATION PRINCIPLES 11-1
11.1   INTRODUCTION 11-1
11.2   ENVIRONMENTAL LIMITS 11-1
11.3   INSTALLATION LIMITS 11-2
   11.3.1 Pulling Cable *11-2*
   11.3.2 Cable Placement *11-7*
11.4   NEC COMPLIANCE 11-8
11.5   END PREPARATION 11-8
   11.5.1 Principle *11-8*
   11.5.2 Cosmetics *11-8*
11.6   REVIEW QUESTIONS 11-8

12 CONNECTOR INSTALLATION PRINCIPLES 12-1
12.1   INTRODUCTION 12-1
12.2   CABLE END PREPARATION 12-1
   12.2.1 Dimensions *12-1*
   12.2.2 Fiber Preparation *12-4*
12.3   ADHESIVES 12-5
   12.3.1 Epoxy *12-5*
   12.3.2 Quick Cure Adhesives *12-8*
   12.3.3 Hot Melt Adhesive *12-9*
   12.3.4 Fiber Insertion *12-9*
12.4   END FINISHING 12-10
   12.4.1 Fiber Removal *12-10*
   12.4.2 Air Polishing *12-11*
   12.4.3 Pad Polishing *12-11*
12.5   CLEAVE AND CRIMP INSTALLATION 12-14
   12.5.1 Cleaving *12-15*
   12.5.2 Insertion *12-15*
   12.5.3 Crimping *12-15*
12.6   SUMMARY 12-15
12.7   REVIEW QUESTIONS 12-16

13 SPLICING PRINCIPLES 13-1
13.1 INTRODUCTION 13-1
13.2 CABLE-ENCLOSURE COMPATIBILITY 13-1
13.3 END PREPARATION 13-2
13.4 FIBER GROUPING 13-3
13.5 ATTACHMENT 13-3
13.5.1 Configurations 13-4
13.5.2 Grounding And Bonding 13-4
13.6 TRAY PREPARATION 13-5
13.6.1 Buffer Tube Attachment 13-5
13.7 FIBER PREPARATION 13-6
13.7.1 Splice Cover 13-6
13.7.2 Strip Length 13-6
13.7.3 Cleaver Function 13-6
13.7.4 Cleave Length 13-7
13.7.5 Cleave Quality 13-7
13.8 FUSION SPLICING 13-8
13.8.1 Fiber Placement 13-8
13.8.2 Splicer Operation 13-9
13.9 MECHANICAL SPLICING 13-11
13.10 FIBER PLACEMENT 13-12
13.11 TRAY PLACEMENT 13-12
13.12 TESTING 13-12
13.13 ENCLOSURE CLOSURE 13-13
13.14 SUMMARY 13-13
13.15 REVIEW QUESTIONS 13-13

PART THREE

PRINCIPLES AND METHODS OF TESTING AND INSPECTION

14 INSERTION LOSS PRINCIPLES AND METHODS 14-1
14.1 INTRODUCTION 14-1
14.1.1 Characteristics To Be Tested 14-1
14.1.2 Power Loss Tests 14-1
14.2 TEST PRINCIPLE 14-1
14.3 SINGLEMODE TESTING 14-1
14.4 MULTIMODE TESTING 14-2
14.4.1 Cause Of Differences In Multimode Loss Measurements 14-2
14.4.2 Multimode Test Procedures 14-4
14.5 MULTIPLE WAVELENGTH TESTS 14-6
14.6 BI-DIRECTIONAL TESTING 14-6
14.7 BIMM TESTING 14-8
14.8 RANGE TESTS 14-8
14.9 SUMMARY OF TEST STANDARDS 14-8
14.9.1 TIA/EIA-14-B 14-8
14.9.2 TIA/EIA-7-A 14-8
14.9.3 IEC 61280-4-1 14-9
14.10 EQUIPMENT REQUIREMENTS 14-9
14.10.1 Source 14-9

14.10.2 Meter *14-10*
14.10.3 Reference leads *14-10*
14.10.4 Barrels *14-13*
14.11 ADVANTAGES 14-13
14.12 DISADVANTAGE 14-14
14.13 REVIEW QUESTIONS 14-14

15 OTDR PRINCIPLES AND METHODS 15-1
15.1 INTRODUCTION 15-1
15.2 TYPES 15-1
15.3 PRINCIPLES 15-2
15.4 BLOCK DIAGRAM 15-2
15.5 THEORETICAL TRACE 15-3
15.6 TRACE FEATURES 15-3
15.6.1 Reflectance *15-3*
15.6.2 Dead, Or Blind, Zones *15-4*
15.6.3 Concealed Features *15-4*
15.6.4 Map To Trace *15-5*
15.7 3 BASIC TRACES 15-5
15.7.1 Reflective Loss *15-5*
15.7.2 Non-Reflective Loss *15-7*
15.7.3 Bad Launch *15-8*
15.7.4 Unusual Traces *15-8*
15.7.5 Wavelength Effect *15-12*
15.8 USING THE OTDR 15-12
15.8.1 Set Up *15-12*
15.8.2 Launch Cable *15-13*
15.8.3 Measurements *15-13*
15.9 FTTH/PON LINK TRACES 15-19
15.10 OTDR INSERTION LOSS COMPARISON 15-20
15.11 THE REAL WORLD 15-20
15.12 REVIEW QUESTIONS 15-21

16 REFLECTANCE AND ORL- PRINCIPLES AND METHODS 16-1
16.1 INTRODUCTION 16-1
16.1.1 Reflectance *16-1*
16.1.2 Optical Return Loss *16-1*
16.2 REFLECTANCE PRINCIPLES 16-1
16.3 REFLECTANCE TESTING 16-1
16.3.1 Equipment Required *16-1*
16.3.2 Procedure *16-2*
16.3.3 Set Up *16-2*
16.3.4 Verify Low Reflectance *16-2*
16.3.5 Test Reflectance *16-3*
16.4 ORL TESTING 16-3
16.5 INTERPETATION 16-4
16.6 FOTP-8 16-4
16.7 STANDARDS 16-4
16.8 REVIEW QUESTIONS 16-4

17 DISPERSION TESTING PRINCIPLES AND METHODS 17-1

17.1	Introduction	17-1
17.2	Back To Basics	17-1
17.3	Chromatic Dispersion	17-2
17.3.1	Reason For Test	17-2
17.3.2	Test Methods	17-2
17.3.3	Test Results	17-3
17.4	PMD	17-3
17.4.1	Cause	17-3
17.4.2	Limits	17-4
17.4.3	Test Method	17-5
17.4.4	Test Results	17-5
17.5	Reference	17-5
17.6	Review Questions	17-6

18 OTHER TESTS AND EQUIPMENT 18-1
18.1	Introduction	18-1
18.2	Other Tests	18-1
18.2.1	Protocol Testing	18-1
18.2.2	Back Reflection	18-1
18.3	Other Equipment	18-1
18.3.1	Fiber identifier	18-1
18.3.2	VFL	18-1
18.3.3	Attenuator	18-2
18.3.4	Traffic Detector	18-2
18.3.5	Talk Set	18-2
18.3.6	Microscope	18-2
18.4	Summary	18-2
18.5	Review Questions	18-2

19 CERTIFICATION PRINCIPLES AND METHODS 19-1
19.1	Introduction	19-1
19.2	Required Information	19-1
19.3	Insertion Loss Calculations	19-1
19.4	Development Of A Strategy	19-2
19.4.1	Insertion Loss Certification	19-3
19.4.2	OTDR Certification	19-3
19.5	An Alternative Strategy	19-4
19.6	Summary	19-4
19.7	Review Questions	19-5

20 CONNECTOR INSPECTION 20-1
20.1	Applicability	20-1
20.2	Equipment Required	20-1
20.3	Procedure	20-1
20.3.1	Cleaning	20-1
20.3.2	General Instructions	20-2
20.3.3	Back Light	20-2
20.4	Evaluation Criteria	20-3
20.4.1	Core	20-3
20.4.2	Everywhere Else	20-4
20.4.3	Bad Clad? Be Glad!	20-5

20.4.4 Ferrule Features *20-5*
20.4.5 Connector Disposition *20-6*
20.5 TROUBLESHOOTING 20-6
20.5.1 Dirt On Connector *20-6*
20.5.2 Faint Stains On Connector *20-6*
20.5.3 No Fiber Found *20-6*
20.5.4 Core Does Not Backlight *20-6*
20.6 REVIEW QUESTIONS 20-6

PART FOUR

INSTALLATION PROCEDURES

HOW TO USE THESE PROCEDURES
SAFETY PRECAUTIONS

21 CABLE END PREPARATION 21-1
21.1 INTRODUCTION 21-1
21.2 TOOLS AND SUPPLIES 21-1
21.3 FIBER HANDLING 21-1
21.3.1 Coated Fiber *21-1*
21.3.2 Uncoated Fiber 1 .... *21-2*
21.3.3 Uncoated Fiber 2 .... *21-2*
21.4 PULLING 21-3
21.4.1 Loose Tube *21-3*
21.4.2 Premises Cable *21-4*
21.5 TERMINATION 21-5
21.5.1 Loose Tube *21-5*
21.5.2 Premises *21-8*
21.6 CONTINUITY TEST 21-8
21.7 ONE PAGE SUMMARY 21-8
21.7.1 Pulling *21-8*
21.7.2 Termination *21-8*

22 CONNECTOR INSTALLATION: EPOXY 22-1
22.1 INTRODUCTION 22-1
22.2 MATERIALS AND SUPPLIES 22-1
22.3 PROCEDURE 22-2
22.3.1 Set Up Oven *22-2*
22.3.2 Preinstall Cable *22-2*
22.3.3 Remove Jacket *22-2*
22.3.4 Install Boots And Sleeves *22-2*
22.3.5 Prepare Epoxy *22-2*
22.3.6 Prepare End *22-3*
22.3.7 Clean Fiber *22-5*
22.3.8 Connector Tests *22-5*
22.3.9 Epoxy Injection *22-5*
22.3.10 Fiber Insertion *22-6*
22.3.11 Crimp Sleeve *22-6*
22.3.12 Connector Insertion *22-6*
22.3.13 Remove Excess Fiber *22-7*

- 22.3.14 Air Polish 22-8
- 22.3.15 Multimode Polish 22-8
- 22.3.16 Final Cleaning 22-9
- 22.3.17 Inspect Connector 22-9
- 22.3.18 White Light Test 22-10
- 22.3.19 Final Assembly 22-10
- 22.3.20 Singlemode Polishing 22-10
- 22.3.21 Final Cleaning 22-11
- 22.3.22 Inspection 22-11
- 22.4 TROUBLESHOOTING 22-11
  - 22.4.1 Installation 22-11
  - 22.4.2 Polishing 22-12
  - 22.4.3 Singlemode Polishing 22-13
- 22.5 ONE PAGE SUMMARY 22-13
  - 22.5.1 Installation 22-13
  - 22.5.2 Multimode Polish 22-14
  - 22.5.3 Singlemode Polish 22-14

23 CONNECTOR INSTALLATION: HOT MELT ADHESIVE 23-1

- 23.1 INTRODUCTION 23-1
- 23.2 MATERIALS AND SUPPLIES 23-1
- 23.3 PROCEDURE 23-2
  - 23.3.1 Preinstall Cable 23-2
  - 23.3.2 Set Up Oven 23-2
  - 23.3.3 Load Holders 23-2
  - 23.3.4 Remove Outer Jacket 23-3
  - 23.3.5 Install Boots 23-3
  - 23.3.6 Prepare End 23-3
  - 23.3.7 Clean Fiber 23-4
  - 23.3.8 Fiber Insertion 23-4
  - 23.3.9 Remove Excess Fiber 23-5
  - 23.3.10 Air Polishing 23-6
  - 23.3.11 Multimode Polishing 23-6
- 23.4 SINGLEMODE POLISHING 23-7
- 23.5 FINAL CLEANING 23-7
  - 23.5.1 Best Method 23-7
  - 23.5.2 Method B 23-7
  - 23.5.3 Method C 23-7
- 23.6 CAP INSTALLATION 23-7
- 23.7 INSPECT CONNECTOR 23-8
- 23.8 WHITE LIGHT TEST 23-8
- 23.9 SC FINAL ASSEMBLY 23-8
- 23.10 SALVAGE 23-8
  - 23.10.1 Procedure 23-8
  - 23.10.2 Alternate Procedure 23-8
- 23.11 TROUBLESHOOTING 23-9
  - 23.11.1 Installation 23-9
  - 23.11.2 Polishing 23-9
- 23.12 ONE PAGE SUMMARY 23-10
  - 23.12.1 Installation 23-10

23.12.2 Multimode Polishing *23-10*

24 CONNECTOR INSTALLATION: CLEAVE AND CRIMP 24-1
24.1 INTRODUCTION 24-1
24.2 TOOLS AND SUPPLIES REQUIRED 24-2
24.3 PROCEDURE 24-2
24.3.1 Remove Jacket *24-2*
24.3.2 Load Tool *24-2*
24.3.3 Prepare Fiber End *24-4*
24.3.4 Prepare Cleaver *24-4*
24.3.5 Clean Fiber *24-5*
24.3.6 Cleave Fiber *24-5*
24.3.7 Install Fiber *24-5*
24.3.8 Connector Removal *24-6*
24.3.9 VFL Evaluation *24-6*
24.3.10 Final Assembly *24-6*
24.4 ST-™ COMPATIBLE PROCEDURE 24-7
24.5 TEST LOSS 24-7
24.6 TROUBLESHOOTING 24-7
24.6.1 Blinking Yellow Light *24-7*
24.6.2 High Loss *24-7*
24.7 ONE PAGE SUMMARY 24-7

25 MID-SPAN SPLICING 25-1
25.1 INTRODUCTION 25-1
25.1.1 Tools And Supplies Required *25-1*
25.2 CABLE END PREPARATION 25-2
25.2.1 Dimensions *25-2*
25.2.2 End Preparation *25-2*
25.3 ENCLOSURE PREPARATION 25-2
25.4 CABLE ATTACHMENT 25-3
25.5 BUFFER TUBE ATTACHMENT 25-4
25.6 FIBER LENGTH VERIFICATION 25-4
25.7 OTDR SET UP 25-5
25.8 SPLICING 25-5
25.8.1 Fusion Splicing *25-5*
25.8.2 Mechanical Splice *25-8*
25.9 TEST LOSS 25-10
25.9.1 Fusion Splices *25-10*
25.9.2 Mechanical Splices *25-10*
25.10 FIBER COILING 25-10
25.11 BUFFER TUBE COILING 25-11
25.12 TRAY ATTACHMENT 25-12
25.13 ENCLOSURE FINISHING 25-12
25.14 TROUBLESHOOTING 25-12
25.14.1 Bad Cleaves *25-12*
25.14.2 Fusion Splicing *25-13*
25.14.3 Mechanical Splicing *25-13*
25.15 ONE PAGE SUMMARY 25-14
25.15.1 Cable End Preparation *25-14*
25.15.2 Enclosure Preparation *25-14*

25.15.3 Set Up OTDR *25-14*
25.15.4 Fusion Splicing *25-14*
25.15.5 Mechanical Splice *25-14*
25.15.6 Test Loss *25-14*
25.15.7 Coil Fiber *25-14*
25.15.8 Coil Buffer tube *25-14*
25.15.9 Attach Tray *25-14*
25.15.10 Finish Enclosure *25-14*
25.15.11 Test Loss *25-14*

26 APPENDICES 26-1
26.1 INDICES OF REFRACTION 26-1
26.1.1 Multimode *26-1*
26.1.2 Singlemode *26-1*
26.2 BACKSCATTER COEFFICIENTS 26-1
26.2.1 Multimode *26-1*
26.2.2 Singlemode *26-1*
26.3 RI INACCURACIES 26-1
26.4 DISTANCE INACCURACIES 26-2
26.5 GLOSSARY 26-3
26.6 CHAPTER 13 ANSWERS 26-10
26.7 CHAPTER 19 ANSWERS 26-10
26.8 ADVANTAGES AND DISADVANTAGES OF 1, 2, AND 3 LEAD TEST METHODS ... 26-13
26.9 ALL CONNECTOR TYPES 26-14
26.10 STUDY GUIDE FOR CFOT EXAMINATION, VERSION 11 26-15
26.11 CFOT PREPARATION PUZZLES 26-17
26.12 THE AUTHOR 26-24
26.13 CROSSWORD EXERCISES 26-25
26.14 PEARSON TECHNOLOGIES' SERVICES 26-25

TABLE OF FIGURES

FIGURE 1-1: STAGE 1-FULL DUPLEX TRANSMISSION 1-6
FIGURE 1-2: STAGE 1-EXTENDED DUPLEX DISTANCE TRANSMISSION.. 1-6
FIGURE 1-3: STAGE 2- INCREASED DUPLEX TRANSMISSION DISTANCE WITH OPTICAL AMPLIFIER 1-6
FIGURE 1-4: STAGE 3- DUPLEX WAVELENGTH DIVISION MULTIPLEXING (WDM)......................... 1-6
FIGURE 1-5: STAGE 4- INCREASED DUPLEX CAPACITY WITH DWDM 1-6
FIGURE 1-6: STAGE 4- INCREASED CAPACITY AND TRANSMISSION DISTANCE 1-7
FIGURE 1-7: FULL DUPLEX SINGLE FIBER DWDM 1-7
FIGURE 1-8: OPTICAL LINE TERMINAL (OLT) FOR BI-DIRECTIONAL TRANSMISSION IN A PON 1-7
FIGURE 1-9: MULTIPLE TRANSMISSION PATHS WITH OPTICAL ADM 1-8
FIGURE 1-10: STAGE 8-INCREASED FLEXIBILITY WITH ROADM AND TUNABLE LASER 1-8
FIGURE 1-11: A HIERARCHICAL LOGICAL STAR, PHYSICAL STAR NETWORK............................... 1-9
FIGURE 2-1: CENTRAL WAVELENGTH AND SPECTRAL WIDTH 2-2
FIGURE 2-2: E AND H LIGHT ENERGY FIELDS 2-3
FIGURE 2-3: IN PHASE (TOP) AND OUT OF PHASE WAVES 2-3
FIGURE 2-4: REFLECTION AT AN INTERFACE 2-4
FIGURE 2-5: REFRACTION OF LIGHT 2-4
FIGURE 2-6: NO MODAL DISPERSION 2-5
FIGURE 2-7: MODAL DISPERSION 2-5
FIGURE 2-8: PULSE DISPERSION CAUSES SIGNAL INACCURACY 2-6
FIGURE 2-9: TIME INTERVAL VS. BIT RATE... 2-7
FIGURE 2-10: MULTIMODE 100 GBPS TRANSMISSION.. 2-8
FIGURE 3-1: FIBER STRUCTURE 3-1
FIGURE 3-2: AN OPTICAL FIBER WITH DARKENED PRIMARY COATING 3-1
FIGURE 3-3: CORE OFFSET AND EXCESS LOSS 3-2
FIGURE 3-4: CLADDING NON-CIRCULARITY AND LOSS 3-3
FIGURE 3-5: CORE PROFILE OF STEP INDEX FIBER 3-3
FIGURE 3-6: RAYS REFLECT IN SI FIBER 3-4
FIGURE 3-7: RAYS OUTSIDE CRITICAL ANGLE ESCAPE 3-4
FIGURE 3-8: RAY PATHS IN A SI FIBER 3-4
FIGURE 3-9: GI CORE PROFILE 3-4
FIGURE 3-10: CURVED RAY GI PATHS.. 3-5
FIGURE 3-11: APPARENT 'RAY' PATHS IN SINGLEMODE FIBER 3-6
FIGURE 3-12: SINGLEMODE MFD DIAMETER... 3-6
FIGURE 3-13: DISPERSION RATE VS. WAVELENGTH FOR G.652 SINGLEMODE FIBERS............... 3-7
FIGURE 3-14: TOTAL DISPERSION FOR DS, G.653 FIBER 3-7
FIGURE 3-15: G.655 TOTAL DISPERSION... 3-8
FIGURE 3-16: G.655 DISPERSION 3-8
FIGURE 3-17: RAYLEIGH SCATTERING IN CORE... 3-9
FIGURE 3-18: ATTENUATION RATE VS. WAVELENGTH (MIT OPEN COURSEWARE) 3-10
FIGURE 4-1: LOOSE BUFFER TUBE CROSS SECTION 4-1
FIGURE 4-2: LOOSE BUFFER TUBE CABLE 4-1
FIGURE 4-3: TIGHT BUFFER TUBE CROSS SECTION... 4-1
FIGURE 4-4: TIGHT BUFFER TUBE CABLE 4-1
FIGURE 4-5: THE MECHANICAL DEAD ZONE 4-2
FIGURE 4-6: FURCATION/FAN OUT KIT FOR LOOSE TUBE FIBERS 4-2
FIGURE 4-7: END ALIGNMENT OF A BROKEN FIBER IN A TIGHT BUFFER TUBE 4-3
FIGURE 4-8: 168-FIBER ARMORED CABLE WITHOUT INTERNAL JACKET..................................... 4-4
FIGURE 4-9: 216-FIBER ARMORED CABLE WITH INTERNAL JACKET... 4-4
FIGURE 4-10: MULTIPLE FIBER PER TUBE 4-5
FIGURE 4-11: MFPT CABLE 4-5
FIGURE 4-12: CENTRAL BUFFER TUBE CROSS SECTION.. 4-5
FIGURE 4-13: TWELVE FIBER RIBBON.. 4-6

FIGURE 4-14: RIBBON CABLE CROSS SECTION	4-6
FIGURE 4-15: RIBBON CABLE	4-6
FIGURE 4-16: ALTERNATE RIBBON DESIGN	4-6
FIGURE 4-17: PREMISES CABLE	4-7
FIGURE 4-18: HIGH COUNT PREMISES CABLE	4-7
FIGURE 4-19: BREAK OUT CABLE	4-8
FIGURE 4-20: BREAK OUT CABLE CROSS SECTION	4-8
FIGURE 4-21: BLOWN FIBER SYSTEM COMPONENTS	4-9
FIGURE 4-22: OPERATING TEMPERATURE RANGE	4-12
FIGURE 5-1: FERRULE CONTACT	5-1
FIGURE 5-2: INCREASED LOSS FROM NON-CONTACT CONNECTORS	5-1
FIGURE 5-3: THE SC/APC CONNECTOR	5-2
FIGURE 5-4: SC CONNECTOR DESIGN WITH A LATCHING MECHANISM	5-3
FIGURE 5-5: CONNECTOR WITH LATCHING RING	5-3
FIGURE 5-6: CONNECTORS WITH PLUGS AND BARREL	5-3
FIGURE 5-7: JACK (LEFT) AND CONNECTOR (RIGHT)	5-3
FIGURE 5-8: KEY ON CONNECTOR	5-4
FIGURE 5-9: CONNECTOR BACK SHELL	5-4
FIGURE 5-10: STRAIN RELIEF BOOT	5-5
FIGURE 5-11: CRIMP SLEEVE	5-5
FIGURE 5-12: LO LC DUPLEX CONNECTOR	5-5
FIGURE 5-13: AIR GAPS CREATE REFLECTANCE	5-6
FIGURE 5-14: SIZE COMPARISON OF CONNECTORS: SC (TOP), MU (MIDDLE) AND LC (BOTTOM)	5-7
FIGURE 5-15: LC SIMPLEX CONNECTOR	5-8
FIGURE 5-16: LX.5 CONNECTOR	5-8
FIGURE 5-17: SC AND LX.5 BARRELS	5-8
FIGURE 5-18: MU CONNECTOR	5-8
FIGURE 5-19: VOLITION™ JACK	5-9
FIGURE 5-20: VOLITION™ PLUG AND JACK	5-9
FIGURE 5-21: TYCO/AMP MT-RJ SYSTEM	5-10
FIGURE 5-22: CORNING CABLE SYSTEMS MT-RJ SYSTEM	5-10
FIGURE 5-23: E2000 CONNECTOR	5-10
FIGURE 5-24: MTP® CONNECTOR	5-10
FIGURE 5-25: ST-™ COMPATIBLE CONNECTOR	5-11
FIGURE 5-26: SC SIMPLEX CONNECTOR	5-11
FIGURE 5-27: SC DUPLEX CONNECTOR	5-11
FIGURE 5-28: 905 SMA CONNECTOR	5-12
FIGURE 5-29: 906 SMA CONNECTOR	5-12
FIGURE 5-30: BICONIC CONNECTOR	5-12
FIGURE 5-31: MINI BNC CONNECTOR	5-12
FIGURE 5-32: FC CONNECTOR	5-12
FIGURE 5-33: D4 CONNECTOR	5-12
FIGURE 5-34: FDDI CONNECTOR	5-13
FIGURE 5-35: ESCON CONNECTOR	5-13
FIGURE 5-36: ESCON FERRULES EXPOSED	5-13
FIGURE 5-37: HIGH YIELD FROM LARGE BEAD	5-14
FIGURE 6-1: 'V-GROOVE'	6-2
FIGURE 6-2: PROFILE ALIGNMENT SPLICER	6-2
FIGURE 6-3: MECHANICAL SPLICES	6-4
FIGURE 6-4: FUSION SPLICES IN SPLICE COVERS	6-5
FIGURE 6-5: HALF LENGTH SPLICE TRAY	6-5
FIGURE 6-6: OUTDOOR SPLICE ENCLOSURE	6-5
FIGURE 6-7: INDOOR SPLICE ENCLOSURE	6-5
FIGURE 6-8: SHRINKABLE FUSION SPLICE COVER	6-5
FIGURE 6-9: ADHESIVE SPLICE COVERS	6-5
FIGURE 6-10: FULL LENGTH SPLICE TRAY	6-6

FIGURE 6-11: GRIPPING MECHANISM FOR INDOOR SPLICE ENCLOSURE 6-6
FIGURE 6-12: GRIPPING MECHANISM FOR OUTDOOR SPLICE ENCLOSURE 6-6
FIGURE 6-13: VALVE FOR INTERNAL PRESSURE 6-6
FIGURE 6-14: REVIEW FIGURE 1 6-7
FIGURE 6-15: REVIEW FIGURE 2 6-7
FIGURE 6-16: REVIEW FIGURE 3 6-7
FIGURE 7-1: COUPLER 7-1
FIGURE 7-2: UNIDIRECTIONAL COUPLER FUNCTION 7-1
FIGURE 7-3: BI-DIRECTIONAL COUPLER FUNCTION 7-1
FIGURE 7-4: ALL WAVELENGTHS IN OUTPUT OF SPLITTER 7-2
FIGURE 7-5: A SPLITTER 7-2
FIGURE 7-6: WAVELENGTH DIVISION DE-MULTIPLEXOR FUNCTION 7-2
FIGURE 7-7: CDWM AND DWDM 7-2
FIGURE 7-8: FUNCTION OF OADM 7-3
FIGURE 7-9: EDFA DEMONSTRATION SYSTEM 7-3
FIGURE 7-10: THE EDFA AMPLIFICATION PROCESS 7-3
FIGURE 7-11: MEMS MIRRORS 7-4
FIGURE 7-12: OPTICAL SWITCH MECHANICS 7-4
FIGURE 7-13: FIVE FIBER ROTARY JOINT 7-5
FIGURE 7-14: SPLITTER NETWORK FOR QUESTION 11 7-6
FIGURE 8-1: RELATIVE SIZES OF LD AND SINGLEMODE CORE 8-2
FIGURE 8-2: VCSEL ANNULAR LAUNCH AREA 8-2
FIGURE 8-3: PROFILE DIP IN OLD MULTIMODE FIBERS 8-3
FIGURE 9-1: INDOOR ENCLOSURES 9-1
FIGURE 9-2: OUTDOOR SPLICE ENCLOSURE 9-1
FIGURE 9-3: WALL MOUNTED ENCLOSURE 9-1
FIGURE 9-4: SINGLE PIECE ENCLOSURE SHELL 9-2
FIGURE 9-5: MULTIPLE PIECE ENCLOSURE SHELL 9-2
FIGURE 9-6: SMALL FTTH ENCLOSURE 9-2
FIGURE 9-7: SC PATCH PANEL MODULES 9-2
FIGURE 9-8: INNER DUCT 9-3
FIGURE 9-9: STORAGE LOOP HOLDER 9-3
FIGURE 9-10: ADSS POLE CABLE GRIP 9-4
FIGURE 9-11: MID SPAN TAP CABLE GRIP 9-4
FIGURE 9-12: END CABLE SUPPORT 9-4
FIGURE 9-13: BEND RADIUS CONTROL FROM TROUGH TO RACK 9-4
FIGURE 9-14: BEND RADIUS CONTROL IN TROUGH 9-4
FIGURE 9-15: BEND RADIUS CONTROL IN RACK 9-4
FIGURE 10-1: PULLING EQUIPMENT 10-1
FIGURE 10-2: BARE FIBER ADAPTER 10-3
FIGURE 11-1: SWIVEL WITH SHEAR PINS 11-2
FIGURE 11-2: PULLER WITH SLIP CLUTCH 11-3
FIGURE 11-3: A MID PULL 11-3
FIGURE 11-4: PULLER WITH LOAD GAGE 11-3
FIGURE 11-5: UNDERGROUND INSTALL, STEP 1 11-4
FIGURE 11-6: CABLE STORAGE IN A FIGURE 8 PATTERN 11-5
FIGURE 11-7: UNDERGROUND INSTALL, STEP 2 11-5
FIGURE 11-8: KELLEMS GRIP 11-5
FIGURE 11-9: CABLE WOUND INCORRECTLY 11-6
FIGURE 11-10: MAP FOR QUESTION 5 11-9
FIGURE 12-1: SINGLE FIBER CABLE END PREPARED FOR CONNECTOR INSTALLATION 12-1
FIGURE 12-2: EIGHT FIBER PREMISES CABLE END PREPARED FOR CONNECTOR INSTALLATION 12-1
FIGURE 12-3: END PREPARATION DIMENSIONS FOR FIBERS WITH INDIVIDUAL JACKETS 12-1
FIGURE 12-4: PROPER JACKET REMOVAL LENGTH 12-2
FIGURE 12-5: EXCESSIVE JACKET REMOVAL LENGTH 12-2
FIGURE 12-6: BUFFER TUBE BUTTS AGAINST FERRULE 12-3

FIGURE 12-7: CRIMP NEST IN A CRIMPER	12-3
FIGURE 12-8: INDOOR CABLE ATTACHMENT LOCATION	12-4
FIGURE 12-9: OUTDOOR CABLE ATTACHMENT LOCATION	12-4
FIGURE 12-10: BUFFER TUBE COILED IN INDOOR ENCLOSURE	12-4
FIGURE 12-11: SC INNER TUBE	12-6
FIGURE 12-12: LC INNER TUBE	12-6
FIGURE 12-13: SUFFICIENT EPOXY IN BACK SHELL	12-6
FIGURE 12-14: LARGE BEAD FOR INEXPERIENCED INSTALLERS	12-7
FIGURE 12-15: SMALL BEAD FOR MODERATELY EXPERIENCED INSTALLERS	12-7
FIGURE 12-16: BEAD WITHOUT APPLICATION OF ADDITIONAL ADHESIVE	12-8
FIGURE 12-17: BEAD SIZE WITH THREE ADHESIVE APPLICATIONS	12-9
FIGURE 12-18: FIBER BROKEN BELOW FERRULE	12-10
FIGURE 12-19: BROKEN FIBER APPEARANCE	12-10
FIGURE 12-20: LOCATION OF SCRIBE	12-11
FIGURE 12-21: WEDGE SCRIBERS	12-11
FIGURE 12-22: DESIRED AIR POLISH CONDITION	12-11
FIGURE 12-23: UNDESIRED CONDITION	12-11
FIGURE 12-24: FIBER FLUSH WITH FERRULE	12-12
FIGURE 12-25: UNDERCUT FIBER	12-12
FIGURE 12-26: POLISHING FIXTURES	12-13
FIGURE 12-27: BEVEL ON END OF FIBER	12-13
FIGURE 12-28: INTERNAL STRUCTURE OF THE CLEAVE AND CRIMP CONNECTOR	12-15
FIGURE 13-1: SIX KINKED BUFFER TUBES	13-1
FIGURE 13-2: CABLE END PREPARATION LENGTHS	13-2
FIGURE 13-3: ENCLOSURE ATTACHMENT LOCATIONS	13-2
FIGURE 13-4: STRENGTH MEMBER ATTACHMENT	13-2
FIGURE 13-5: BUFFER TUBES OUT OF ENCLOSURE	13-2
FIGURE 13-6: MECHANICAL SPLICE TOOL	13-3
FIGURE 13-7: CABLE ATTACHMENT	13-3
FIGURE 13-8: STRENGTH MEMBER ATTACHMENT	13-3
FIGURE 13-9: FIBERS GROUPED AND ROUTED WITH TUBING	13-3
FIGURE 13-10: INDOOR ENCLOSURE CABLE ATTACHMENT	13-4
FIGURE 13-11: BUTT CONFIGURATION	13-4
FIGURE 13-12: IN-LINE CONFIGURATION	13-4
FIGURE 13-13: BUFFER TUBE ATTACHMENT TO SPLICE TRAY	13-4
FIGURE 13-14: GROUNDING AND BONDING HARDWARE	13-4
FIGURE 13-15: ENCLOSURE GROUNDING LOCATION	13-5
FIGURE 13-16: SPLICE HOLDER INSERTS	13-5
FIGURE 13-17: INCORRECT SPLICE HOLDERS	13-5
FIGURE 13-18: HEAT SHRINK SPLICE COVER	13-6
FIGURE 13-19: ADHESIVE SPLICE COVER	13-6
FIGURE 13-20: PRECISION V-GROOVES	13-7
FIGURE 13-21: ACCEPTABLE CLEAVE	13-7
FIGURE 13-22: CLEAVES WITH MISSING GLASS	13-7
FIGURE 13-23: UNACCEPTABLE CLEAVE DUE TO EXTRA GLASS	13-7
FIGURE 13-24: UNACCEPTABLE HIGH ANGLE CLEAVE	13-8
FIGURE 13-25: FIBERS CORRECTLY POSITIONED	13-8
FIGURE 13-26: FIBERS INCORRECTLY POSITIONED	13-8
FIGURE 13-27: EXCESSIVE FIBER SEPARATION	13-8
FIGURE 13-28: FIBERS MISALIGNED IN SPLICER	13-9
FIGURE 13-29: OVERRUN	13-9
FIGURE 13-30: BULGED SPLICE DUE EXCESSIVE OVERRUN	13-9
FIGURE 13-31: HOURGLASS SPLICE DUE INSUFFICIENT OVERRUN	13-9
FIGURE 13-32: RAMPED (LEFT) AND NON-RAMPED (RIGHT) SPLICING CURRENTS	13-10
FIGURE 13-33: EQUAL BOW RESULTS IN LOW LOSS 3M FIBRLOK™ SPLICE	13-11
FIGURE 13-34: COILED FIBER	13-12

FIGURE 13-35: PIGTAILS AND TUBING ATTACHMENT LOCATIONS 13-12
FIGURE 13-36: VELCRO STRAP ATTACHMENT 13-12
FIGURE 13-37: MOISTURE SEALS AND GASKET GROOVES 13-13
FIGURE 13-38: PRESSURIZATION VALVE 13-13
FIGURE 13-39: ENCLOSURE PLACEMENT 13-13
FIGURE 14-1: INPUT POWER MEASUREMENT 14-1
FIGURE 14-2: OUTPUT/RECEIVER POWER MEASUREMENT 14-2
FIGURE 14-3: LOSS IN LINK 14-2
FIGURE 14-4: INPUT POWER MEASUREMENT WITH TWO LEADS 14-2
FIGURE 14-5: LED POWER DISTRIBUTION 14-3
FIGURE 14-6: VCSEL LAUNCH POWER DISTRIBUTION 14-3
FIGURE 14-7: INCREASED POWER LOSS AT CONNECTION WITH OVERFILLED CORE 14-3
FIGURE 14-8: REDUCED POWER LOSS AT CONNECTION WITH UNDER FILLED CORE 14-4
FIGURE 14-9: ENCIRCLED FLUX TEMPLATE FOR CORE POWER DISTRIBUTION 14-4
FIGURE 14-10: MEASUREMENT OF INPUT POWER WITH EF COMPLIANT SOURCE 14-5
FIGURE 14-11: LOSS MEASUREMENT WITH EF COMPLIANT SOURCE 14-5
FIGURE 14-12: EF MODE CONTROLLER (COURTESY ARDEN PHOTONICS) 14-5
FIGURE 14-13: MEASUREMENT OF INPUT POWER WITH EF COMPLIANT TEST LEAD 14-5
FIGURE 14-14: LOSS MEASUREMENT WITH EF COMPLIANT TEST LEAD 14-5
FIGURE 14-15: MEASUREMENT OF INPUT POWER FOR HOML QUALIFICATION 14-5
FIGURE 14-16: MEASUREMENT OF OUTPUT POWER FOR HOML QUALIFICATION 14-5
FIGURE 14-17: MEASUREMENT OF INPUT POWER, HOML>0.6 DB 14-5
FIGURE 14-18: MEASUREMENT OF INSERTION LOSS 14-6
FIGURE 14-19: MEASUREMENT OF INPUT POWER, 0.1 DB<HOML<0.5 DB 14-6
FIGURE 14-20: MEASUREMENT OF INSERTION LOSS 14-6
FIGURE 14-21: CONNECTORS AT UNEQUAL DISTANCES FROM LINK ENDS 14-7
FIGURE 14-22: MINIMUM CONFIGURATION OF FTTH/PON NETWORK 14-7
FIGURE 14-23: CATEGORY 1 MULTIMODE SOURCE WITH MANDREL 14-9
FIGURE 14-24: THREE WAVELENGTH SINGLEMODE SOURCE 14-9
FIGURE 14-25: MODE CONDITIONER FOR MULTIMODE TESTING 14-10
FIGURE 14-26: METER WITH ADAPTERS 14-10
FIGURE 14-27: LOCATION OF POWER MEASUREMENT 14-10
FIGURE 14-28: SINGLEMODE INPUT POWER MEASUREMENT 14-11
FIGURE 14-29: FIRST REFERENCE LEAD TEST 14-11
FIGURE 14-30: SECOND REFERENCE LEAD TEST 14-11
FIGURE 14-31: LEAD QUALIFICATION WITH EF COMPLIANT SOURCE 14-12
FIGURE 14-32: LEAD QUALIFICATION WITH EF COMPLIANT TEST LEAD 14-12
FIGURE 14-33: LEAD QUALIFICATION WITH HOML BETWEEN 0.1 DB AND 0.5 DB 14-12
FIGURE 14-34: LEAD QUALIFICATION WITH HOML>0.5 DB 14-13
FIGURE 14-35: BARRELS 14-13
FIGURE 15-1: MAINFRAME OTDR 15-1
FIGURE 15-2: TEKTRONIX MINI-OTDR 15-2
FIGURE 15-3: EXFO MINI-OTDR 15-2
FIGURE 15-4: USB-POWERED OTDR MODULE 15-2
FIGURE 15-5: RAYLEIGH BACK SCATTERING 15-2
FIGURE 15-6: FUNCTIONAL DIAGRAM OF OTDR 15-2
FIGURE 15-7: THEORETICAL BACKSCATTER TRACE 15-3
FIGURE 15-8: SYMBOLS USED 15-3
FIGURE 15-9: TRACE WITH BACKSCATTER AND REFLECTIONS 15-4
FIGURE 15-10: BASIC TRACE WITH MODIFIED PEAKS 15-4
FIGURE 15-11: TWO CLOSELY-SPACED REFLECTIVE EVENTS 15-5
FIGURE 15-12: TRACE OF CONNECTORS, SINGLE PAIR 15-5
FIGURE 15-13: REFLECTIVE LOSS 15-5
FIGURE 15-14: REFLECTIVE LOSS TRACE FROM RADIUS CONNECTORS 15-6
FIGURE 15-15: REFLECTIVE LOSS FROM A MULTIMODE MECHANICAL SPLICE 15-6
FIGURE 15-16: REFLECTIVE LOSS TRACE FROM A SINGLEMODE MECHANICAL SPLICE 15-6

FIGURE 15-17: REFLECTIVE LOSS TRACE FROM BROKEN TIGHT TUBE CABLE	15-6
FIGURE 15-18: REFLECTIVE EVENT FROM MULTIPLE REFLECTION	15-6
FIGURE 15-19: NON-REFLECTIVE FUSION SPLICE	15-7
FIGURE 15-20: TRACE FOR 0 dB FUSION SPLICE	15-7
FIGURE 15-21: NON-REFLECTIVE LOSS FROM SINGLEMODE MECHANICAL SPLICE	15-8
FIGURE 15-22: NON-REFLECTIVE LOSS FROM APC CONNECTORS	15-8
FIGURE 15-23: NON-REFLECTIVE LOSS FROM CABLE PARAMETER VIOLATION	15-8
FIGURE 15-24: BAD LAUNCH	15-8
FIGURE 15-25: TRACE WITHOUT END REFLECTION	15-9
FIGURE 15-26: GAINER	15-9
FIGURE 15-27: MULTIPLE SEGMENTS APPEAR FROM A SINGLE SEGMENT	15-10
FIGURE 15-28: SINGLE SEGMENT TRACE EXPECTED	15-10
FIGURE 15-29: TRACE WITH GHOST	15-11
FIGURE 15-30: GHOST REFLECTION IN MIDDLE	15-11
FIGURE 15-31: HIGH REFLECTANCE CONNECTOR	15-11
FIGURE 15-32: OVERSHOOTING AFTER HIGH REFLECTANCE CONNECTOR	15-11
FIGURE 15-33: TRACE WITH LAUNCH CABLE	15-13
FIGURE 15-34: FIRST SEGMENT LENGTH MEASUREMENT, REFLECTIVE CONNECTION	15-14
FIGURE 15-35: FIRST SEGMENT LENGTH MEASUREMENT, NON-REFLECTIVE EVENT	15-14
FIGURE 15-36: SEGMENT LENGTH WITH REFLECTIVE ENDS	15-15
FIGURE 15-37: SEGMENT WITH REFLECTIVE AND NON-REFLECTIVE ENDS	15-15
FIGURE 15-38: SEGMENT LENGTH WITH NON-REFLECTIVE CONNECTIONS AT BOTH ENDS	15-15
FIGURE 15-39: CURSOR PLACEMENT FOR ESTIMATED REFLECTIVE CONNECTION LOSS	15-16
FIGURE 15-40: INCORRECT CURSOR PLACEMENT	15-16
FIGURE 15-41: INCORRECT CURSOR PLACEMENT	15-16
FIGURE 15-42: INCORRECT CURSOR PLACEMENT	15-16
FIGURE 15-43: CURSOR PLACEMENT, NON-REFLECTIVE CONNECTION LOSS	15-16
FIGURE 15-44: WIDE SPACING OF CURSORS	15-16
FIGURE 15-45: CURSORS IN DROP OFF	15-17
FIGURE 15-46: ACCURATE REFLECTIVE LOSS MEASUREMENT	15-17
FIGURE 15-47: ACCURATE NON-REFLECTIVE LOSS MEASUREMENT	15-17
FIGURE 15-48: ATTENUATION RATE MEASUREMENT	15-18
FIGURE 15-49: IMPROPER CURSOR PLACEMENT FOR ATTENUATION RATE MEASUREMENT	15-18
FIGURE 15-50: IMPROPER CURSOR PLACEMENT FOR ATTENUATION RATE MEASUREMENT	15-18
FIGURE 15-51: CORRECT ATTENUATION RATE MEASUREMENT WITH NON-UNIFORMITY	15-18
FIGURE 15-52: ATTENUATION RATE MEASUREMENT WITH REDUCED CURSOR SEPARATION	15-18
FIGURE 15-53: FTTH/PON LINK WITH 1X2 SPLITTER	15-19
FIGURE 15-54: DOWNSTREAM TRACE THROUGH 1X2 SPLITTER	15-20
FIGURE 15-55: SINGLEMODE LINK AT 1310 NM	15-20
FIGURE 15-56: THE SAME SINGLEMODE LINK AT 1550 NM	15-20
FIGURE 15-57: FTTH/PON FOR QUESTION 37	15-27
FIGURE 15-58: BLANK TRACE FOR QUESTION 38, 2200 M LEG	15-27
FIGURE 15-59: BLANK TRACE FOR QUESTION 38, 2200 M AND 1650 M LEGS	15-27
FIGURE 15-60: BLANK TRACE FOR QUESTION 38, 2200 M, 1650 M AND 1100 M LEGS	15-27
FIGURE 15-61: BLANK TRACE FOR QUESTION 38, 2200 M, 1650 M, 1100 M AND 550 M LEGS	15-28
FIGURE 16-1: REFLECTANCE TEST SET (COURTESY OF NETTEST)	16-1
FIGURE 16-2: STRUCTURE OF REFLECTANCE TEST SET	16-2
FIGURE 16-3: REFLECTANCE REFERENCE CONFIGURATION	16-2
FIGURE 17-1: TIME INTERVAL FOR OC-48	17-1
FIGURE 17-2: TOTAL DISPERSION, (CD AND PMD) FOR OC-48	17-2
FIGURE 17-3: TOTAL DISPERSION, (CD AND PMD) AT OC-192	17-2
FIGURE 17-4: ORTHOGONAL STRESS STATE IN IDEAL FIBER	17-4
FIGURE 17-5: ORTHOGONAL STRESS STATE IN REAL FIBER	17-4
FIGURE 18-1: VFL TEST ON CLEAVE AND CRIMP CONNECTOR (BOTTOM, GLOWING)	18-2
FIGURE 19-1: MAP 1	19-2
FIGURE 19-2: MAP FOR QUESTION 1	19-5

Figure	Page
FIGURE 19-3: MAP FOR QUESTION 2	19-5
FIGURE 19-4: MAP FOR QUESTION 3	19-6
FIGURE 19-5: MAP FOR QUESTION 4	19-6
FIGURE 19-6: MAP FOR QUESTION 5	19-6
FIGURE 20-1: USE OF INSPECTION MICROSCOPE	20-2
FIGURE 20-2: CONNECTOR WITH BACK LIGHT	20-2
FIGURE 20-3: SAME CONNECTOR NO BACK LIGHT	20-2
FIGURE 20-4: SHATTERED END WITH BACK LIGHT	20-3
FIGURE 20-5: SHATTERED END	20-3
FIGURE 20-6: GOOD CONNECTOR- BACK LIGHT	20-3
FIGURE 20-7: CLEANING RESIDUE	20-4
FIGURE 20-8: SCRATCH IN CORE	20-4
FIGURE 20-9: CRACK AND NON-ROUND CORE	20-4
FIGURE 20-10: CRACK IN CORE	20-4
FIGURE 20-11: DIRT ON CONNECTOR	20-5
FIGURE 20-12: DIRT ON FERRULE	20-5
FIGURE 20-13: CLEANING LIQUID RESIDUE	20-5
FIGURE 20-14: ACCEPTABLE, IMPERFECT CLAD	20-5
FIGURE 20-15: FERRULE FEATURES	20-5
FIGURE 21-1: BREAK AWAY SWIVEL WITH PULL ROPE	21-3
FIGURE 21-2: SETTING SLITTER DEPTH	21-3
FIGURE 21-3: PROPER SLIT DEPTH	21-3
FIGURE 21-4: EXCESSIVE SLIT DEPTH	21-4
FIGURE 21-5: STRENGTH MEMBERS ATTACHED TO SWIVEL	21-4
FIGURE 21-6: CUTTING OF PREMISES CABLE JACKET	21-5
FIGURE 21-7: PREMISES CABLE ATTACHED TO PULL ROPE	21-5
FIGURE 21-8: END PREPARATION DIMENSIONS	21-5
FIGURE 21-9: CUTTER ON BUFFER TUBE	21-7
FIGURE 22-1: TEMPLATE FOR LENGTH BARE FIBER	22-2
FIGURE 22-2: CLEANING OF CLAUSS STRIPPER	22-3
FIGURE 22-3: FIBER STRAIGHT IN STRIPPER	22-3
FIGURE 22-4: TEMPLATE OF STRIP LENGTHS	22-4
FIGURE 22-5: TUBING CUTTER ON CABLE	22-4
FIGURE 22-6: CUTTING ARAMID YARN TO LENGTH	22-4
FIGURE 22-7: FIBER CLEANING	22-5
FIGURE 22-8: CENTRAL TUBE OF SC CONNECTOR	22-5
FIGURE 22-9: CRIMPING CRIMP SLEEVE	22-6
FIGURE 22-10: CONNECTORS IN CURING OVEN	22-7
FIGURE 22-11: SCRIBING THE EXCESS FIBER	22-7
FIGURE 22-12: AIR POLISHING OF FIBER	22-8
FIGURE 22-13: POLISHING	22-8
FIGURE 22-14: WHITE LIGHT TEST	22-10
FIGURE 22-15: ALIGNMENT HOUSINGS	22-10
FIGURE 22-16: INSERTION OF INNER HOUSING	22-10
FIGURE 22-17: INNER HOUSING FULLY INSERTED	22-10
FIGURE 23-1: HOT MELT COOLING STAND	23-2
FIGURE 23-2: LOADED CONNECTOR HOLDER	23-2
FIGURE 23-3: SC AND HOLDER PRIOR TO ROTATION	23-2
FIGURE 23-4: SC AND HOLDER AFTER ROTATION	23-2
FIGURE 23-5: TEMPLATE OF STRIP LENGTH FOR PREMISES CABLE (TO SCALE)	23-3
FIGURE 23-6: CLEANING OF CLAUSS STRIPPER	23-3
FIGURE 23-7: FIBER STRAIGHT IN STRIPPER	23-3
FIGURE 23-8: CLEANING FIBER WITH MOIST TISSUE	23-4
FIGURE 23-9: CONNECTORS IN OVEN	23-4
FIGURE 23-10: CABLE INSERTED	23-5
FIGURE 23-11: CABLE IN CLIP	23-5

FIGURE 23-12: SCRIBING THE EXCESS FIBER	23-5
FIGURE 23-13: AIR POLISHING OF FIBER	23-6
FIGURE 23-14: POLISHING	23-6
FIGURE 23-15: WHITE LIGHT TEST	23-8
FIGURE 24-1: UNICAM™ VFL-CRIMPER TOOL	24-1
FIGURE 24-2: PRIOR GENERATION UNICAM™ CLEAVER	24-1
FIGURE 24-3: LATEST UNICAM CLEAVER, OPEN	24-1
FIGURE 24-4: OPEN UNICAM INSTALLATION TOOL, CRIMP JAW CLOSED	24-2
FIGURE 24-5: OPEN CRIMP JAW INCORRECT	24-2
FIGURE 24-6: CAPS REMOVED, SC AND LC	24-3
FIGURE 24-7: CAMS IN CORRECT POSITION, SC AND ST™-COMPATIBLE	24-3
FIGURE 24-8: DATE CODE ON TOP SURFACE	24-3
FIGURE 24-9: ADAPTER NOT ON FERRULE	24-3
FIGURE 24-10: INCORRECT CONNECTOR LOCATION	24-4
FIGURE 24-11: ADAPTER FULLY ON FERRULE	24-4
FIGURE 24-12: ADAPTER NOT FULL ON FERRULE	24-4
FIGURE 24-13: IMPROPER LEAD IN TUBE POSITION	24-4
FIGURE 24-14: CLEANING FIBER WITH MOIST TISSUE	24-5
FIGURE 24-15: CLEAVER WITH FIBER INSTALLED	24-5
FIGURE 24-16: FIBER SLIGHTLY BENT	24-5
FIGURE 24-17: SC BOOT INSTALLED	24-6
FIGURE 24-18: VFL TEST	24-6
FIGURE 24-19: LC WITHOUT INSTALLATION HOUSING	24-6
FIGURE 24-20: PROPER ALIGNMENT OF DATE CODE AND KEY	24-7
FIGURE 24-21: INNER HOUSING FULLY INSERTED	24-7
FIGURE 25-1: CABLE END PREPARATION DIMENSIONS	25-2
FIGURE 25-2: PROPER BUFFER TUBE LENGTH	25-2
FIGURE 25-3: 7 AND 12 MM PLUGS	25-3
FIGURE 25-4: MOISTURE SEAL COMPONENTS	25-3
FIGURE 25-5: END CAP REQUIRING DRILLING	25-3
FIGURE 25-6: ENCLOSURE PLUGS	25-3
FIGURE 25-7: GROUNDING AND BONDING HARDWARE	25-3
FIGURE 25-8: STRENGTH MEMBER ATTACHMENT HARDWARE	25-3
FIGURE 25-9: ATTACHMENT MECHANISM	25-4
FIGURE 25-10: BUFFER TUBES ATTACHED TO TRAY	25-4
FIGURE 25-11: PROPER FIBER LENGTH	25-5
FIGURE 25-12: POTENTIAL BEND RADIUS VIOLATION	25-5
FIGURE 25-13: SPLICER GROOVES	25-6
FIGURE 25-14: ACTIVE FIBER MENU	25-6
FIGURE 25-15: CLEAVE ANGLE SETTING	25-6
FIGURE 25-16: SPLICE COVER LENGTH SETTING	25-6
FIGURE 25-17: PROPER FIBER LOCATION FOR SPLICING	25-7
FIGURE 25-18: SPLICE COVER IN HEATING OVEN	25-7
FIGURE 25-19: ADHESIVE SPLICE COVER	25-8
FIGURE 25-20: SPLICE TOOL FEED-IN GROOVES	25-8
FIGURE 25-21: POSITION OF HOLDERS FOR 250μ PRIMARY COATING	25-8
FIGURE 25-22: OPEN SPLICE IN TOOL	25-8
FIGURE 25-23: POSITION OF HOLDERS FOR 900μ TIGHT TUBE	25-9
FIGURE 25-24: FIRST FIBER IN TOOL	25-9
FIGURE 25-25: SPLICE WITH EVEN BOWS	25-9
FIGURE 25-26: ARM IN CLOSING POSITION	25-10
FIGURE 25-27: CLOSED FIBRLOK II™ SPLICE	25-10
FIGURE 25-28: FIBERS PARALLEL TO SIDES	25-11
FIGURE 25-29: TWISTED FIBERS AT SECOND END	25-11
FIGURE 25-30: FIBERS REVERSED TWISTED AT INPUT END	25-11
FIGURE 25-31: FIRST SPLICE HOLDER POSITION	25-11

© PEARSON TECHNOLOGIES INC.

FIGURE 25-32: STARTING POSITIONS OF SPLICE HOLDER 25-11
FIGURE 25-33: TRAY WITH COVER 25-11
FIGURE 25-34: STARTING POSITION 25-11
FIGURE 25-35: TWISTED BUFFER TUBES 25-12
FIGURE 25-36: TRAY SUPPORT STRUCTURE 25-12
FIGURE 25-37: BOLT HOLE FOR TRAY ATTACHMENT 25-12
FIGURE 25-38: ATTACHMENT WITH VELCRO BANDS 25-12
FIGURE 25-39: TRAY ATTACHMENT WITH RUBBER HOLDERS 25-12

TABLE OF TABLES

TABLE 2-1: COMMON COMMUNICATION WAVELENGTHS	2-1
TABLE 2-2: TELEPHONE WAVELENGTH BANDS	2-2
TABLE 2-3: NOMINAL NA VALUES	2-4
TABLE 2-4: FAST ETHERNET DISTANCES	2-7
TABLE 2-5: 850 NM GIGABIT ETHERNET (1000BASE-SX) DISTANCES	2-7
TABLE 2-6: 1300 GIGABIT ETHERNET (1000BASE-LX) DISTANCES	2-7
TABLE 2-7: 850 NM 10 GIGABIT ETHERNET (10GBASE-SX) DISTANCES	2-7
TABLE 2-8: 1300 NM 10 GIGABIT ETHERNET DISTANCES	2-8
TABLE 2-9: 850 NM, 40 AND 100 GIGABIT DISTANCES	2-8
TABLE 2-10: 850 NM FIBER CHANNEL DISTANCES	2-8
TABLE 3-1: NOMINAL FIBER NA VALUES	3-2
TABLE 3-2: FIBER DIAMETER TOLERANCES	3-2
TABLE 3-3: OFFSET AND OVALITY VALUES	3-3
TABLE 3-4: BANDWIDTH DISTANCE PRODUCTS (MHZ-KM)) FOR OM-X DESIGNATIONS	3-5
TABLE 3-5: MODE FIELD DIAMETERS	3-6
TABLE 3-6: MAXIMUM CABLE ATTENUATION RATES	3-9
TABLE 3-7: TYPICAL CABLE ATTENUATION RATES	3-9
TABLE 4-1: NEC FIBER CABLE RATINGS	4-9
TABLE 4-2: CABLE JACKET COLOR CODES	4-12
TABLE 4-3: COLOR CODING SEQUENCE	4-12
TABLE 5-1: CONNECTOR COLOR CODES	5-5
TABLE 8-1: OPBA FOR FIBER STANDARDS	8-3
TABLE 8-2: 850 NM 10 GIGABIT DISTANCES VS. NUMBER OF CONNECTOR PAIRS	8-4
TABLE 14-1: MULTIMODE INSERTION LOSS SITUATIONS	14-4
TABLE 14-2: HOML MANDREL DIAMETERS	14-5
TABLE 17-1: CHROMATIC DISPERSION TEST RESULTS	17-3
TABLE 17-2: SUMMARY OF PMD LIMITS FOR OC-192 AND 10 GB ETHERNET	17-5
TABLE 17-3: PMD TEST RESULTS	17-5
TABLE 19-1: THE OPTICAL POWER LOSS CALCULATION	19-2
TABLE 19-2: 850 NM, 62.5µ MULTIMODE VALUES	19-2
TABLE 19-3: CALCULATION, MAXIMUM LOSS	19-2
TABLE 19-4: CALCULATION, TYPICAL LOSS	19-2
TABLE 19-5: 1300 NM MULTIMODE VALUES	19-5
TABLE 19-6: 1310 NM SINGLEMODE VALUES	19-5
TABLE 26-1: CALCULATION, MAXIMUM LOSS	26-10
TABLE 26-2: CALCULATION, TYPICAL LOSS	26-10
TABLE 26-3: CALCULATION, MAXIMUM LOSS	26-10
TABLE 26-4: CALCULATION, TYPICAL LOSS	26-10
TABLE 26-5: CALCULATION, MAXIMUM LOSS	26-11
TABLE 26-6: CALCULATION, TYPICAL LOSS	26-11
TABLE 26-7: CALCULATION, MAXIMUM LOSS	26-11
TABLE 26-8: CALCULATION, TYPICAL LOSS	26-11
TABLE 26-9: CALCULATION, MAXIMUM LOSS	26-11
TABLE 26-10: CALCULATION, TYPICAL LOSS	26-11
TABLE 26-11: CALCULATION, MAXIMUM LOSS	26-11
TABLE 26-12: CALCULATION, TYPICAL LOSS	26-11
TABLE 26-13: CALCULATION, MAXIMUM LOSS	26-12
TABLE 26-14: CALCULATION, TYPICAL LOSS	26-12
TABLE 26-15: CALCULATION, MAXIMUM LOSS	26-12
TABLE 26-16: CALCULATION, TYPICAL LOSS	26-12

AUTHOR'S PREFACE

Successful fiber optic network installation requires:

- Low optical power loss
- Low reflectance
- Low installation cost
- Low installation time
- High reliability

This text and handbook guides the reader to:

- Successful installation of fiber optic cables, connectors, and splices
- Proper testing and interpretation of test results
- Confidence in his ability to be successful
- Becoming certified by the Fiber Optic Association (www.thefoa.org) and other certifying organizations

To lead to these four goals, this book includes four parts:

- Essential Information
- Principles Of Installation
- Principles And Methods Of Testing And Inspection
- Procedures

Essential Information

Chapters 1-9 present the language of fiber optic networks and their components. This language includes the products with which the installer works, the types of products, their advantages, and the performance numbers that the installer can expect. This first part acts as a comprehensive textbook for those studying fiber optic technology with the intent of involvement in field installation. With the foundation created by the first part, the installer can understand the principles in the second and third parts. In addition, he can follow the procedures in the fourth part.

Principles Of Installation

Chapters 10-13 present the principles, or basic characteristics, that give rise to the procedures for installing cables, connectors, and splices. These principles are rarely included in installation procedures from manufacturers. Such inclusion would complicate those procedures and distract the installer from his focus on installation. Knowledge of these principles makes the installer sensitive to the importance of the procedures. With this sensitivity, the installer is able to follow the procedures and achieve the four requirements of successful installation.

Principles And Methods Of Testing And Inspection

Chapters 14-20 present the principles and methods of testing, inspection, and interpretation of results, which I call 'network certification'. These chapters address testing by insertion loss, OTDR, reflectance, ORL, and dispersion, and microscopic inspection. Chapter 19 enables the installer to interpret the test results in a manner that resuls in high reliability.

Procedures

Chapters 21-25 present installation procedures that, when followed, result in successful installation. These chapters can be used as training procedures and as field procedures. In addition, when there is a long time between fiber activities, these procedures can be used as a tool for review.

Certification

Use of this book enables the reader to become certified by the Fiber Optic Association (FOA, www.thefoa.org). Chapters 1-11, 14-15, and 20 include most of the information that the installer needs to become certified as a Certified Fiber Optic Technician (CFOT). Chapters 12-25 include most of the information required to be certified as a Certified Fiber Optic Specialist in connector installation (CFOS/C), in splicing (CFOS/S), and in testing (CFOS/T). Much of this same information will enable the reader to pass other fiber optic certification examinations. Appendices 26.5, 26.9, and 26.10 provide tools to assist the reader in passing the FOA certification examinations. Finally, this text will enable the reader to pass the FOA instructor certification, CFOS/I.

Training Benefits

For training, this text includes 432 review questions and exercises designed to reinforce understanding. For hands-on training, Chapters 14, 15, and 21-25 contain detailed installation procedures to make hands-on training easy. These procedures include troubleshooting sections and one-page review sections that help identify and avoid problems during training. In Chapters 11-20, key points are highlighted with the symbols ▶ for a principle and ▶▶ for a method. Trainers can request anwers to the review questions by email. Note that a pdf of answers to the review questions is available upon request. To receive this pdf, send a request to: fiberguru@ptnowire.com. In addition, a power point presentation of Chapters 1-20 is available for a fee.

Applicability

The knowledge information in the text applies equally well to data networks, telephone networks, and CATV networks, CCTV and process control links.

Evolution

In 1990, the predecessor to this book started as a training manual. That predecessor evolved into a text I finished in 1995. Immediately after finishing that text, I began this text, with frequent revisions and improvements.

This version, Version 9, brings the text up to date with the different types of links (Chapters 1 and 8), new insertion loss test procedures (Chapter 14), and with a new cleave-and-crimp connector installation tool (Chapter 24).

This book has developed from 36 years of work with fiber optic products and training people in installation and testing of these products. The elements of this book have come from: manufacturers' instructions; professional associates, for whose continuous generosity of time and information I am grateful; and from my efforts to find ways to avoid the errors made by the more than 8700 trainees who have attended our 530 presentations.

Some elements have come from trainees, who have asked practical questions and shared their observations. Finally, some have come from the testing we have performed in our laboratory to identify methods to improve, simplify, and speed up the installation and learning processes. We include much information in this book because its lack results in increases in power loss and installation cost or reduced reliability. This aspect of development resulted in the subtitle of this book "The Essentials For Success."

The result of this evolution is a series of procedures that work well for almost all people. 'Working well' means the procedures result in low power loss, the prime concern of the installer, low installation cost, through reduced rework, and maximum reliability, through avoidance of both common and subtle errors.

Whether you are studying fiber optics for the first time or you are a field installer, you will find this book highly useful and an investment that will pay back many times its cost. Finally, this book will help you set up field installation and cable assembly operations. As this book remains a work in progress, I encourage your comments and observations.

In Closing

For 36 years, I have been blessed to work in fiber optic communications and witness the amazing creativity of fiber professionals. I have worked with many brilliant people, witnessed genius-class solutions, and trained many who have learned what they needed, and by their questions, motivated me to expand my understanding of this powerful technology. To all those, I am grateful.

Best Regards,

Eric R. Pearson, CFOS/C/T/S/I
Pearson Technologies Inc.
770-490-9991
www.ptnowire.com
fiberguru@ptnowire.com
September 2014

PART ONE
ESSENTIAL INFORMATION

Essential:

"Absolutely necessary; extremely important". In the context of this text, essential means information and understanding without which the installer would not achieve the four goals of installation.

1 INTRODUCTION TO FIBER NETWORKS

Chapter Objectives: you will learn the basics of fiber networks. These basics will allow you to put fiber transmission in its proper context.

1.1 NINE ADVANTAGES

In general, optical fiber is used when its capabilities exceed the limitations of other communication media. Optical fiber is the medium of choice when its nine characteristics favor its use. These characteristics are:

- Nearly unlimited bandwidth
- Long transmission distance
- Low cost per bit
- EMI and RFI immunity
- Dielectric construction
- Small size
- Light weight
- Ease of installation
- Intrinsically secure transmission

1.1.1 UNLIMITED BANDWIDTH

Optical fiber has essentially unlimited bandwidth. While 'unlimited bandwidth' may sound like an exaggeration, it is a realistic and reasonable description of fiber capacity. A study by Lucent Technologies indicated the theoretical capacity of one singlemode fiber is on the order of 200 Tbps, or 200 million Mbps. Such a capacity deserves the term 'essentially unlimited'.

One part of this high capacity results from the ability to modulate light sources at very high data rates. At the time of this writing, a practical maximum, single wavelength data rate is 40 Gbps. This rate is 4,000 times the rate of the early, Ethernet fiber optic transceivers.

A second part of high capacity results from the ability to transmit multiple wavelengths on one fiber simultaneously. Such transmission is called wavelength division multiplexing (WDM), coarse wavelength division multiplexing (CWDM), and dense wavelength division multiplexing (DWDM).

WDM refers to simultaneous transmission of two widely separated wavelengths on the same fiber. For multimode fibers, widely separated means 850 nm and 1300 nm. For singlemode fibers, widely separated means 1310 nm and 1550 nm.

CWDM refers to simultaneous transmission of up to 16 wavelengths with separations of 20 nm on the same fiber. DWDM refers to simultaneous transmission of up to 200 wavelengths with separations of as little as 0.4 nm on the same fiber (G.692).

While the G.692 standard allows 200 wavelengths, Lucent Technologies demonstrated the ability to transmit 1000 wavelengths.

A third part of high capacity results from the ability of an optical pulse to transmit more than one data bit. At this time, this technology, coherent transmission, enables transmission of eight electrical bits with one optical pulse.

This 'essentially unlimited' bandwidth means simply replacing the optoelectronics on the fiber ends can increase that fiber bandwidth. This upgradeability results in low life cycle and low total cost ownership.

1.1.2 TRANSMISSION DISTANCE

Optical fiber allows extremely long transmission distance without return to the electrical regime. Long transmission distance results in a reduction in the cost of mid-span signal repeaters and regenerators. Elimination of these electronics reduces hardware and maintenance costs significantly.

While not a current 'champion' result, Williams Communications demonstrated the ability to transmit 5000 km (3100 miles) in the optical regime. 5000 km is approximately the distance between Boston MA and Los Angeles CA!

In the early 1990s, long distance capability provided two major benefits to the

CATV industry. This long transmission distance enabled a major cost reduction through a reduction in the number of satellite 'farms' necessary to support a service area. In addition, long transmission distance enabled CATV companies to reduce the number of coax amplifiers between a satellite down link and set top boxes. This reduction resulted in reduced equipment and maintenance costs, and improved signal quality.

1.1.3 EMI AND RFI

Long transmission distance capability results primarily from low attenuation rate (3.5.2) and low pulse dispersion (2.3.3). However, a third property of optical fiber communications is equally important. This property, immunity to electromagnetic interference (EMI) and radio frequency interference (RFI) enables optical signals to travel long distances without the need for signal correction due to interference from EM and RF signals in the environment.

1.1.4 LOW BIT COST

The combination of multiple wavelength transmission, low power loss, low pulse dispersion, and EM and RF immunity results in low cost per bit. This low cost has made fiber the medium of choice for long distance, high capacity communication. This low cost has resulted in the displacement of satellites as the 'king' of long distance communication. Now satellites are a back up for optical fiber transmission, a reverse of the original relationship!

At the present time, some local area networks (LANs) with centralized backbones have a total initial installed cost that is lower than that of traditional horizontal UTP, vertical fiber networks. Such networks are known as fiber to the desk (FTTD) and collapsed backbone networks.[1] This cost advantage results from a reduction in the costs of telecommunication rooms and switches. Use of fiber significantly reduces the cost of such rooms and the cost of support for such rooms.[2]

Finally, the low cost of optical fiber has led to increased implementation of fiber-to-the-home (FTTH) networks. The characteristics of high bandwidth and multiple wavelength transmission enable cost effective delivery of multiple services, also known as 'triple play' services.

1.1.5 DIELECTRIC CONSTRUCTION

Conductive cables must be grounded and bonded to prevent induced currents from entering a building. Such currents are caused by lightning and ground potential rise. Such currents can injure people and damage electronics. Optical fiber cables can be made without conductive elements. Such **dielectric** construction eliminates both the initial installed cost and the maintenance cost of grounds and bonds. After all, grounds and bonds do not last forever.

1.1.6 SMALL SIZE

The small size of optical fibers and their cables results in reduced system cost. For example, large cities with filled underground conduit systems have two methods of increasing telephone capacity: 1) dig up the streets to install more conduits or 2) replace large copper cables with small fiber cables.

This author's rough estimate of the replacement ratio is 37,305, 3" diameter, 900 pair cables to one 1" diameter fiber cable with single wavelength transmission. With dense wavelength division multiplexing (DWDM) allowing 200 wavelengths per fiber, this replacement ratio is increased 200 times. The cost advantage of using fiber instead of digging up streets is significant.

1.1.7 LIGHT WEIGHT

Optical fiber cables are significantly lighter than copper cables. As a result, fiber finds

---

[1] See the cost model offered by the Fiber Optic LAN Section (FOLS) of the TIA at www.fols.org. The FOLS and Pearson Technologies co-developed this model.

[2] See the fiber to the desk cost comparison at www.fols.org. Pearson Technologies Inc. and the FOLS developed this cost comparison.

use in field tactical military, shipboard, and aircraft applications. In field tactical applications, reduced weight enables soldiers to carry increased cable lengths. Such increased lengths enable placement of electronic monitoring equipment at the front line while the monitoring personnel are in a safe location away from that line. In addition, the non-radiating nature of optical fibers prevents the enemy from detecting the equipment location.

In shipboard applications, the light weight of optical cables increases the stability of the ships by reducing the weight above the waterline. Finally, in aircraft applications, the light weight increases mission endurance.

1.1.8 INSTALLATION EASE

Because of their lightweight and small size, fiber cables are easy to install. In addition, fiber connector installation methods have advanced sufficiently to enable installation by junior and senior high school students with minimal training!

1.1.9 SECURITY

Fiber transmission is nearly completely secure. Its security comes from two aspects.

- No radiation
- Difficulty in tapping signal without such tapping being detected

An optical signal travels in the center of the fiber, known as the core or the mode field diameter (3). As such, no signal is radiated externally to the fiber or its cable.

In addition, any effort to tap optical power from the fiber reduces the power delivered to the receiver. With the addition of a simple signal intensity monitoring circuit to the receiver, a transmission system becomes secure.

Of course, the desire for security can be expanded beyond these two aspects. For example, a fiber transmission system developed in the 1980s required six fibers. The system monitored the signal intensity on all fibers. If the signal level dropped by as little as 0.2 dB, the system sounded alarms.

Two active fibers carried the encoded signal. The other fibers carried garbage signals. Periodically, the signal switched from the active fibers to those that carried the garbage signals. Any attempt to capture and decode the signal would be useless, as there would be no discernable difference between the signal and the garbage.

Some DWDM telephone links use a similar approach to monitor link status. One of the wavelengths, usually 1625 nm, is an OTDR signal. The OTDR monitors the loss along the entire link continuously. When there is an increase in the power loss anywhere along the link, the monitoring equipment activates an alarm.

1.2 NETWORKS AND FULL DUPLEX LINKS

In the most general sense, a network is a series of electronic devices that communicate **digital** information with each other. In fiber optic networks, these electronic devices are **optoelectronic** devices, as they process both electrical and optical signals.

The network can be:

- A local area network (LAN)
- A data network of LANs
- A storage area network (SAN)
- A wide area network (WAN)
- A long distance, or long haul, or telephone network
- A CATV network
- A process control network.

This communication is via connections. A point-to-point **link** forms this connection. 'Point-to-point' means that electronic conversion and distribution of the electrical data /signals occur at the link ends only.

- From the installer's point of view, **link** is the most important term, as all installation actions are performed on the components of each of the links.

- **Incorrect installation procedures reduce the optical power** delivered to the receiver on a link and **reduce the reliability** of the components of a link. Thus,

power loss is the installer's most important concern.

Transceivers perform this conversion. Transceivers are optoelectronic devices that provide simultaneous optical to electrical and electrical to optical conversion; i.e., simultaneous transmission and reception. Thus, transceivers enable **full duplex** communication. Usually, duplex transmission requires two fibers, one transmit and one receive. As we will see, the requirement for two fibers is not absolute.

- Installers need remember that link communication requires two fibers.
- **A common problem is crossed fibers.**

With crossed fibers, some test results can be proper. However, communication will not occur. For the obvious reason, transmitters do not communicate properly with transmitters!

The fiber patch cables most commonly used for duplex transmission are two simplex cables or a single 'zip cord' duplex cable. In contrast, UTP networks require two pairs of conductors. Thus, a fiber pair is not the same as UTP pair.

1.3 FIBER LINK EVOLUTION

Fiber optic networks have evolved from a series of simple links to highly complex links. We describe this evolution in nine stages. Note that these stages simplify their descriptions. They are not necessarily sequential.

1.3.1 STAGE 1

In 'Stage 1,' fiber networks consisted of a series of point-to-point links (Figure 1-1) with a single wavelength and one optical bit per electrical symbol (i.e., 0 or 1). Starting with and continuing after the 100 Mbps protocols, FDDI and Fast Ethernet, four electrical bits were encoded into five optical bits.

The components of such links were optoelectronics, fiber in cable, connectors, and splices. The transmission distances of these links were limited by one of two causes of signal distortion:

- Power loss in the fiber (attenuation, 3.5.2).
- Pulse spreading from dispersion (dispersion, 3.5.1)

To extend the distance to greater than allowed by these two causes of distortion, the optical signal was converted to an electrical signal and regenerated (Figure 1-2-). Such regeneration was expensive.

After installation, testing of Stage 1 links is simple: connect test equipment to the ends. Measurement of the power loss at the wavelength of operation is sufficient to verify proper operation of the optoelectronics.

1.3.2 STAGE 2

In Stage 2, designers overcame the distance limitation by adding optical amplifiers (7.2.4) in the middle of the link (Figure 1-3). These devices amplify the optical signal to compensate for the power loss from attenuation in the fiber.

Although such amplifiers require power for operation, they are considered passive devices because they handle the optical signal as an optical signal without conversion of the optical signal to an electrical signal. Examples of optical amplifiers are erbium-doped fiber amplifiers (EDFA) and Raman amplifiers (7).

After installation, testing of Stage 2 links is simple: connect test equipment to the ends of the segments limited by the optical amplifiers. Measurement of the power loss at the wavelength of operation is sufficient to verify proper operation of the optoelectronics.

The addition of optical amplifiers compensated for power loss, but did nothing for increasing dispersion. To compensate for dispersion, network designers added dispersion compensating modules. These devices reduced the width of the optical pulses.

The first generation of dispersion compensation was based on a specialized optical fiber. The wavelengths that moved fast in the transmission fiber moved slowly in the compensation fiber; those that moved slowly in the trans-mission fiber, moved fast in the compensation fiber.

The second generation of dispersion compensation was based on Bragg gratings. These gratings forced the different wavelengths within the optical pulse (spectral width, 2.2.2) to travel different lengths in the grating, resulting in a reduction in pulse width. Both gratings and compensating fibers are passive devices (7), which increase power loss.

1.3.3 STAGE 3

In Stage 3, links took advantage of wavelength division multiplexing (WDM), with two widely separated wavelengths on the same fiber pair. Widely separated means 850 and 1300 nm wavelengths on multimode fibers and 1310 nm and 1550 nm on singlemode fibers.

These wavelengths are combined and separated at the ends of the link. Such multiplexing is possible because widely separated wavelengths do not interfere with one another.

To enable such multiplexing, designers added passive devices prior to the optoelectronics at the ends of link (Figure 1-4). Such devices were wavelength division multiplexors and wavelength division de-multiplexors (7). Except for optical amplifiers, all passive devices exhibit power loss.

In WDM, each wavelength carries a unique data stream, resulting in an increase in capacity per fiber. This increase is less than double, as dispersion (2.3.3, 3.4.3.3) limits the capacity of one wavelength to less than that of the second.

1.3.4 STAGE 4

In Stage 4, the concept of multiple wavelengths was expanded to simultaneous transmission of many, closely spaced wavelengths on the same fiber (Figure 1-5). Of course, DWDM transmission distance can be increased through the use of optical amplifiers (Figure 1-6).

This technology, dense wavelength division multiplexing (DWDM) enabled transmission of wavelengths separated by as little as 0.4 nm. According to the ITU standard, G. 692, up to 200 wavelengths can be carried on the same fiber. This number is not the limit, as Bell Labs demonstrated in 2000 by transmitting 1000 wavelengths on a single fiber.[3] Such links require sophisticated and precise lasers, wavelength division multiplexors, and wavelength division de-multiplexors.

Closely spaced wavelengths can be carried by a single fiber as long as four conditions are met:

- Stable wavelength
- Narrow spectral width lasers
- A limitation on the total power in the core of the fiber,
- A requirement for a minimum amount of dispersion.

These expensive DWDM lasers required precise temperature control to avoid wavelength drift. Such drift would result in interference between the data streams carried by adjacent wavelengths.

Optical amplifiers extended the distance by simultaneous amplification of multiple wavelengths. However, their limited amplifier wavelength range required multiple optical amplifiers.

1.3.5 STAGE 5

In Stage 5, the concept of DWDM was modified to enable transmission of multiple wavelengths in both directions on one fiber (Figure 1-7). This approach reduces the cost of the transmission path. Such transmission requires the same conditions as in Stage 4.

An example of such bi-directional transmission is the FTTH passive optical network (PON). In this network, three wavelengths are transmitted on a single fiber, two downstream to the customer and one upstream from the customer (Figure 1-8). The 1550 nm wavelength carries the CATV signals; the 1490 nm wavelength, digital voice and Internet signals; the 1310 nm, voice and Internet signals from the customer. Provision of

---

[3] (http://gcn.com/articles/2000/03/15/bell-labs-expands-fiberoptic-capacity-to-1000-channels.aspx)

these three services is known as 'triple play'.

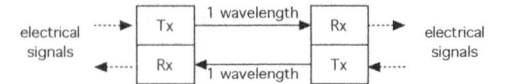

Figure 1-1: Stage 1-Full Duplex Transmission

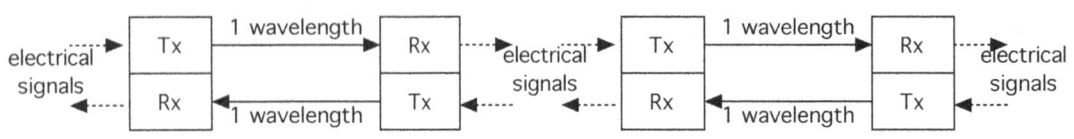

Figure 1-2: Stage 1-Extended Duplex Distance Transmission

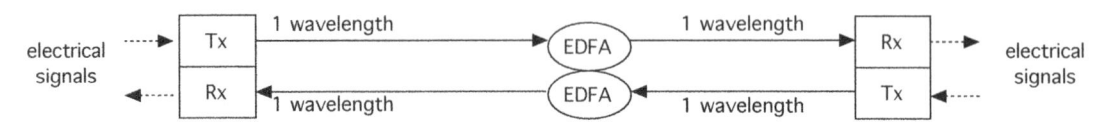

Figure 1-3: Stage 2- Increased Duplex Transmission Distance With Optical Amplifier

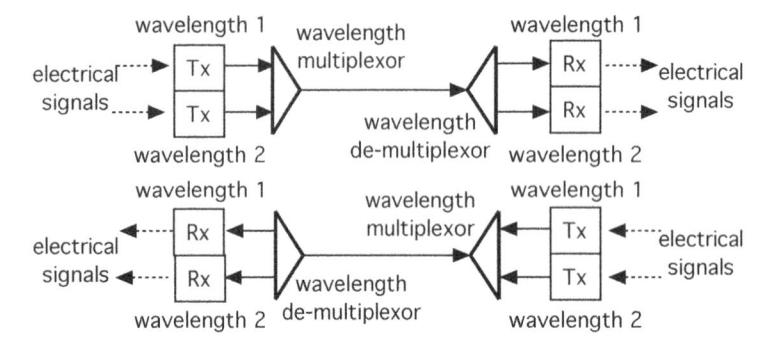

Figure 1-4: Stage 3- Duplex Wavelength Division Multiplexing (WDM)

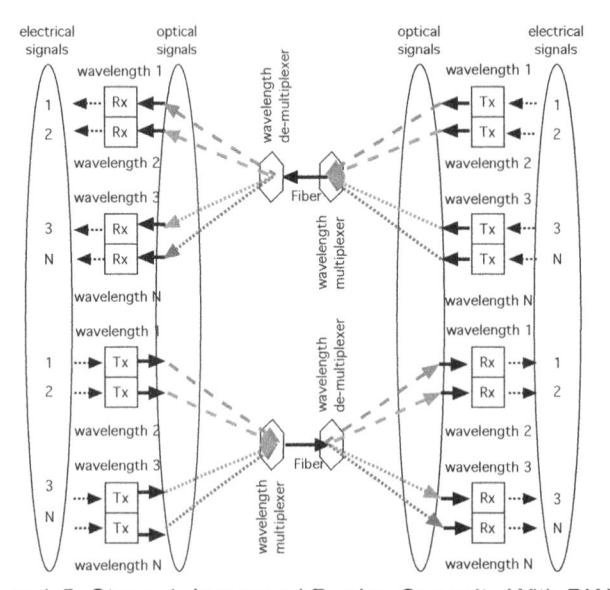

Figure 1-5: Stage 4- Increased Duplex Capacity With DWDM

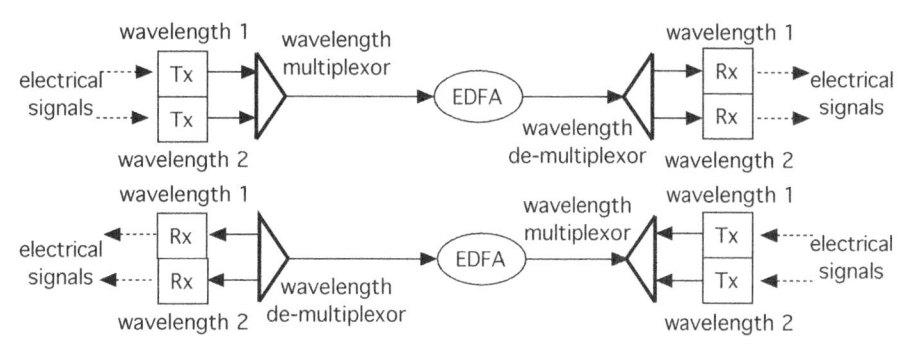

Figure 1-6: Stage 4- Increased Capacity And Transmission Distance

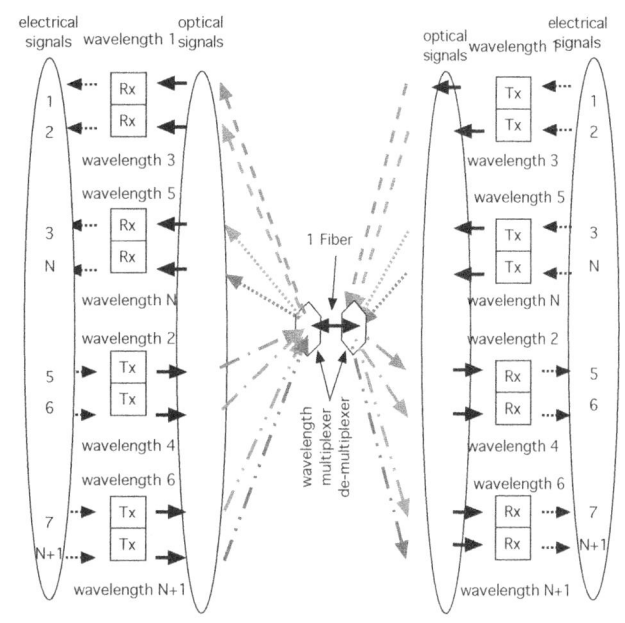

Figure 1-7: Full Duplex Single Fiber DWDM

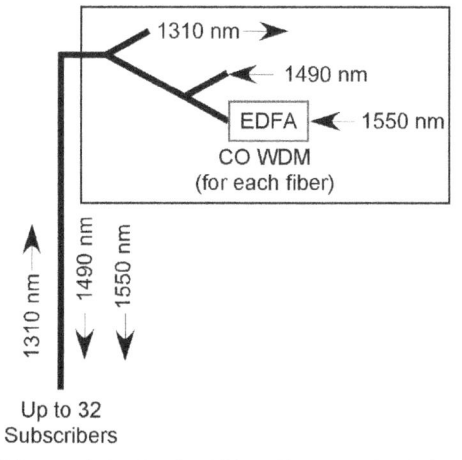

Figure 1-8: Optical Line Terminal (OLT) For Bi-Directional Transmission In A PON

1.3.6 STAGE 6

In Stage 6, the spacing of adjacent wavelengths was increased to 20 nm to allow use of reduced cost, uncooled lasers. This technology, known as coarse wavelength division multiplexing (CWDM) is used in metropolitan area networks.

After installation, testing of Stage 3, 4, 5, and 6 links is simple: connect test equipment to the ends. Measurement of the power loss at the wavelengths of operation is sufficient to verify proper operation of the optoelectronics.

In the first six stages, optoelectronics are restricted to the ends of the optical path. With such links, all electrical signals were restricted to transmission between the same two end points. Such transmission made testing simple: test equipment placement was at the end points. Testing of links in stages after Stage 6 becomes increasingly complicated.

This restriction required network planners to know the bandwidths required between the end points. As we know, accurate prediction of future bandwidth requirements is nearly impossible.

Such links were inflexible. They did not allow for mid span addition or delivery of signals. The lack of flexibility of point-to-point links led to the development of the next three stages.

1.3.7 STAGE 7

In Stage 7, mid-span, fixed wavelength division add/drop multiplexors (ADMs) enable a single optical path to have multiple end points (Figure 1-9). Such configurations increase design flexibility. However, fixed ADMs still required the planners to know the future bandwidth requirements of each wavelength path.

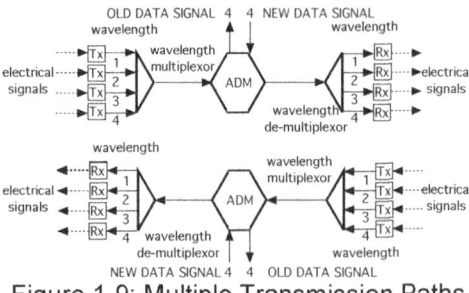

Figure 1-9: Multiple Transmission Paths With Optical ADM

1.3.8 STAGE 8

In Stage 8, the reconfigurable optical add/drop multiplexor (ROADM) enables changing the capacity to meet the demand. The use of tunable wavelength lasers in combination with ROADMs increased this flexibility (Figure 1-10). In the extreme, this combination enables transmission of (almost) any bandwidth between any two locations in a network! This is an ideal goal, to be certain.

After installation, testing of Stage 3, 4, and 5 links is simple: connect test equipment to the ends at the optoelectronics. Measurement of the power loss at the wavelengths of operation is sufficient to verify proper operation of the optoelectronics.

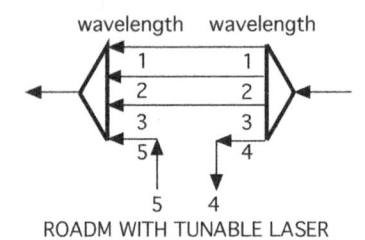

Figure 1-10: Stage 8-Increased Flexibility With ROADM And Tunable Laser

Up through Stage 8, optical transmission had the same two characteristics:

➢ Method of detection
➢ Number of electrical signal bits for each optical bit

The method of detection used in Stages 1-8 was direct detection: at the receiver, the optical signal was directly detected. A 'one' electrical bit corresponded to detection of a 'high' optical power level. A 'zero' electrical bit corresponded to a low optical power level. In other words, the amplitude of the optical signal determines the information. This method is known as intensity modulation (IM or AM), return to zero (RZ), non-return to zero (NRZ), and on/off keying (OOK).

Direct detection was sufficient up to 25 Gbps.[4] Above this optical data rate, direction detection becomes difficult due to the increased signal distortion that results from chromatic dispersion and polarization mode dispersion (2.3.3). These dispersions limit transmission distance and/or bandwidth.[5]

To meet the need for increased bandwidth over existing link lengths, the fiber optic industry developed a second method of detection, coherent detection. This method was common in the radio industry but too expensive for use in fiber optic communication below 25 Gbps.

---

[4] Some fiber optic systems use direct detection at 40 Gbps.

[5] To avoid confusion, we use the commonly-used term 'bandwidth'. However, the correct term is the digital term 'bit-rate'.

1.3.9 STAGE 9

In Stage 9, coherent detection, the transmitted optical signal is combined with an optical signal created by the receiver. The receiver recovers the transmitted optical signal by analyzing the combined signal.

The second characteristic was the number of electrical signal bits transmitted by an optical pulse. Up through Stage 8, five optical bits translated to four electrical signal bits. After Stage 8, the optical signal was modulated such that each optical pulse carried more than one electrical signal bit. With such modulation, each optical bit carries up to 8 electrical data bits. By this method, the optical pulse rate was reduced by the number of bits transmitted by a single optical pulse. This reduction in optical pulse rate reduced the amount of optical dispersion. If effect, this method compensates for the signal distortion that results from dispersion.

For example, a coherent 100 Gbps link with each optical pulse transmitting 8 electrical bits can operate at an optical pulse rate of 12.5 Gbps. The amount of dispersion allowed at this optical rate is greater than that allowed at an optical rate of 100 Gbps. This reduction in the level of dispersion allows an increase in transmission distance.

In Stage 9, the intensity of the optical pulse is not modulated. Instead the polarization and phase of the pulse are modulated. Recovery of the electrical signals requires high-speed digital signal processing (DSP) of the characteristics of the optical pulse at the receiver. Thus, a single wavelength can carry 100 Gbps. When a link transports multiple wavelengths, the capacity of a single fiber extends to more than one terabit per second (Tbps).

1.4 TOPOLOGIES

The links in Figure 1-1 to Figure 1-7 are arranged to form a network. The physical arrangement of the links defines the **physical topology**. The manner in which the data move through the network defines the **logical topology**. These two topologies need not be the same.

There are four topologies:

- Linear, or bus
- Ring
- Star
- Mesh

By considering convenience or cost, the network designer chooses the physical topology. The protocol, or standard, such as Ethernet, the dominant data network standard, Synchronous Optical Network (SONET), a telephone network standard, determines the logical topology.

Some standards allow multiple logical topologies. SONET, SDH and Ethernet are examples. In data networks, **the star physical topology** is most common. When multiple stars exist in the same network, the network forms a **hierarchical star topology** (Figure 1-11).

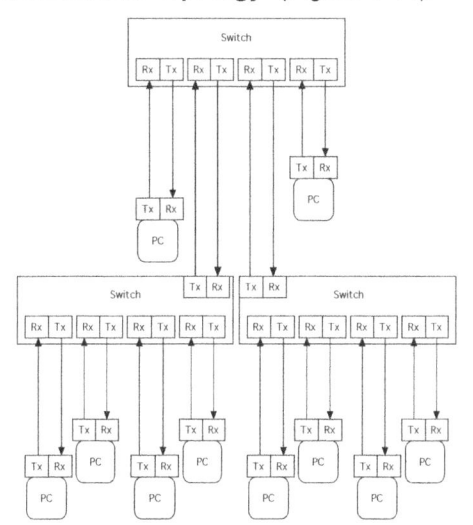

Figure 1-11: A Hierarchical Logical Star, Physical Star Network

Topologies are concerns of the designer. As long as the designer defines the cable arrangement, topologies are of little concern to the installer.

1.5 LINK COMPONENTS

As indicated in previous sections of this chapter, links contain the following components, which we present in the chapters indicated:

- Fiber (Chapter 3)
- Cable (Chapters 4, 11, 21)
- Connectors (Chapters 5, 12, 20, 22-25)

- Splices (Chapter 6, 13, 26)
- Passive devices (Chapter 7)
- Optoelectronics (Chapter 8)

In addition, all networks contain hardware, which we present in Chapter 9. The installer needs to plan the installation, which we present in Chapter 10.

The installed links need testing, which we present in Chapters 14-18. Finally, the installer needs to interpret the test results, which we present in Chapter 19.

Of course, none of these chapters would make sense without a fundamental understanding of light, which we present in Chapter 2.

1.6 SUMMARY

Networks enable communication between computers and peripherals. Optical networks have advantages over networks using other media. The optical network consists of a series of links. Each link consists of fibers enclosed in a protective cable structure, connectors, splices, hardware, and optoelectronics. In some networks, the links include passive devices. The links are arranged to create a physical topology and to support a logical topology.

1.7 REVIEW QUESTIONS

1. To the installer, what is the most important term?

2. What does 'full duplex' mean?

3. How may advantages does optical fiber transmission offer?

4. What are these advantages?

5. What three characteristics enable long distance transmission?

6. Define WDM

7. True of false: data transmission standards require analog transmission.

8. In a fiber data communication network, each node is serviced with ___ fiber(s).

9. What is a SAN?

10. For what does SONET stand?

11. What is the designation for a single delivery of voice, video and data services?

12. What does PON mean?

13. How many wavelengths are used in a FTTH/PON network?

14. What are the FTTH/PON wavelengths?

15. What is a physical topology?

16. What is a logical topology?

17. What does an optical amplifier do?

18. Why is an optical amplifier considered a passive device?

2 A LIGHT OVERVIEW

Chapter Objectives: you learn the language of light. This language consists of properties, behaviors, and numbers. With this language, you will understand the behavior of light in fibers and connectors and the significance of the different types of fiber (3).

2.1 INTRODUCTION

As it relates to fiber optic communications, light has properties and behaviors, both of which properties determine the capabilities and limitations of fiber transmission.

Light can be described by at least four different concepts:

- Rays
- Particles
- Waves
- Energy fields

Light is described as a 'ray', since it travels in a straight line. Light is described as a particle, since light reflects in a manner similar to that in which a particle bounces. Light is described as a wave in that it exhibits a periodic nature like that of ocean waves.

Waves carry energy, as we all know from being in the water during a windy day. This energy exists within a volume. These concepts of light as a wave and as an energy field become important in understanding the behavior of light in single-mode fibers.

2.2 PROPERTIES

In this section, we examine eight properties of light. These properties are:

- Wavelength
- Spectral width
- Speed
- Power
- Volume
- Pulse width
- Polarization
- Phase

2.2.1 WAVELENGTH

When we think about light, we use the term 'color'. We translate 'color' into the technical term, which is 'wavelength.'

A stone dropped in a pond creates ripples with a circular form. The distance from peak to peak, or from trough to trough, is the 'wavelength' of the ripples. Similarly, light exhibits a periodic wave-like nature. The wavelength of light, λ, is the measure, in nanometers (nm), of the distance between successive peaks, or between successive troughs.

The wavelength of determines the light behavior in a fiber. Specifically, the wavelength of light determines the two fiber behaviors of dispersion (3.5.1) and attenuation (3.5.2). In addition, the wavelength determines the mode of transmission.

Common wavelengths range from 780 nm to 1625 nm (Table 2-1). While some multimode Fiber Channel links operate at 780 nm, this wavelength is not common in other applications. 1625 nm is used as a testing wavelength and for out-of-band status monitoring. As a practical matter, common transmission wavelengths range from 850 nm to above 1550 nm.

	Multi-	mode	Single-	mode
Data	850 780	1300	1310	
CATV			1310	1550
Telephone			1310	1550
DWDM CWDM			1310-	1550 1625
WDM	850	1300	1310	1550
FTTH/PON			1310	1490 1550

Table 2-1: Common Communication Wavelengths

Because wavelength determines behavior, the installer must know the wavelength at which the link will operate. The installer will test at this wavelength so that testing simulates, as closely as possible, the operation of the optoelectronics on the cable ends. In the telephone industry, wavelength ranges are referred to by the term 'band' (Table 2-2).

© PEARSON TECHNOLOGIES INC.

Band	Descriptions	Wavelength Range, nm
O	Original	1260-1360
E	Extended	1360-1460
S	Short	1460-1530
C	Conventional	1530-1565
L	Long	1565-1625
U	Ultra long	1625-1675

Table 2-2: Telephone Wavelength Bands

2.2.2 SPECTRAL WIDTH

When we use the term 'wavelength', we imply, incorrectly, that light has a single wavelength. In almost all cases, the opposite is true: the light in fiber communication systems consists of more than one wavelength. These wavelengths occupy a range centered on a 'central', or 'peak' wavelength (Figure 2-1). The measure, in nanometers (nm), of this range is the 'spectral width' of the light.

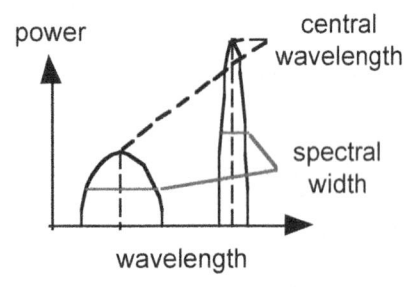

Figure 2-1: Central Wavelength and Spectral Width

Because each wavelength travels at a slightly different speed, spectral width is one factor that determines the accuracy with which the fiber will transmit or the maximum distance to which a fiber can transmit an accurate signal. Spectral width is specified as a 'full width, half maximum (FWHM)' value. This value is the range of wavelengths at 50% of the maximum power of the wavelength-power chart (Figure 2-1).

2.2.3 LIGHT SPEED

In school, we learned about the speed of light: the speed of light, c, is the speed at which light travels in a vacuum.[6] However, the speed of light in any material is less than that in a vacuum. The technical term for speed of light in a material is 'index of refraction,' the 'refractive index', and group index. This term is abbreviated as RI, IR, and η.[7]

The RI is defined in Equation 2-1.

RI= (speed of light in vacuum/speed of light in a material)

Equation 2-1

With this definition, the RI is always greater than 1. The RI of optical fibers ranges from approximately 1.46 to 1.52 (26.1). The installer uses the RI to calibrate the OTDR. With calibration, the OTDR provides accurate fiber length and attenuation rate measurements (15.8.1).

2.2.4 OPTICAL POWER

Light has intensity, also known as power. For example, when a strobe light is flashed at us, we blink because of the high intensity. However, when a traffic signal changes color, we do not blink, because the low intensity does not overload the receptors in our eyes.

In fiber optics, we measure two types of power:

➢ Absolute power
➢ Relative power

When we measure absolute optical power, we use two terms: milliwatts and dBm. A milliwatt is one one-thousandth of a watt. A dBm is a power level relative to one milliwatt.

dBm is defined in Equation 2-2.

dBm= 10 log (power level/one milliwatt)

Equation 2-2

For example:

1 milliwatt = 0 dBm
10 milliwatt = +10 dBm
One tenth of a milliwatt = –10 dBm

We use the terms milliwatt or dBm in three ways:

➢ As the power level launched into a fiber by a transmitter
➢ As the power at the receiver
➢ As the power level required by the receiver for proper functioning

---

[6] c= 2.994 x10^8 m/sec.

[7] η is the Greek letter 'eta'.'

When we measure relative optical power, we use the term 'dB'. A dB is defined in Equation 2-3:

dB= 10 log (new power level/arbitrary reference power level)

Equation 2-3

We use the term dB to indicate:

- Power loss of a component in a link
- Total loss in a fiber link
- Maximum loss that can occur between a properly functioning transmitter-receiver pair

For example:

- Connectors are specified with a maximum loss of 0.75 dB/pair (5.4.2).
- Transmitter-receiver pairs can be specified with a maximum allowable loss of 12 dB (Table 8-1).

2.2.5 VOLUME

In the case of singlemode transmission (3.4.3), light is best described as an energy field. This field occupies a volume. This field is known as the 'mode field'. This volume defines the region within which *most* of the energy travels in a singlemode fiber.

2.2.6 PULSE WIDTH

In a transmission system, each optical pulse in a fiber has width. This width results from the time required to turn a light source on and off. In a communication system, this width is less than the time interval of each bit. For example, in a one-gigabit transmission system, the pulse at the receiver must have a width of less than 10^{-9} seconds.

In OTDR testing, the term 'pulse width' has a second meaning. It is the length of time that the OTDR laser transmits power into the fiber (15.8.1). During testing with an OTDR, the pulse width determines the length of fiber that can be accurately tested. For example, testing of a short fiber requires a short pulse width; that of a long fiber requires a long pulse width.

2.2.7 POLARIZATION

The energy field of light has two axes, the E, or electric field, and the H, or magnetic field. These fields are perpendicular to each other (Figure 2-2). These two fields can have different field intensities and can move at different speeds. When these two field intensities are independently modulated, the optical pulse can transmit increased capacity. When the two fields travel at different speeds, polarization mode dispersion (PMD) results (2.3.3.5).

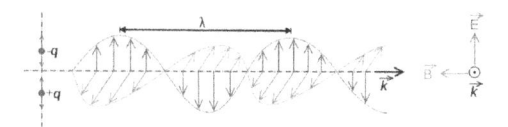

Figure 2-2: E And H Light Energy Fields[8]

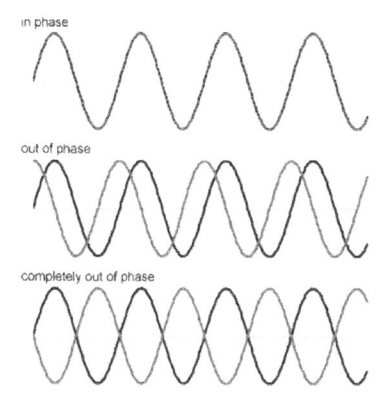

Figure 2-3: In Phase (Top) And Out Of Phase Waves[9]

2.2.8 PHASE

Because light has a wave-like nature, it has a phase. An optical signal can be 'in phase' or 'out of phase' (Figure 2-3) with itself. The phase of a signal can be modulated. When modulated, the phase can carry information, enabling the optical pulse to transmit increased capacity.

2.3 BEHAVIOR

Light has five behaviors:

- Reflection
- Refraction
- Dispersion

---

[8] (Courtesy www.qed.co.uk)
[9] (Courtesy http://mcat-review.org)

- Attenuation
- Skew

2.3.1 REFLECTION

Two types of reflections occur in our daily life:

- We see the sky reflected in a lake
- We can see ourselves dimly reflected when we look through a closed window

In optical links, these two reflections enable communication and can cause signal inaccuracy. The reflection of the sky in a lake results from a change in the speed of light, or RI, at the air-water boundary. This change results in a total reflection. This reflection occurs as long as the angle of reflection is proper.

2.3.1.1 TOTAL REFLECTION

In the terms of physics, we see the sky as long as the rays of light strike the water within a 'critical angle' (Snell's Law, Figure 2-4). If we look at the water closer and closer to our feet, we stop seeing the sky at some angle.

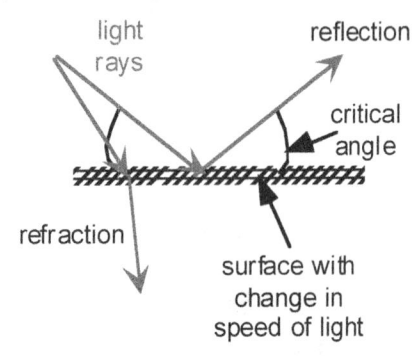

Figure 2-4: Reflection At An Interface

Instead, we see into the water. Temporarily, we call the maximum angle at which the reflection occurs the 'critical angle'. This reflection within a critical angle enables transmission through an optical fiber (3.4.1).

We might be tempted to use the term 'critical angle'. However, the technical term for critical angle is 'numerical aperture' (NA). The NA, a measure of the critical angle, is defined in Equation 2-4.

NA= sine (critical angle)

Equation 2-4

Since the sine of an angle, in degrees, is a dimensionless number, the NA is dimensionless (Table 2-3).

Critical Angle, °	NA
8.05	0.14
11.53	0.20
15.96	0.275

Table 2-3: Nominal NA Values

2.3.1.2 PARTIAL REFLECTION

When we see a dim reflection on the surface of a closed window, we are seeing a partial, or 'Fresnel' reflection. A Fresnel reflection can occur any time light moves from one medium to another. More precisely, such a reflection will occur if the RI in the two media, air and glass, are different. Such reflections are important in fiber communication systems because connectors and splices create locations at which the speed of light can change. Excessive connector and splice reflections can occur, resulting in signal transmission errors (5.4.5).

2.3.2 REFRACTION

We have all seen refraction, or bending, of light. When we look at a pen in a glass of water from the side, the pen appears to be bent (Figure 2-5). This bending occurs whenever light moves from a material with one RI to a material a different RI. This bending becomes important in understanding the functioning of graded index multimode fibers (3.4.2).

Refraction obeys a simple rule: the bending will be towards the lower speed region. Stated differently, the bending will be towards the higher RI region.

Figure 2-5: Refraction of Light

2.3.3 DISPERSION

Dispersion is also known as 'pulse spreading' and 'pulse broadening. Dispersion is the delay of some of the optical power in a pulse at the output end of the fiber. Dispersion refers to the fact that optical power that enters the fiber at the same time does not exit the fiber at the same time.

Dispersion results in an optical pulse width that increases continuously through the fiber.

> ➢ When dispersion exceeds the maximum amount allowed by the data rate, the output signal is different from the input signal.

This difference creates signal inaccuracy. Dispersion is one of the two mechanisms that limit transmission distance. Dispersion limits distance because it determines the accuracy with which a fiber transmits data.

Five types of dispersion occur in data transmission systems:

- Modal dispersion
- Chromatic dispersion
- Material dispersion
- Waveguide dispersion
- Polarization Mode Dispersion

2.3.3.1 MODAL DISPERSION

In a fiber, light can take two types of paths:

- All rays can travel the same path and have same path length (Figure 2-6)
- Rays can travel different paths and have differing path lengths (Figure 2-7)

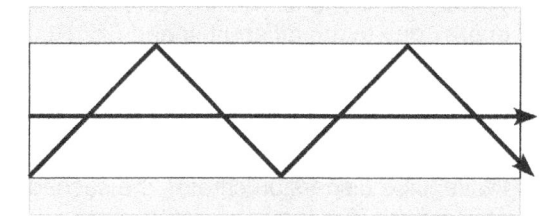

Modal dispersion occurs when the rays travel multiple paths and have multiple path lengths. Since the rays do not travel the same path, different rays will arrive at the fiber end at different times. In short, **modal dispersion is path length dispersion**. Modal dispersion is the largest cause of dispersion. This dispersion occurs in multimode fibers, but not in singlemode fibers.

Figure 2-6: No Modal Dispersion

Figure 2-7: Modal Dispersion

2.3.3.2 CHROMATIC DISPERSION

Chromatic dispersion is the second largest of the five types. Chromatic dispersion occurs because all commercial fiber optic light sources emit a range of wavelengths, known as the spectral width (2.2.2). Each of these wavelengths travels at a slightly different speed. Even if all the rays in a pulse are traveling the same path length, as occurs in a singlemode fiber, rays with different wavelengths arrive at different times at the output end of the fiber. This form of dispersion is active in both multimode and singlemode fibers.

2.3.3.3 MATERIAL DISPERSION

Material dispersion occurs in the core because of the structure of that core. Optical fibers are glass fibers. Glass fibers are amorphous materials. Amorphous materials have no crystal structure, i.e., no constant and repeating atomic level structure. Two rays of light having the same wavelength and traveling the same path length can travel in differ-ent regions of the core. Because of this amorphous structure, these different regions will not have the same exact atomic level composition and the same RI. Atomic level differences in composition in the core result in differences in speed, just as the differences between the compositions in the GI core result in differences in speeds. This form of dispersion occurs in both multimode and singlemode fibers.

Material dispersion creates one of the three fundamental limits on bandwidth and transmission distance. We cannot eliminate or compensate for material dispersion, as we can for modal and chromatic dispersions.

2.3.3.4 WAVEGUIDE DISPERSION

Waveguide dispersion is the spreading of energy due to the different speeds, or RIs, in the core and the cladding. This form of dispersion is most important in single-mode fibers, since a significant amount of energy travels in the cladding (3.4.3). Waveguide dispersion creates the second of the three fundamental limits on bandwidth and transmission distance. We cannot eliminate or compensate for waveguide dispersion, as we can for modal and chromatic dispersions.

2.3.3.5 POLARIZATION MODE DISPERSION

When light is described as an energy wave or field, it has two orthogonal axes, the E, or electromotive force, and the H, or magneto motive force. When the energy in these two axes travels at different speeds in a fiber, polarization mode dispersion (PMD) results. PMD becomes important above 2.5 Gbps at long transmission distances.

Since these rates and distances are well above those in data networks, PMD is not a concern in such networks. However, PMD can be critical for telephone, CATV, and long distance networks. We cannot eliminate or compensate for PMD, as we can for modal and chromatic dispersions. As a practical matter, PMD is more important on 'old' fibers that were installed 15 or more years ago than on fibers installed after approximately 2000.

2.3.3.6 DISPERSION AND SIGNAL ACCURACY

These five types of dispersion create the total pulse dispersion. If we send a single pulse down the fiber, dispersion would be irrelevant to accuracy. However, we send multiple pulses, each within its own time interval. Each of these pulses spreads. If multiple pulses spread sufficiently so that sufficient energy from a pulse arrives at the fiber end in a time interval not its own, the input and output signals will not be the same. In other words, we will experience signal inaccuracy (Figure 2-8).

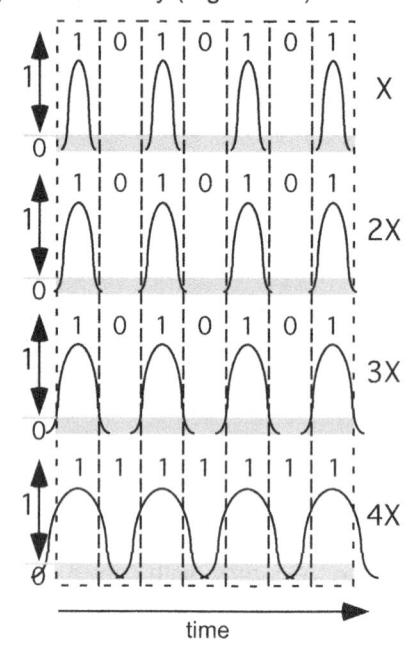

Figure 2-8: Pulse Dispersion Causes Signal Inaccuracy

In Figure 2-8, we examine pulses digital transmission. To simplify this figure, we ignore attenuation. In digital transmission systems, we have a threshold power level. For power levels below the threshold, we call the signal 'zero'. For power levels above the threshold, we call the signal 'one'. This threshold may be at zero power or at some level above zero. In systems that include optical amplifiers, the '0' power level is always above zero.

In Figure 2-8, we transmit an alternating string of ones and zeros. Each of these ones (or zeroes) occurs within a time interval. The size of this interval is determined by factors that are too complex for this book. Each of these ones is comprised of multiple rays of light that enter the fiber at the same time. Each of these ones has a range of wavelengths. Each of these ones will experience material and waveguide dispersion. In multimode fibers, the pulses experience modal dispersion.

Due to dispersion, the rays exit the fiber over a time range. The first rays to arrive at the fiber output end are represented by the leading edge of each pulse; the last rays, by the trailing edge (Figure 2-8). The time difference between the leading and trailing edges is the pulse width.

As the transmission distance increases, so does the pulse width (Figure 2-8, 2X). As pulse width increases, energy from one pulse arrives at the fiber end in the time interval of an adjacent pulse (Figure 2-8, 3X). With excessive width, an input zero becomes an output one (Figure 2-8, 4X).

The time interval for each bit at the end and the amount of dispersion that can occur both shrink as the bit rate increases (Figure 2-9).

In spite of the importance of pulse dispersion, the installer need not be concerned. The installer can affect none of the seven factors that control dispersion:

- the size of the core
- the manner in which light enters the core
- the fiber NA
- the spectral width of the transmitter
- atomic level variation of the composition of the core
- waveguide dispersion
- the length of the link.

However, he needs to recognize that dispersion limits distance. This concern occurs during troubleshooting of high bit error rates (BER). If the power loss tests indicate proper installation and operation, the problem may be due to excessive distance.

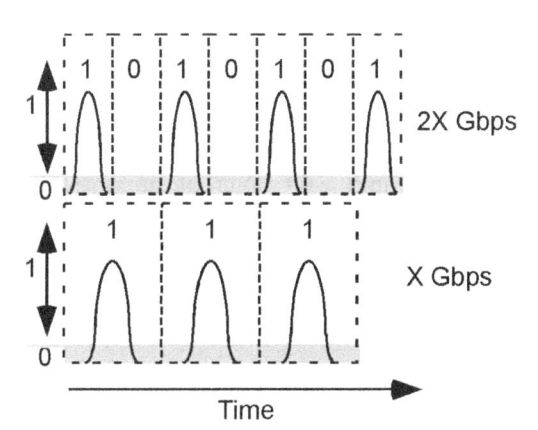

Figure 2-9: Time Interval Vs. Bit Rate

As an example, imagine links from two locations to the same location (A to C and B to C). Image that someone decides to transmit from location A to location C by cross connection at location B. While the indivi-dual links, A to B and C to B, may be sufficiently short , the link A to C may have excessive dispersion.

2.3.3.7 DISTANCE LIMITATION

In general, dispersion limits the transmission distance on multimode fibers before power loss does. Again, in general, attenuation limits the distance of single-mode fibers before dispersion does. Table 2-5 to Table 2-10 have distance limits.

	Wavelength	OM-x	Distance, m
100BASE-F	1300	OM1	2000
100BASE-SX	850	OM1	300

Table 2-4: Fast Ethernet Distances

OM-x	Distance, m
OM1	275
OM2	550
OM3	800
OM4	1100

Table 2-5: 850 nm Gigabit Ethernet (1000BASE-SX) Distances

Wavelength	OM-x	Distance, m
1300	OM1	550
1300	OM2	550
1300	OM3	550
1300	OM4	550
1310	SM	5000
1310	SM	10000
1550	SM	~70000

Table 2-6: 1300 Gigabit Ethernet (1000BASE-LX) Distances

OM-x	Distance, m
OM1	33
50μ, 400 MHz-km	66
OM2	82
OM3	300
OM4	550

Table 2-7: 850 nm 10 Gigabit Ethernet (10GBASE-SX) Distances[10]

---

[10] Based on link with two connector pairs.

Designation	OM-x	Distance, m
10GBASE-LRM	OM1	220
10GBASE-LRM	OM2	220
10GBASE-LRM	OM3	220
10GBASE-LRM	OM4	220
10GBASE-LX4	OM1	300
10GBASE-LX4	OM2	300
10GBASE-LX4	OM3	300
10GBASE-LX4	OM4	300

Table 2-8: 1300 nm 10 Gigabit Ethernet Distances

Designation	OM-x[11]	Distance
40GBASE-SR4	OM3	100
40GBASE-SR4	OM4	125
100GBASE-SR10	OM3	100
100GBASE-SR10	OM4	125

Table 2-9: 850 nm, 40 and 100 Gigabit Distances

Data Rate	OM-x	Distance, m
4 Gbps	OM1	70
	OM2	150
	OM3	380
	OM4	400
8 Gbps	OM1	21
	OM2	50
	OM3	150
	OM4	200
16 Gbps	OM1	15
	OM2	35
	OM3	100
	OM4	130

Table 2-10: 850 nm Fiber Channel Distances

2.3.4 ATTENUATION

Attenuation is the power loss that occurs when an optical signal travels through a fiber. Attenuation is the second of the two mechanisms that limits the transmission distance: if attenuation in the link is excessive, the power level at the receiver is insufficient. With insufficient power, the receiver cannot accurately convert the optical signal to an electrical signal. Attenuation is stated in units of dB. Attenuation rate of fiber is stated in units of dB/km. When referring to fiber, it is common to use dB/km.

2.3.5 SKEW

Skew can be compared to dispersion, as it represents differences in travel time. However, skew is the difference in travel time, in nanoseconds (ns), of optical signals traveling on different fibers. This difference is important in multimode fiber transmitting 40 and 100 Gbps Ethernet.

In these two Ethernets, the electrical signal is de-multiplexed into four or ten, respectively, 10 Gbps data streams (Figure 2-10). Each stream travels on a separate fiber. Skew must be limited so that the de-multiplexed data streams can be accurately multiplexed at the receiver. For these two data rates, the maximum skew is 79 ns. Such multimode links are used in data centers, in which the transmission distance is short.

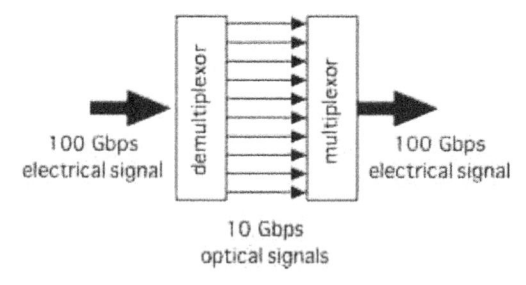

Figure 2-10: Multimode 100 Gbps Transmission

2.4 SUMMARY

Light has characteristics, with terms and units of measure. We use wavelength and spectral width with units of nanometers (nm); power, with units of dBm or dB; speed of light in an optical material with the dimensionless refractive index (RI); and critical angle with the dimensionless numerical aperture (NA).

We know that light reflects and refracts, has behavior that can be described with the terms rays, particles, waves, and energy fields; and that optical power

---

[11] See Section 3.4.2.2

disperses and is reduced as it travels through a fiber.

With these basic concepts, you will easily understand the behavior of light in optical fibers (3).

2.5 REVIEW QUESTIONS

1. What are the eight properties of light?
2. What are common multimode wavelengths?
3. What are common singlemode wavelengths?
4. What are the units of wavelength?
5. How many wavelengths does the typical transmitter emit?
6. What are the six wavelength designations used for telephone systems?
7. Why does the installer need to know the wavelength?
8. What are the five behaviors of light?
9. What are the units of absolute power measurement?
10. What are the units of relative power measurement?
11. How many types of reflection are important in a fiber optic link?
12. What are the types in Question 11?
13. Where does each type in Question 12 occur?
14. Express the simple rule of refraction in terms of speed.
15. Express the simple rule of refraction in terms of RI.
16. What is dispersion?
17. Why is dispersion important?
18. How many forms of dispersion are there?
19. What is the largest form of dispersion?
20. What is the second largest form of dispersion?
21. If dispersion is excessive, what happens to the signal?
22. What can result in excessive dispersion?
23. What is the technical term for critical angle?
24. What are the units of measure for this term for critical angle?
25. What are the units of measure for the index of refraction?
26. Where is a Fresnel reflection important?
27. Express 100 mW in terms of dBm.
28. Express 1000 μW in terms of dBm.
29. The transmitter of a link launches 1 mW into a fiber. The receiver requires at least 100 μW for accurate operation. Express, in dB, the optical power budget for this transmitter-receiver pair.
30. What is skew?
31. Where is skew important?
32. On what type of fiber is skew important?
33. At what data rates is skew important?
34. What is the benefit of modulating the phase of the optical signal?
35. What is the benefit of modulating the polarization of the optical signal?

3 FIBERS

Chapter Objectives: you learn the types of fiber, the advantages of these types, the language of fibers, and the numbers that describe fibers and their performance.

3.1 INTRODUCTION

In this chapter, we examine the following aspects of optical fibers :

- Function
- Structure
- Types
- Performance

3.2 FUNCTION

The fiber is the medium through which light travels. The function of the fiber is guidance of the light between the transmitter and receiver with minimum signal distortion. Minimum signal distortion means no difference between the input and output signals and minimum power loss.

3.3 STRUCTURE

The fiber provides this function through its structure. The structure consists of at least two, but usually three, regions. These regions are the core, the cladding, and the primary coating (Figure 3-1 and Figure 3-2).

A= core= 8.2-62.5 µm

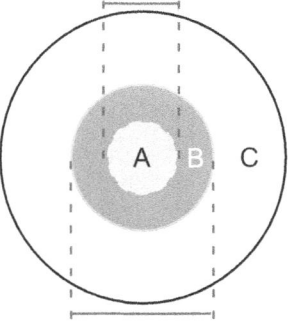

B=cladding= 125 µm
C= primary coating

Figure 3-1: Fiber Structure

The core is the central region of the fiber in which *most* of the light energy travels. The cladding surrounds the core, confines the light to the core, and increases the fiber size so that the fiber is easily handled. The primary coating, formerly called the buffer coating, protects the cladding from mechanical and chemical surface damage. The installer removes, or strips, this coating for connector installation and splicing.

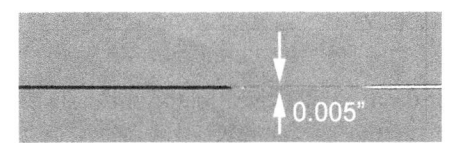

Figure 3-2: An Optical Fiber With Darkened Primary Coating

A common misconception is that the coating provides strength to the fiber. The reality is that the coating protects the cladding from damage, thus allowing the fiber to retain its intrinsic high strength. Any strength the primary coating provides is so low as to be insignificant.

3.3.1 DIAMETERS

These three regions are characterized by their diameters, stated in microns (µ). As a matter of designation, fibers are referred to by their core diameter, cladding diameter, and, occasionally, by their coating diameter. For instance, a fiber might be called a 50/125 fiber or a 50/125/250 fiber. Occasionally, the core and cladding diameters are reversed, as in 125/50.

In telephone and CATV networks, the core diameter is 8-10µ. In data networks, the core diameter can range from 8-10 µ to 50 µ.

Beginning in 1985, data network, or short-haul LAN, standards specified the fiber core diameter of 62.5 µ. As a result, this fiber is found in many data networks operating at data rates below 1 Gbps.

The last four of the multimode data standards indicate a preferred use of 50µ fibers. This preference is due to its increased transmission distance. These standards are:

- 1000BASE-SX
- 10GBASE-SX
- 40GBASE-SX
- 100GBASE-SX

Each fiber has an NA. As a practical matter, the core diameter and the NA are related (Table 3-1).

Core Diameter, μ	NA
8.2-10	0.14
50	0.20
62.5	0.275

Table 3-1: Nominal Fiber NA Values

For data, CATV, and telephone networks, the cladding diameter is always 125 μ. Fibers for other applications have cladding diameters between 140 and 1,000 μ.

The primary coating diameter is approximately 245μ but can be as high as 500μ. It is common to refer to the primary coating diameter as 250μ, which is its diameter when it is color-coded. As a practical matter, the installer rarely need concern himself with the diameter of the primary coating.

At the time of this writing, there are expectations that fibers with a reduced primary coating diameter will become available. The motivation for this reduction is a doubling of the number of fibers in the same cable diameter. This reduced diameter is expected to be 200μ.

Most fibers have an ultra violet (UV)-cured acrylate primary coating. This coating has two layers, a soft, low modulus inner layer and a hard, high modulus outer layer.

3.3.2 TOLERANCES

The terms 'core diameter' and 'cladding diameter' imply a constant value along the fiber. It is unrealistic to expect such constancy. Excessive variation in the core diameters of two connected fibers can result in excess power optical loss.

Excessive variation in the cladding diameters of two connected fibers can result in either excess loss or in inability to fit the fiber into a connection. If one of the connected fibers is has a small cladding, the cores may not align. In this case, connected fibers will exhibit high power loss. If one of the fibers has an excessively large cladding diameter, it may not fit into a connector or mechanical splice.

Knowledge of the core and cladding diameter tolerances can be useful during troubleshooting lack of fit or excessive connection loss. The Building Wiring Standard, TIA/EIA-568-C, provides these tolerances though reference to ICEA S-83-596-2001, an international cable standard (Table 3-2).

Diameter	Singlemode	Multimode
Core	± 0.5, 1310 nm ± 0.7, 1550 nm	± 3, 850 nm ± 3, 1300 nm
Cladding	± 1 μ	± 2 μ

Table 3-2: Fiber Diameter Tolerances

3.3.3 OFFSET AND NON CIRCULARITY

The terms 'diameter' and 'core centered in cladding' imply a perfectly circular fiber and perfectly centered core. Reality is not 'perfect'. The fiber can be slightly non-circular and the core can be slightly off center in the cladding. Both lack of perfections result in the possibility of excess power loss at splices and connectors. This excess power loss can occur due to imperfect alignment of cores (Figure 3-3 and Figure 3-4).

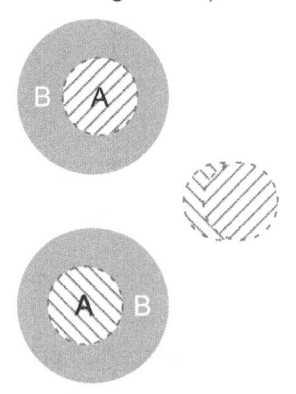

Figure 3-3: Core Offset And Excess Loss

To limit excess power loss, the fiber needs to meet limitations on:

- Core offset, also called core-cladding non-concentricity

➢ Cladding non-circularity, aka 'ovality'

Knowledge of these two characteristics may become important during troubleshooting of excess connection loss that cannot be explained by any other reason. Fibers with low offset and non-circularity (Table 3-3) do not exhibit excess loss. These values come from ICEA S-83-596-2001.

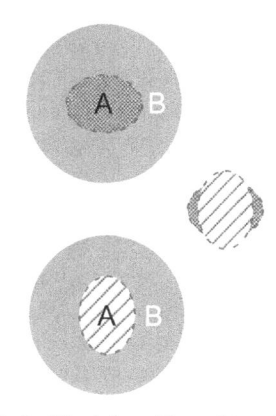

Figure 3-4: Cladding Non-Circularity and Loss

Fiber	Wavelength	Ovality	Offset, µ
Singlemode	1310 nm	≤ 1%	≤ 0.6
Singlemode	1550 nm	≤ 1%	≤ 0.6
Multimode		≤ 2%	≤ 3.0

Table 3-3: Offset And Ovality Values

3.4 TYPES

The core and cladding can be either glass or plastic. Fibers can have:

➢ Glass cores and glass cladding
➢ Plastic cores and plastic cladding
➢ Glass cores and plastic cladding.

Since fibers used in data, CATV, and telephone networks have glass cores and glass claddings, we shall focus this book on these 'all-glass' fibers. For all-glass fibers, the core and cladding are fused and inseparable.

➢ In other words, **you cannot strip the cladding from the fiber**. You can strip the primary coating!

All glass communication fibers are of three types. The type depends upon the core diameter and structure. The three types are:

➢ Multimode step index
➢ Multimode graded index
➢ Singlemode

The term 'mode' can be crudely translated to mean 'path'. The term 'multimode' indicates that rays of light can travel multiple paths in the core (Figure 2-7). The term 'singlemode' indicates that all rays of light behave as though they travel a single path (Figure 2-6).

3.4.1 MULTIMODE SI

The first optical fiber was a large core, multimode, step index (SI), plastic optical fiber (POF). To the time of this writing, most POF fibers are step index. Such fibers are used in consumer devices, such as digital audio links, and in some automotive applications.

Step index fibers have a core with a single chemical composition. Because of this single composition, the speed of light and RI, are constant (Figure 3-5). Because the composition changes at the core-cladding boundary, rays of light can be reflected (2.3.1).

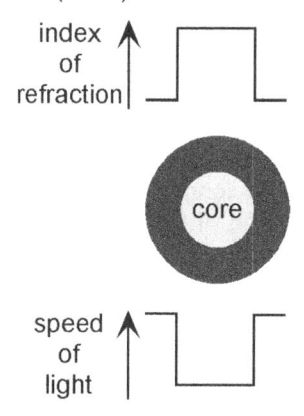

Figure 3-5: Core Profile Of Step Index Fiber

Rays of light that enter the core at an angle less than or equal to the critical angle, (2.3.1), will be reflected at the core boundary and remain in the core (Figure 3-6). Rays of light that enter the core at an angle greater than the critical angle escape into the cladding and do not reach the receiver (Figure 3-7).

Rays of light reflect from at one side of the core to the opposite side. As long as the rays remain within the critical angle defined by the NA, the rays reflect to the opposite side (Figure 3-8). The name of this process is 'total internal reflection.'

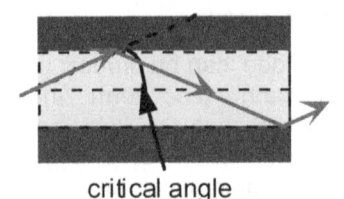

Figure 3-6: Rays Reflect In SI Fiber

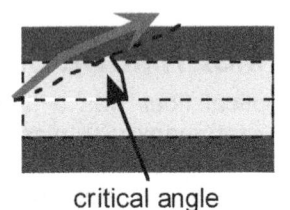

Figure 3-7: Rays Outside Critical Angle Escape

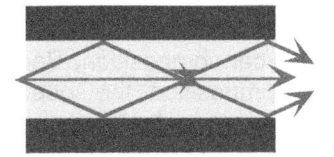

Figure 3-8: Ray Paths In A SI Fiber

In a SI fiber, rays of light in the same pulse can travel both parallel to the axis and at any angle to the axis up to the maximum angle defined by the NA. While these two rays enter the fiber at the same time, they travel different paths, have different path lengths and must arrive at the output end of the fiber at significantly different times. This difference in time of arrival results in a significant amount of modal dispersion (2.3.3). This dispersion results in a SI bandwidth[12] that is too low to be of much use. For this reason, the SI fibers are not used in data networks.

3.4.2 MULTIMODE GI

3.4.2.1 STANDARD GI MULTIMODE

The second type of multimode fiber, a graded index (GI) fiber, has a relatively large core with up to 2500 different chemical compositions. It is difficult and expensive to make, because of the different chemical compositions.

This GI core structure creates two mechanisms that reduce dispersion. The first mechanism is the compensation for the different path lengths. These compositions are chosen so that the speed of light in the center of the core is lowest (Figure 3-9). This speed increases as the distance from the axis increases. With this structure, axial rays at the center travel the shortest path length at the slowest speed. Critical angle rays travel the longest length at the highest average speed. This structure provides compensation for the different path lengths that different rays travel in the multimode core.

This compensation allows rays that travel different path lengths to arrive at approximately the same time. This compensation reduces the dispersion and increases the bandwidth to a level acceptable for use in data networks. For this reason, all the data transmission standards for local area networks (LANs) allow use of GI multimode fibers.

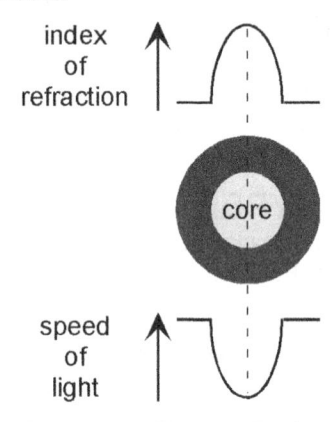

Figure 3-9: GI Core Profile

The second method by which GI fibers reduce dispersion is refraction. At each boundary between the layers, the rays bend back towards the axis (2.3.2). This bending reduces the difference between the path lengths. Thus, light in a graded index multimode fiber tends to travel in what appear to be 'curved' paths (Figure 3-10).

---

[12] The correct technical term is bit-rate, since data systems transmit digital signals. However, the analog term 'bandwidth' is commonly used to mean digital bit-rate. Reluctantly, we use this incorrect term to avoid confusing our readers.

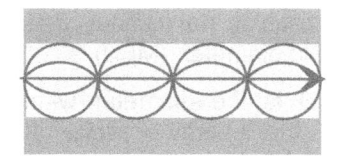

Figure 3-10: Curved Ray GI Paths

The bandwidth, or information carrying capacity, of multimode fibers is indicated by four terms:

- the 'bandwidth distance product' (BWDP)
- 'differential modal dispersion' (DMD)
- the transmission distance possible transmission of 1 Gbps and/or 10 Gbps Ethernet signals
- The OM-x designation

The first term, bandwidth distance product, is measured in units of MHz-km, with typical values from 160 MHz-km to 3700 MHz-km. This term is applicable to fibers that carry light from LEDs (8.2.3). As a complication, we measure BWDP by two methods, the overfilled launch (OFL) and the equilibrium modal bandwidth (EMB).

The BWDP does not accurately correlate with transmission distance for 1 Gbps and/or 10 Gbps Ethernet signals. The differential modal dispersion (DMD) measurement method does. While not commonly stated, DMD is the method by which fiber manufacturers guarantee gigabit transmission distances. More commonly, the transmission distance for gigabit transmission for the fiber is stated.

3.4.2.2 LO GI MULTIMODE

The second term, differential modal dispersion, is applicable to fibers that carry light from VCSELs (8.2.5). Fibers designed for use with VCSELs are 'laser optimized' (LO) fibers. Such fibers have transmission distances greater than those stated in the data standards.

Laser optimized fibers are multimode fibers designed to match the characteristics of VCSELs (8.2.5). VCSELs are the light source used in multimode data networks for data rates from 1 to 100 Gbps.

		OFL		EMB
		850	1300	850
OM1	62.5/125	200	500	
OM2	50/125	500	500	
OM3	50/125	1500	500	2000
OM4	50/125	3500	500	4700

Table 3-4: Bandwidth Distance Products (MHz-km)) For OM-x Designations

LO fibers are identified in two ways:

- By their OMx (optical fiber, multimode) designations (Table 3-4)
- By their transmission distances at 1 and/or 10 Gbps

The designations OM-1 and OM-2 are now standard performance fibers. Such fibers provide for transmission distances as stated in the data standards. OM-3 and OM-4 are laser optimized fibers. Other LO fibers have transmission distances that exceed those stated in the standards.

- Of specific importance to installers is the need to use LO patch cords with LO cables. Connecting standard multimode patch cords to LO cables increases dispersion and reduces transmission distance.

3.4.3 SINGLEMODE

3.4.3.1 WHY SINGLEMODE?

While the GI multimode fiber provides increased bandwidth and transmission distance, it does so at a significant manufacturing cost. This cost results from the 2500 different chemical compositions in 0.001" and the need to maintain those compositions constant along of the fiber. This length can be more than 600 miles! The manufacturing difficulty and the limitations on maximum possible GI BWDP limited the capacity of this fiber type and led to a search for a fiber with increased capacity.

In the 1970's, the clever scientists and engineers working on fiber development found such a fiber. The design of this fiber arose from consideration of the quantum mechanics of light. This consideration indicated that all rays of light would behave as though they were traveling parallel to the axis of the fiber (Figure 3-11). Such propagation resulted when the core was sufficiently small and the wavelength sufficiently long. With all rays traveling the same path, the modal dispersion of multimode fibers would be eliminated. Since light 'rays' in such a fiber would follow a single path, this fiber became known as 'singlemode' in North America and 'monomode', elsewhere.[13]

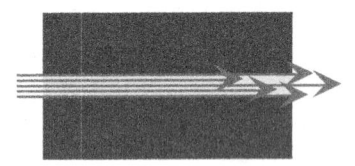

Figure 3-11: Apparent 'Ray' Paths In Singlemode Fiber

3.4.3.2 SINGLEMODE CHARACTERISTICS

Singlemode fibers have seven characteristics:

- Long wavelength operation
- A 'cut off' wavelength
- Very small core diameter
- A 'mode field diameter' (MFD)
- Essentially unlimited bandwidth
- A 'dispersion rate' specification
- A zero dispersion wavelength

Since singlemode operation requires a long wavelength, singlemode communication fibers function at a wavelength of least 1310 nm. Singlemode fibers transmit wavelengths ranging from 1310 to 1625 nm.

Should the wavelength be too low, the singlemode fiber will behave as a multimode fiber. The minimum wave-length at which a fiber behaves as a singlemode fiber is the 'cut-off wave-length'.

The most common singlemode core diameters are 8.2-9.4 µ. However, this characteristic is not critical. Instead, the 'mode field diameter' (MFD) is.

The MFD is the diameter within which most of the energy travels. The MFD (Figure 3-12) is larger than the core by approximately 1 µ. Thus, some of the energy travels in the singlemode cladding. MFDs vary with fiber type and manufacturer (Table 3-5).

core≈ 8.2-10 µm

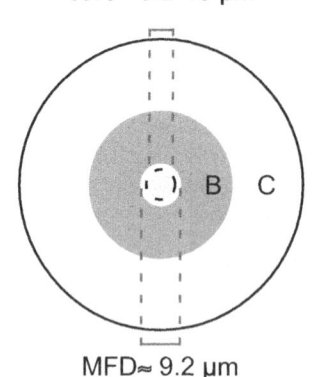

MFD≈ 9.2 µm
B=cladding= 125 µm
C= primary coating

Figure 3-12: Singlemode MFD Diameter

Fiber Type	MFD, µ
OFS Depressed Clad	8.8 ± 0.5 @1310 nm
OFS Matched Clad	9.3 ± 0.5 @1310 nm
OFS True Wave	8.4 ± 0.6 @1550 nm
Corning LEAF ®:	9.2-10.0 @ 1550 nm
Corning SMF-28	9.2 ± 0.6 @ 1310 nm
Corning SMF-28	10.4 ± 0.8 @ 1550 nm
Corning MetroCor	8.1± 0.5 @ 1550

Table 3-5: Mode Field Diameters

To understand this phenomenon, we step away from fiber optics for a minute. Imagine that you are working with a son or daughter on a high school science fair project. The project is to build a model of a singlemode fiber. You have decided to use a piece of pipe to simulate the fiber.

The center of the pipe simulates the core. The wall of the pipe simulates the cladding. You decide to use a ping-pong ball to simulate the photon traveling down the core. You intend to shoot the ping-pong ball straight down the pipe. However, light has properties of an

---

[13] Light in a singlemode fiber does not behave as a ray. Instead, it behaves as an energy field. For this reason, we use the phrase 'behave as though' and put the word 'rays' in quotes.

energy field. At this time, you have not modeled these properties. After additional consideration, you decide to rub the ping-pong ball with a piece of fur or silk in order to create a static charge on the ball. The static charge creates an energy field.

When the ball travels down the pipe close to the inside wall, some of the static field travels in the wall. Analogously, some of the optical energy in a singlemode fiber travels in the cladding of the fiber.

The theoretical bandwidth of a singlemode fiber is 200 Tbps (1.1.1). This high bandwidth makes the singlemode fiber ideal for long haul, or inter-exchange, telephone networks, CATV networks, and fiber to the home (FTTH) networks.

We specify singlemode capacity with the term 'dispersion rate' in units of pico-seconds/nanometer/kilometer (ps/(nm-km)). A maximum dispersion rate for a singlemode fiber at its optimum wavelength is 2.8 ps/(nm-km).

Singlemode fibers have an optimum wavelength for dispersion, known as the 'zero dispersion wavelength' (ZDW). At this wavelength, the waveguide and material dispersions cancel out (Figure 3-13). With such cancellation, the fiber exhibits its minimum dispersion and maximum bandwidth or bit rate. The further the transmitter wavelength is from the ZDW, the larger will be the dispersion rate and the lower will be the bandwidth. Multimode fibers have a ZDW of approximately 1300 nm.

Singlemode fibers have evolved into a series of four types. These types have different dispersion performances and ITU G.65x designations:

> Non Dispersion Shifted (G.652)
> Dispersion Shifted (DS, G.653)
> Dispersion Shifted, Non-Zero Dispersion (DS-NZD, G.655)
> Large Effective Area (LEAF™)

3.4.3.3 SINGLEMODE TYPES

Standard singlemode fibers, known as 'non-dispersion shifted' fibers and G.652, have a ZDW of approximately 1310 nm (Figure 3-13). These fibers will transmit signals at wavelengths above 1310 nm, but the dispersion rate will be higher than that at 1310 nm. As a result, the maximum bandwidth at increased wavelengths will be less than that at 1310 nm. This standard singlemode fiber is used in data communication, telephone, CATV, and FTTH/PON networks. This fiber can be used to transmit two or more wavelengths.

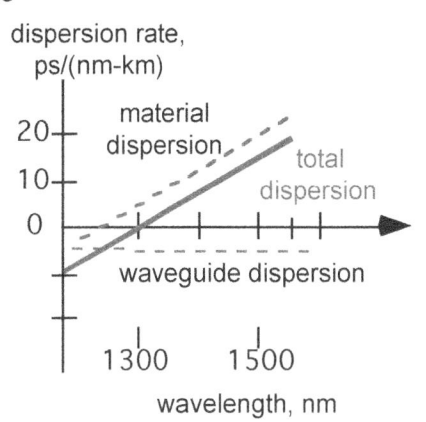

Figure 3-13: Dispersion Rate Vs. Wavelength For G.652 Singlemode Fibers[14]

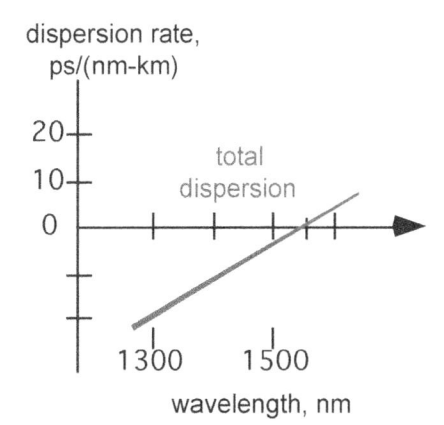

Figure 3-14: Total Dispersion For DS, G.653 Fiber

To provide both low attenuation rate and low dispersion rate, fiber designers created a second-generation singlemode fiber, the 'dispersion shifted' (DS) or G.653 fiber. This fiber exhibits both low attenuation rate and low dispersion rate at 1550 nm through a design change that shifted the ZDW from 1310 nm to 1550

---

[14] After Fiber Optics Handbook For Engineers and Scientists, (1990), Frederick C. Allard, p. 1.38.

nm (Figure 3-14). This fiber behaves in a manner symmetrical to the NDS fiber: it exhibits low dispersion rate at 1550 nm and increased dispersion rate at 1310 nm.

While the DS fiber extended transmission distances, it was best suited for use with only one or two wavelengths. To obtain extremely high aggregate capacity on a single fiber, network designers wanted to transmit multiple, closely-spaced wavelengths. This process is known as dense wavelength division multiplexing (DWDM, 7.2.3.3). When multiple, closely spaced, wavelengths travel on a DS fiber, they interact to create new wavelengths. This interaction creates noise and signal errors in wavelengths other than the two wavelengths that are interacting.

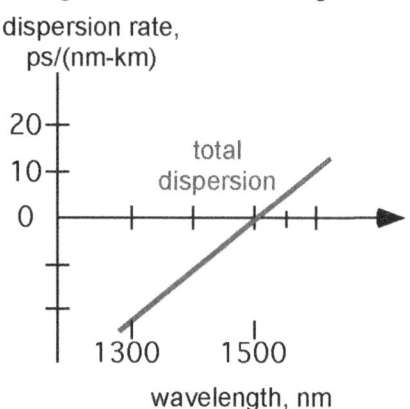

Figure 3-15: G.655 Total Dispersion

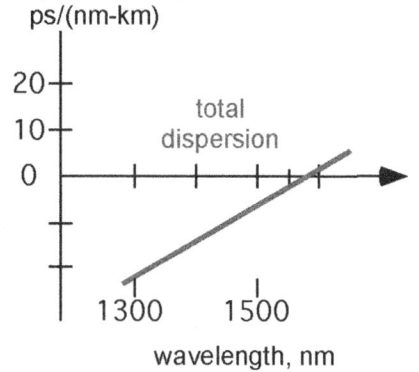

Figure 3-16: G.655 Dispersion

The cause for this interaction is the extremely low dispersion rate at 1550 nm. If the dispersion is low, this interaction becomes significant. The clever fiber engineers developed a solution in the form of a third generation fiber: the dispersion shifted, non zero dispersion fiber (DS-NZD). This fiber has a ZDW close to, but not at, 1550 nm (Figure 3-15 and Figure 3-16). Fibers are available with ZDWs between 1405 and 1560 nm. With this low level of dispersion, the interaction between multiple wavelengths is essentially eliminated. Such a singlemode fiber transmits 200 different wavelengths. Although 200 is the number of wavelengths specified by the ITU, Lucent Technologies has demonstrated the ability to transmit 1000 wavelengths in a single singlemode fiber.

While this new fiber allowed for many wavelengths, each of approximately 10 Gbps and 1 mW of power, it created a new problem. The process of shifting the ZDW reduces the MFD (3.4.3). This reduction, in turn, results in an increased power density, which, in turn, results in non-linear optical effects that create transmission errors.

For example, if 56 wavelengths, each of 1 mW, are launched into a 9.2 µ MFD fiber, and if 90% of the power is within the MFD, the power density is 489,000 watts /square inch. Extremely high power density!

To reduce these non-linear effects, Corning Inc. created the LEAF™ fiber, a DS-NZD fiber. In the LEAF™ fiber, the effective area in which the power is confined is larger than that in standard DS-NZD fiber. This increased MFD results in reductions in both power density and transmission errors. With reduced power density, total power in the core can be increased. Such an increase increases transmission distance.

3.5 PERFORMANCE

Two characteristics of an optical pulse change as the pulse travels through a fiber:

- The width of the pulse increases
- The height of the pulse reduces

These two changes give rise to the two main optical performance characteristics of the fiber, dispersion and attenuation.

3.5.1 DISPERSION

In multimode fibers, dispersion is indicated by one of four fiber performance parameters:

> - 'Bandwidth distance product' (BWDP)
> - 'differential modal dispersion' (DMD)
> - Transmission distance of 1 Gbps and/or 10 Gbps Ethernet signals
> - The OM-x designation

In singlemode fibers, dispersion is determined by the dispersion rate and the transmitter wavelength. As dispersion is controlled by factors that the installer cannot degrade, dispersion is not an installer concern. However, the installer need be aware of excessive dispersion as a potential cause of link malfunction. If transmission distance exceeds the maximum distance of the protocol (Table 2-5 to Table 2-10), excessive dispersion can occur.

3.5.2 ATTENUATION

In 2.3.3, you learned that dispersion limits the transmission distance or the bit rate. Attenuation, or power loss, the second major fiber performance characteristic, limits transmission distance also. If the power loss through a fiber is excessive, there is insufficient power at the receiver.

3.5.2.1 CAUSE

Most power loss results from 'Rayleigh scattering'. Imagine that you are in a dark dusty room with a high intensity flashlight. A friend stands outside of the path of the beam. Your friend can see the dust because the light scatters from its path.

In an analogous manner, light traveling in the core is scattered by the atoms to the core boundary at angles greater than the critical angle. Such light escapes the core, reducing the receiver power (Figure 3-17).

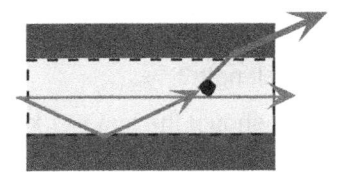

Figure 3-17: Rayleigh Scattering In Core

We refer to this power loss as "attenuation rate", in units of dB/km. The maximum and typical rates decrease as the core diameter decreases (Table 3-6 and Table 3-7) and as wavelength increases (Figure 3-18). For this reason, operation of singlemode fibers at a wavelength of 1550 nm is desirable to achieve increased transmission distance.

The network designer uses the maximum attenuation rate as a specification for cables. The typical values are those that the installer will see on properly installed cables. The network designer and installer use both the maximum and typical attenuation rates to calculate the acceptance value, which is the power loss value that the installer must not exceed for maximum reliability (19).

Wavelength, nm	Core Diameter, µ	Attenuation Rate, dB/km
850	62.5	3.5
850	50	3.5
1300	62.5	1.5
1300	50	1.5
1310	8.2	0.5, 1.0
1550	8.2	0.5, 1.0

Table 3-6: Maximum Cable Attenuation Rates

Wavelength, nm	Core Diameter, µ	Attenuation Rate, dB/km
850	62.5	2.8-3.0
850	50	2.5-2.7
1300	62.5	0.7
1300	50	0.7
1310	8.2	0.30-0.35
1550	8.2	0.20

Table 3-7: Typical Cable Attenuation Rates

3.5.2.2 ZERO WATER FIBER

One class of singlemode fibers is 'zero water' or 'no water' fiber. These fibers have no water peaks. Water peaks occur at specific wavelengths, such as 1380 nm (

Figure 3-18). At this wavelength, water in the core absorbs light, causing an increase in attenuation rate. These fibers are treated to remove the water. This type of fiber enables transmission at all wavelengths from 1380 to 1625, nm without a reduction in transmission distance due to the increased attenuation at the water peak wavelength.

> Power loss is the most important characteristic, since installation mistakes can result in increased power loss.

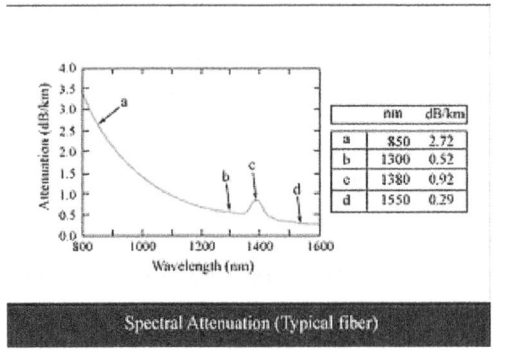

Figure 3-18: Attenuation Rate Vs. Wavelength (MIT Open Courseware)

3.5.3 BEND INSENSITIVE FIBER

Bend insensitive (BI) fibers are those that can be bent to radii smaller than that of standard fibers while exhibiting power loss lower than that of standard fibers. Both 50μ multimode and singlemode BI fibers are available. The good news is that BI fibers are available from multiple manufacturers. The bad news is that not all BI fibers are the same.

BI singlemode fibers, G.657, are compatible with standard, G.652, fibers. Future versions of PONs may use 1625 nm, at which wavelength a standard singlemode fiber is highly sensitive to power loss due to stress.

However, such compatibility may not exist for 50μ multimode fibers. While standard and BI fibers from the same manufacturer may be compatible, BI fibers from one manufacturer may not be compatible with standard fibers from another. These differences may result in changes to the insertion loss measurement procedures (14.7). At the time of this writing, there is no industry approved standard insertion loss measurement procedure. The BI manufacturers have created proprietary test procedures.

A search has revealed no BI 62.5μ fibers. Thus, compatibility and testing are not issues.

3.6 SUMMARY

The installer needs the following fiber information:

> Wavelength(s), nm
> Core diameter, μ
> Mode field diameter, μ
> Cladding, μ
> NA, dimensionless
> Attenuation rate, dB/km, maximum
> Attenuation rate, dB/km, typical
> RI, dimensionless

The installer matches the wavelength of his testing light source to the wavelength in the transmitter. With such matching, the power loss measurement will simulate the loss that the transmitter-receiver pair will experience.

The installer matches the core diameter and the NA of the test leads used in testing (14.10.3) to those characteristics of the fiber to be tested. Without such matching, the loss measurements will be higher than reality.

The installer uses the maximum and typical attenuation rates to calculate the acceptance value. This value is the maximum value at which the link can be accepted. This value indicates that the link is correctly and reliably installed (19).

The installer uses the RI to calibrate the OTDR. With such calibration, the OTDR length and attenuation rate measurements will be accurate.

3.7 REVIEW QUESTIONS

1. Why should the installer know the core diameter?
2. Why should the installer know the NA?
3. Why should the installer know the wavelength?
4. Why should the installer know the attenuation rate?
5. Why should the installer know the RI?

6. What might happen if the installer does not know the core diameter?

7. What might happen if the installer does not know the IR?

8. What might happen if the installer does not know the wavelength?

9. Provide a second answer to Question 8.

10. Are the compositions of the core and cladding the same?

11. Explain your answer to Question 10.

12. Rate the three fiber types from low to high pulse spreading.

13. List common fiber core diameters from high pulse spreading to low pulse spreading.

14. Why should the installer know about pulse spreading?

15. Can the installer increase pulse spreading of an installed network?

16. What of the units of fiber diameters?

17. What are typical multimode core/cladding diameters?

18. What are typical singlemode core/cladding diameters?

19. What are the units of measurement for attenuation rate?

20. List the maximum attenuation rates by wavelength and core diameter.

21. What are typical wavelengths for multimode fibers?

22. What are typical wavelengths for singlemode fibers?

23. For multimode fibers, what are the two terms that indicate capacity?

24. For singlemode fibers, what are the units of measurement of pulse spreading?

25. According to the text, what are typical values of capacity for multimode fibers?

26. What does SI mean?

27. What does mean GI mean?

28. What does LO mean?

29. What does BI mean?

30. True or false: you can strip the cladding from the fiber.

31. True or false: the speed of light is higher in the core than in the cladding.

32. True or false: the RI is higher in the core than in the cladding.

33. True or false: all the light energy travels in the core of all fibers.

34. True or false: the standards recommend the use of SI fibers.

35. True or false: in a SI fiber, there is a gradual change of RI from the center of the fiber to the cladding.

36. True or false: MFD is a multimode term.

37. True or false: BWDP is a singlemode term.

38. True or false: the bandwidth or capacity of a fiber is the same at all wavelengths.

39. True or false: the bandwidth or capacity of a singlemode fiber is less than that of a multimode fiber.

40. True or false: G.652 is a multimode fiber.

41. True or false: a G.652 fiber has a higher bandwidth at 1550 nm than at 1310 nm.

42. True or false: data networks use G.655 fiber.

43. True or false: Raleigh scattering results in dispersion.

44. True or false: chromatic dispersion is a cause of attenuation.

45. True or false: attenuation rate is higher at longer wavelengths.

46. True or false: a typical multimode attenuation rate is 10 dB/km.

47. True or false: the attenuation rate of singlemode fiber is lower than that of multimode fiber.

48. What type of fiber is used in FTTH networks?

49. What is the ITU designation of the fiber in Question 53?

50. What is the ITU designation for bend insensitive singlemode fiber?

51. What type of SI fiber is used?

52. In what type of application is the answer to Question 55 used?

53. As BI singlemode fibers compatible with G.652 fibers?

54. What does OM mean?

55. True or false: bandwidth of standard fiber is less than that of OM fiber.

56. What is the function of the cladding?

4 CABLES

Chapter Objectives: you learn the types of cables, the advantages of these types, the language of cables, and the numbers with which you will work.

4.1 INTRODUCTION

While the fiber can easily perform the task of optical communication, it cannot, in and of itself, survive installation and use. To survive, the fiber needs protection from installation and environmental conditions. The cable is the package that provides this protection.

4.2 STRUCTURE

The cable structure, with its six elements, provides this protection. These elements are:

- Buffer tubes
- Water blocking materials
- Strength members
- Binding tapes
- Jacket(s)
- Armor

4.2.1 BUFFER TUBES

The buffer tube is the first layer of plastic placed around a fiber by the cable manufacturer. This buffer tube can be a loose tube, with a diameter of 2-3 mm (Figure 4-1 and Figure 4-2), or a tight tube, with a diameter of 0.9 mm (900 µ) (Figure 4-3 and Figure 4-4).

The inner diameter of the loose buffer tube is larger than the outside diameter of the fiber. As such, a loose buffer tube can contain one or more fibers. Usually, a loose buffer tube has 12 fibers. However, it can contain 6 or as many as 400.

4.2.1.1 LOOSE BUFFER TUBE

Loose buffer tube designs were those used at the beginning of the development of fiber optic cables. The loose buffer tube designs have four advantages:

- High reliability due to a 'mechanical dead zone'
- Relatively low cost
- Relatively small size
- Ease of achieving moisture resistance

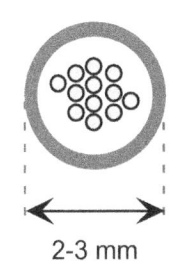

Figure 4-1: Loose Buffer Tube Cross Section

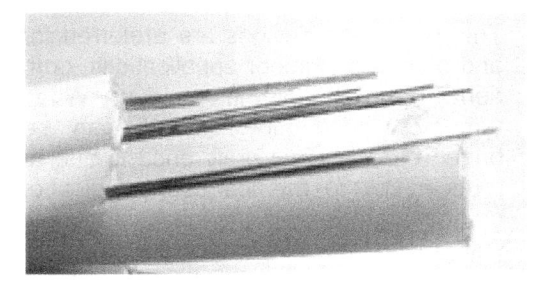

Figure 4-2: Loose Buffer Tube Cable

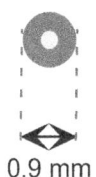

Figure 4-3: Tight Buffer Tube Cross Section

Figure 4-4: Tight Buffer Tube Cable

In these designs, the fiber is a longer than the buffer tube. This excess fiber length and the space between the fiber and the buffer tube create a 'mechanical dead zone' (Figure 4-5). This dead zone improves the reliability of the loose buffer tube over that of the tight tube design.

The mechanical dead zone is a range of force that can be applied to the cable without imposing force on the fiber. This zone exists in tension, bending and compression.

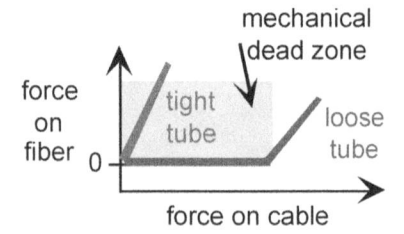

Figure 4-5: The Mechanical Dead Zone

Loose tube designs tend to have reduced costs because they have smaller sizes for a given fiber count than do tight buffer tube cables. This cost advantage increases as the fiber count increases.

The loose tube designs are preferred for and dominate outdoor appplications, both aerial, and buried in conduit. When armored, loose tube designs can be buried directly. This preference is due to reduced cost for the fiber counts typical of outdoor cables and the ability to place stiff, tough jacketing materials over the loose buffer tubes. It is difficult, but not impossible, to put such stiff materials around fibers in tight buffer tubes.

The loose buffer tube designs can achieve moisture resistance easily. Water blocking gels and greases can be placed inside of the buffer tubes and in all unfilled spaces outside the buffer tubes. Alternatively, super-absorbent polymer (SAP) tapes, yarns and powders can be placed inside the buffer tubes and under the jacket.

The loose buffer tube designs have three disadvantages:

> Reduced flexibility
> Increased cost for termination
> Increased cost for end preparation, when gel filled and grease blocked

The reduced flexibility results from the stiff, tough jackets commonly placed on outdoor cables. The increased cost for termination results from the need to install a furcation, also known as a fan out, tube on each fiber (Figure 4-6).

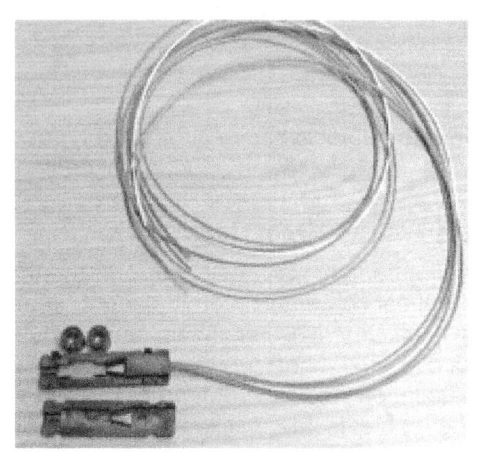

Figure 4-6: Furcation/Fan Out Kit For Loose Tube Fibers

This author has measured the time to prepare gel-filled, grease-blocked cables. A 24-fiber cable requires 2.5 man-hours per end to prepare, for a rate of 0.11 man-hours per fiber per end.

4.2.1.2 TIGHT BUFFER TUBE

In a tight buffer tube, the inner diameter of the buffer tube is the same as the outer diameter of the fiber (Figure 4-3). As such, a tight buffer tube contains a single fiber.

Tight tube cable designs have two advantages:

> End alignment of a broken fiber
> Sufficient strengthening of the fiber for connectorization

The primary advantage, end alignment of a broken fiber (Figure 4-7), occurs because there is no space between the outside of the fiber in the inside of the tight buffer tube. Broken fiber ends cannot move laterally to create core offset, the major source of power loss in all connections. Thus, a broken fiber in a tight tube cable design can provide continuity of signal transmission. In the early 1980's, this advantage provided the incentive for use of tight tube designs in the field tactical, military applications.

The second advantage is the strengthening provided by the tight buffer tube. A 250μ primary coated fiber is not sufficiently rugged to allow repeated handling. As such, the individual fibers in loose buffer tubes must be strengthened with a furcation kit (Figure 4-6) before connectors can be installed. A furcation kit to adds $1 to $2 per fiber end to the installation cost.

The buffer tube strengthens the fiber sufficiently for connector installation. The typical tight buffer tube diameter is 900 µ (0.036"). This diameter creates a structure that is sufficiently rugged to allow repeated handling without fiber damage. A second diameter, 600µ, is used in telephone networks.

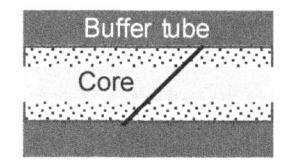

Figure 4-7: End Alignment Of A Broken Fiber In A Tight Buffer Tube

4.2.2 WATER BLOCKING MATERIALS

Fiber optic cables require moisture resistance. This resistance prevents

- Cables channeling water into the electronics
- Fiber breakage from freezing water
- Degradation of fiber strength from ground water

When they are gel filled and grease blocked ('filled and blocked'), loose tube designs have a relatively high cost for end preparation. Because of this disadvantage, most current moisture-resistant cables are made with the super-absorbent polymer (SAP) materials. These materials convert water to a gel. In conversion, the gel swells to many times the original volume of the SAP. Such swelling blocks further moisture ingress.

SAPs exist as tapes, threads, and yarns. The cost of SAP cable is approximately the same as that of gel-filled, grease-blocked cables. However, the labor cost for SAP cables is significantly less than that for gel-filled, grease blocked cables.

4.2.3 STRENGTH MEMBER MATERIALS

All fiber optic cables contain strength member materials. These materials prevent excessive stretching of fibers during cable installation. During this installation, the installer indirectly attaches these strength members to the pulling rope. In all dielectric, self-support cables (ADSS), the strength members allow the cables to withstand high long-term loads without fiber breakage (4.6.1).

These strength members can be located in the center of the cable, as a central strength member, outside the buffer tubes and inside the jacket, or between jackets. Strength member materials include aramid yarns, such as Kevlar®, flexible fiberglass rovings, rigid fiberglass epoxy rods, and steel wires.

4.2.4 BINDING TAPES

Binding tapes, often Mylar or SAP, are placed outside loose buffer tubes to hold the buffer tubes together after stranding and during jacketing. In addition, such tapes provide a heat barrier so that the heat from jacket extrusion does not damage the buffer tubes. Finally, if the tape is a SAP, the tape provides moisture resistance.

4.2.5 JACKETS

All cables have one or more jackets. The jacket, the outer-most layer of the cable structure, protects the cable core[15] and provides compliance with the National Electrical Code (NEC, 4.4). The jacket materials provide this protection through resistance to the conditions imposed on the cable by the environment in which the cable is installed. Such conditions include exposure to UV light, abrasion, water, and chemicals.

In the case of indoor-outdoor cables, a single jacket may meet both the requirements of the environment and the NEC code. In cables with multiple jackets, an inner jacket provides compliance with the NEC; an outer jacket provides protection

---

[15] The buffer tubes and the binding tape form the 'core' of the cable.

against degradation due to exposure to the outdoor environment.

Strength members may be placed between multiple jackets. Such a configuration is a 'strengthened' jacket. Such a configuration is used for the all-dielectric, self-support (ADSS) cables used by power utilities between widely spaced power transmission towers. Such cables are designed to withstand high long-term loads, up to 2000 pounds force, for a twenty year lifetime.

Finally, jacket colors of indoor cables can indicate the type of fiber in the cable. According to the Building Wiring Standard, TIA/EIA-568-C, a yellow indoor fiber cable contains singlemode fiber. Other colors, such as orange, gray, and purple are used for multimode cables (4.7.2.1).

4.2.6   ARMOR

Corrugated, plastic-coated, stainless steel armor provides both high crush resistance and rodent resistance (Figure 4-8 and Figure 4-9). Both resistances are required when the cable is directly buried. The corrugation provides a limited improvement in flexibility. The plastic coating allows the armor to be heat sealed to itself, providing a measure of moisture resistance. The stainless steel material does not corrode, so that the crush and rodent resistances remain constant. Armor always has an external jacket. Armor may have an internal jacket (Figure 4-9).

Figure 4-8: 168-Fiber Armored Cable Without Internal Jacket

Figure 4-9: 216-Fiber Armored Cable With Internal Jacket

4.3   DESIGNS

All fiber optic cables are one of seven designs. Four of these designs are common:

- Multiple fiber per tube
- Central loose tube
- Ribbon
- Premises, aka distribution

4.3.1   MFPT DESIGN

The multiple fiber per tube design (MFPT) consists of multiple loose buffer tubes around a central strength member (Figure 4-10, Figure 4-11, Figure 4-8, and Figure 4-9). Buffer tubes contain 6 or 12 fibers.[16] A binding tape surrounds the buffer tubes. Flexible strength members, either aramid yarn or flexible fiberglass rovings may be over the binding tape and under the jacket.

A ripcord is under the jacket or armor. The installer uses the ripcord to cut through a jacket or armor. The ripcord eases jacket removal.

Optional layers include additional flexible strength members, armor, and a second jacket. When present, the armor and additional jackets include a ripcord for ease of removal.

The MFPT design is one of the two most commonly used designs. The MFPT design has three advantages:

- Relatively low cost
- Relatively small size
- Easy mid span access

---

[16] Other counts are possible but not common.

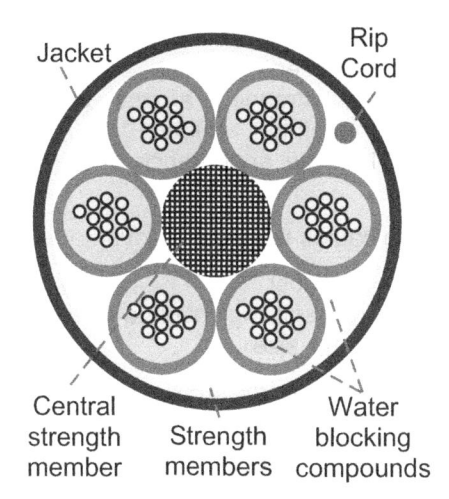

Figure 4-10: Multiple Fiber Per Tube

Figure 4-11: MFPT Cable

While not always true, this design has one of the lowest costs per fiber, particularly in high fiber counts. The small size of this design results in relatively low-cost.

Mid span access refers to the situation in which some of the fibers are dropped off between the cable ends. In this situation, the installer can remove the jacket from the cable while still leaving some protection, the buffer tubes, on the fibers. Because of the buffer tubes, it will be difficult to damage the fibers which are not terminated at the mid span location.

This design has two disadvantages:

> Relatively high labor cost for end preparation
> A requirement for enclosures on the ends

The relatively high cost for end preparation results from the removal of water blocking gels and greases. While not difficult, removal of these gels and greases is time consuming. In addition, removal of MFPT jackets is more difficult than those of tight tube cable designs.

The need for end enclosures increases typical link costs by $3-$6 per fiber per end. For a 24-fiber cable, the enclosures add $144-$288 per link.

4.3.2 CENTRAL LOOSE TUBE DESIGN

There is no reason to limit the number of fibers in a loose buffer tube to 6 or 12. We can place all the fibers in a single loose buffer tube. In this case, the buffer tube will reside in the center of the cable, resulting in a 'central loose tube' or 'central buffer tube' design.

In this design, fibers are arranged in groups of 6 or 12. A color-coded thread or yarn binds together the fibers in each group. Up to 216 fibers can reside in the buffer tube. The buffer tube can contain water blocking gel or SAP material.

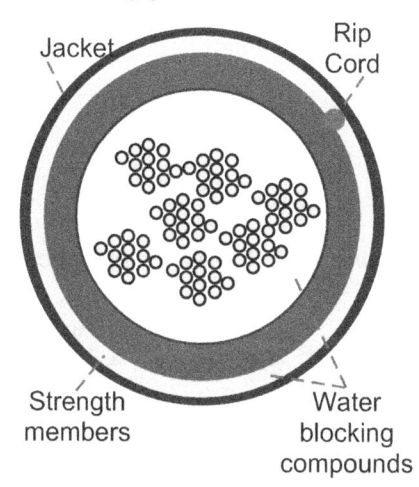

Figure 4-12: Central Buffer Tube Cross Section

Strength members surround the buffer tube, often flexible aramid yarns or fiberglass rovings (Figure 4-12). The empty spaces outside the buffer tube can contain grease or SAP materials to provide moisture resistance. An outer jacket surrounds the strength members. As in the MFPT design, optional additional layers include additional flexible strength members, armor, and a second jacket.

This design has two advantages:

> Low-cost
> Reduced cable size

A major disadvantage of this design is inconvenient mid-span access. Imagine this design with 216 fibers. You wish to drop off 12 fibers at a mid span location. When you remove the jacket and the buffer tube, you expose all 216 fibers. Murphy's Law states that the fibers to drop at the mid-span location will not be on the outside of the bundle. While digging through the bundle to access the desired fibers, the installer may break a fiber. Of course, Murphy's Law also states that the broken fiber will not be the one to terminate at the mid span location!

4.3.3 RIBBON DESIGN

The third commonly used, loose tube design, is the ribbon cable, a cousin to the central loose tube design. A ribbon is a series of 4-24 fibers precisely aligned and glued to a thin tape substrate (Figure 4-13).

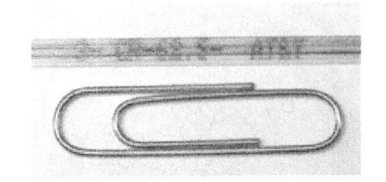

Figure 4-13: Twelve Fiber Ribbon

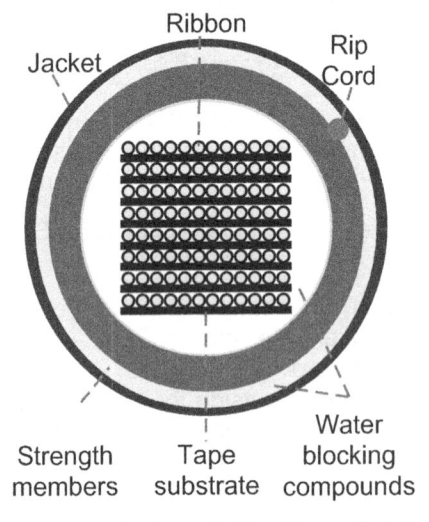

Figure 4-14: Ribbon Cable Cross Section

Ribbons can be stacked on one another to create a design with 416 fibers (Figure 4-15 and Figure 4-16). The ribbons are enclosed in the central loose buffer tube. The buffer tube and the rest of the structure is the same as that of the central loose tube design (Figure 4-14). An alternative ribbon cable design consists of ribbons stacked in loose buffer tubes, which are stranded around a central strength member (Figure 4-16). This design has the appearance of an MFPT design.

Ribbon cables have three advantages:

- Low cost
- Low splicing cost and time
- Small size

When filled with the maximum number of fibers, ribbon cables have low cost. The pre-alignment of fibers on the tape allows simultaneous splicing of all fibers on the ribbon. This procedure, mass fusion splicing and ribbon splicing, results in reduced splicing time and cost. Since all fibers are not individually aligned, ribbon splicing can result in unacceptably high losses on some fibers. Re-splicing the ribbon is the solution. Finally, when the fiber count is very high, hundreds of fibers, ribbon cables have the smallest diameter for a specified fiber count and the highest fiber density possible.

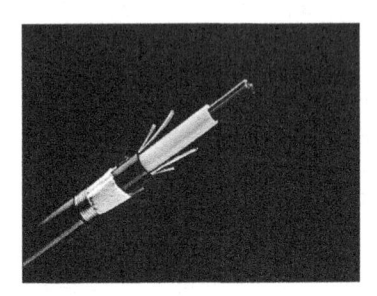

Figure 4-15: Ribbon Cable (Corning Cable Systems Inc.)

A modification to the basic ribbon cable design combines the aspects of a MFPT design with those of a ribbon design (Figure 4-16). This modified design can contain thousands of fibers.

Figure 4-16: Alternate Ribbon Design (Corning Cable Systems Inc.)

4.3.4 PREMISES DESIGN

The premises cable, or 'distribution' and the 'tight pack' cable, is the most-common design in indoor networks. While commonly used indoors, this design can be used outdoors as long as it has the characteristics appropriate for the environment in which it is installed.

The premises cable structure consists of a central strength member surrounded by multiple, tightly buffered fibers (Figure 4-17 and Figure 4-18). Flexible strength members, usually aramid yarns, surround the fibers. A jacket covers the aramid yarns. For moisture resistance, SAP tape or yarn is under the jacket.

This basic structure (Figure 4-17) can be constructed with up to 24 fibers. For higher fiber counts, multiple structures are stranded and covered with an outer jacket (Figure 4-18). Thus, a high fiber count premises design may have an appearance similar to that of the MFPT design.

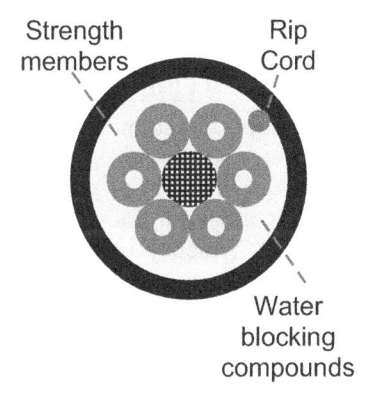

Figure 4-17: Premises Cable

The low-cost for end preparation results from the lack of gel and grease and relatively soft jackets. In 15 minutes, an installer can prepare the end of a 24-fiber premises cable. In comparison, an installer could take two man-hours to prepare the end of a 24 fiber, gel filled, grease blocked loose tube cable.

Figure 4-18: High Count Premises Cable

The premises cable design has four primary advantages:

- Low-cost for low to moderate fiber counts
- Low labor cost and ease of end preparation
- Low-cost per end for termination
- High flexibility

Low cost per end results from the 900 μ buffer tube. This diameter provides strength sufficient to allow repeated handling without damage to the fiber. In comparison, the loose tube designs require a furcation kit (Figure 4-6) to increase the ruggedness of the 250 μ primary coated fibers.

The high flexibility of the premises cable simplifies installation in buildings, short dry conduit runs, and in riser and plenum runs. This high flexibility can work against the installer, in that it allows the cable to experience small bend radii and fiber breakage.

In all cases, the premises cable is allowed when the cable is terminated within a junction box, patch panel, or enclosure. This requirement results from the inability of the tight buffer tubes to survive exposure to the working environment. Finally, this flexibility makes the premises design preferred for field tactical, military applications, electronic news, and broadcasting applications.

Single fiber, tight-tube cables, or 'simplex' cables are used as jumpers, aka patch cords. Two fiber, or 'duplex', cables are used similarly. Some duplex cables have two channels that are attached by a thin web of jacket material. Such cables are 'zip-cord' cables, so named because of their similarity to lamp cord.

4.3.5 BREAK OUT DESIGN

The final design, the breakout cable design (Figure 4-19 and Figure 4-20) was the design used in indoor networks until the mid 1980s. Break out cables were used in conduits, in risers and in air handling plenums. Today, their use is rare except in situations in which junction boxes or enclosures are not used.

The breakout structure starts with a tight buffer tube. Flexible strength members, usually aramid yarns, surround the buffer tube. An inner jacket surrounds the aramid yarns. This structure forms a 'sub cable' or 'sub element.' Multiple sub cables are stranded around a central strength member to form a cable core. Mylar tape and an outer jacket surround the cable core (Figure 4-20).

Figure 4-19: Break Out Cable

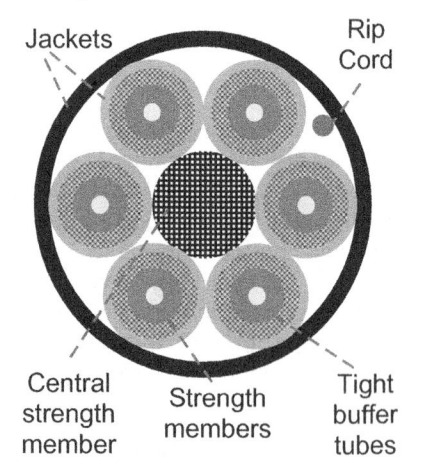

Figure 4-20: Break Out Cable Cross Section

The ruggedness results from strength members associated with every fiber. The additional association of an inner jacket with each fiber results in a sub cable that can be exposed to the working environment.

The break out cable can be used for building cable runs, conduit runs, riser runs, and plenum runs. Since the individual fibers are fully protected by the inner sub cables, the break out cable need not be terminated inside an enclosure. Elimination of the end enclosures reduces the cost per link by $144-$288 per link. In contrast, without an enclosure on the cable ends, all the other designs expose fiber or tight buffer tube to the working environment.

The breakout cable design has two advantages:

> Ruggedness
> Elimination of end enclosures.

The breakout cable has two major disadvantages:

> Large size
> High cost

In small conduits, the large size can restrict the use of break out cables to low fiber counts. The high cost results from both the amount of materials in the cable and from the increased number of processing steps required by this design. A zip-cord duplex, which has the appearance of lamp cord, is a relative of the break-out design. The zip cord duplex and simplex cables are often used for patch cords and back planes

4.3.6 BLOWN PRODUCTS

4.3.6.1 BLOWN FIBER

A blown fiber system consists of proprietary fibers in a small diameter tube and a plumbing system of plastic tubes, elbows and couplings. Blown fiber is a group of fibers, typically up to 12, that is blown by air pressure into an airtight system of plastic tubes (Figure 4-21). The surface of the group is dimpled in the manner of a golf ball. This surface allows the fiber group to float in the plastic tubes, with a significant reduction in friction. This reduction in friction allows elimination of traditional strength members. Blown fiber systems allow installation to lengths of 1000- 2000 m of fiber, removal of fibers for reuse in a new location, rapid installation of fibers, rapid restoration of damaged tubes, and low cost installation of additional fibers as needed.

A primary advantage of this system is a reduction in future installation cost: fibers can be added to the system as needed. In contrast, a traditional fiber optic cable system requires an initial investment in all the fibers needed for the lifetime of the network. This investment eliminates future installation costs.

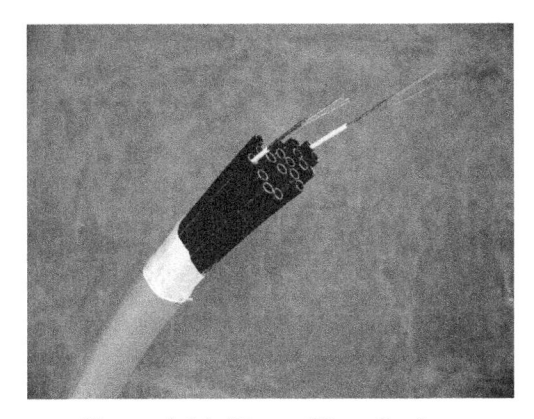

Figure 4-21: Blown Fiber System Components (Courtesy Sumitomo Electric Lightwave)

A primary disadvantage is the cost of the plastic tube system. This cost can make the cost of the first installation higher than the cost of the first traditional cable system.

4.3.6.2 BLOWN CABLE

A blown cable is a small diameter, light duty, cable similar to the cables described in this chapter. Light duty indicates low installation load rating. The installation process allows this low rating.

This process involves attachment of a plug to the end of the cable and use of airflow and air pressure to blow the cable into an airtight conduit. The airflow allows the cable to float in the conduit, resulting in a significant reduction in friction and installation load. This reduction in friction allows reduction in the amount of strength member materials and the size of the cable.

4.4 NEC COMPLIANCE

The National Electrical Code (NEC) influences two aspects of fiber optics, cable requirements and terminology. In the first aspect, the NEC imposes flame spread and smoke generation requirements on all cables placed inside buildings.

Rating	Can Replace
OFNG, OFCG	---
OFN, OFC	---
OFNR, OFCR	OFN, OFC
OFNP, OFCP	OFN, OFC, OFNR, OFCR

Table 4-1: NEC Fiber Cable Ratings

In Article 770, the National Electrical Code identifies two groups of fiber optic cables and four ratings for each group (Table 4-1). The two groups are: OFN and OFC. OFN stands for optical fiber cable non-conductive; OFC, for optical fiber cable, conductive. An OFC cable has at least one conductive element in its structure. A cable with a steel strength member, armor or conductor is conductive and would result in a cable designated as OFC.

Each group has four ratings, listed in order of increasing requirements: two horizontal ratings, riser rated (R) and plenum-rated (P). Each rating has a different test procedure and must meet different requirements.

The first two ratings (Table 4-1) are horizontal. OFNG and OCCG are Canadian ratings. OFN and OFC are US ratings. Such cables can be used for links on a single floor.

A riser rated cable (OFNR, OFCR) can be used in for a link between floors. A plenum rated cable (OFNP, OFCP) is required for a cable installed in an air handling plenum.

The NEC allows use of a higher rated cable in a lower rated location. However, the reverse substitution does not comply with NEC requirements.

As a practical matter, OFNG and OFCG cables are not available in the US, as they are qualified with a Canadian test procedure. As a second practical matter, many manufacturers do not make OFN or OFC cables, as the cost differences between these two ratings and the next higher ratings, OFNR and OFCR, are small enough to eliminate the need to maintain two separate inventories.

As a third practical matter, most indoor fiber optic cables are of the OFN series, since dielectric designs are easier to install in compliance with the NEC than are the OFC cables.

In the second aspect, the NEC defines the language used to describe two types of fiber optic cables. As is to be expected, a cable containing only multimode fibers is a multimode cable; only singlemode, singlemode. The NEC defines a cable containing both multimode and single-mode fibers as a 'hybrid' cable. The NEC defines a cable containing both fibers and conductors as a 'composite' cable.

With some exceptions, outdoor cables do not comply with the requirements of the NEC because their jackets are HDPE, a member of the paraffin family. However, some indoor-outdoor cables are UV-resistant, moisture resistant and are rated either OFNR or OFNP.[17]

4.5 DIELECTRIC DESIGN

The OFN series is an example of an indoor dielectric design. Outdoor cables may be dielectric. Dielectric outdoor cables provide increased safety, since they cannot conduct lightning current or ground potential rise into a building.

Ground potential rise occurs when a conductive path connects two widely spaced locations. 0 volts in Chicago is not the same voltage as 0 volts in Atlanta. Because of a small difference in the voltage, current can flow, resulting in a potentially hazardous condition.

Grounds and bonds are required on cables that are not dielectric. Since such grounds and bonds do not last forever, grounding and bonding increases both initial installation cost and maintenance cost. Because of these safety and cost advantages, outdoor designs tend to be dielectric.

4.6 CHARACTERISTICS

Fiber optic cables are designed to meet two groups of requirements: environmental and installation. The system designer defines both. He defines the environmental requirements so that the cable survives the environment in which the cable is installed. In addition, he defines the installation requirements so that the cable survives the installation process. Because the installer installs the cable, he needs to know the installation limits within which he must handle and

---

[17] Nexans, formerly Berk-Tek, was the first company to offer such products.

install the cable. Without such knowledge, the installer can, and often does, damage the cable.

As presented (4.4), the installer must install a cable with an NEC rating appropriate to the location in which the cable is to be installed. In addition, the installer needs to know ten cable specifications:

- Installation load
- Use load
- Short-term bend radius
- Long term bend radius
- Outer jacket diameter
- Inner jacket diameter
- Buffer tube diameter
- Storage temperature range
- Installation temperature range
- Operating temperature range

4.6.1 LOADS

As a material, tensile loads can damage glass. Because of this characteristic, the cable is designed to limit the elongation of the fiber when a load is applied to the cable.

There are three kinds of damage:

- Breakage of fiber
- Increased attenuation
- Delayed breakage of fiber

Loads are of two types:

- Short-term, or installation
- Long-term, or use

The installation load is the maximum load that the installer can use while pulling the cable into location without damaging the cable. The installation load is important in all installations in which the cable will be pulled into its final location. When the cable is pulled into underground conduits or building conduits, the load imposed on the cable can be high. Because of this characteristic, the installer pulls in the cable in a manner that limits the load applied to the cable (11.3.1.2).

Installation loads range from 110 pounds-force (490N) for single and dual fiber cables to 600 pounds-force (2700 N) for outdoor, high-fiber count cables. Use loads range widely.

If the cable is laid into a cable trough, cable tray, or raceway, the installation load applied to the cable will not be high. In these situations, the installation load is of reduced concern.

The use load is the maximum load that the installer can impose on the cable for the entire life of the cable without damage. The long-term load is important in installations in which the cable is strung, without external support, between widely spaced telephone poles, power transmission towers or buildings.

Another form of the use load is the maximum vertical rise distance. This is the distance to which a cable can be installed without mid span support. The installer must comply with this distance when installing cables in vertical risers and on tall transmission towers.

4.6.2 BEND RADII

As stated earlier, glass does not support tension loads well. Bending produces a tensile stress on the surface of the fiber. This stress can produce three problems: increased attenuation rate, immediate fiber breakage, and delayed fiber breakage.

Because of this property, all fiber optic cables are designed with a limit on the radius to which the cable can be bent. Both short-term and long term loads are specified in the Building Wiring Standard, TIA/EIA-568-C.

The cable is designed to withstand two bend radii: the short-term, or installation, bend radius and the long-term, or unloaded, bend radius. The short term, or installation, bend radius is the minimum radius to which the cable can be bent while the cable is under the maximum installation load. While defined precisely on cable specification sheets, the installation bend radius is specified as:

Short term bend radius = 20 x cable diameter

Equation 4-1

The long-term bend radius is the minimum bend radius to which the cable can be bent while the cable is under no load. This radius is limited more by the cable materials than by the fibers. While defined precisely on the cable specification sheet, the installation bend radius is specified as:

Long term bend radius = 10 x cable diameter

Equation 4-2

While the two rules (Equations 4-1 and 4-2) may appear to be in reverse, they are not. When the cable is bent while under the installation load, the fiber surface is under two forms of tension: axial tension and bending tension. When the cable is under no installation load, the fiber surface can withstand the same level of tension. Since there is no axial tension, the fiber can withstand a higher level of bending tension, which results from a bend radius smaller for the unloaded state than for the loaded state.

4.6.3 DIMENSIONS

Three cable dimensions are important to the installer:

- Outer jacket diameter
- Inner jacket diameter
- Buffer tube diameter

The outer jacket is important whenever the cable is installed into a conduit or inner duct, since the cable must be smaller than the conduit or inner duct. The outer jacket of one-fiber cables and zip-cord duplex cables must fit into the boot of the connector. In addition, the inner jacket of a break out cable must fit into the boot of the connector. The buffer tube must fit into the connector back shell.

4.6.4 TEMPERATURE RANGES

Three temperature ranges can be important to the installer:

- Storage - range
- Operating temperature range

The cable is designed to withstand storage and installation within defined temperature ranges. If the cable is stored at an excessively high temperature, the cable materials can soften, degrade and shrink. Shrinkage of the cable materials can result in an increased attenuation rate. If the cable is stored at an excessively low temperature, the cable jacket can crack. Cracks may expose the fibers to water and chemicals in the environment.

If the cable is installed at excessively low or high temperatures, the materials can be damaged. Flexing of the cable at an excessively low temperature can result in cracking of the jacket. Cracks in the jacket can expose the fibers to water and chemicals in the environment. Flexing of the cable at an excessively high temperature can result in a permanent stretching of the jacket. Such stretching can increase the attenuation rate of the fibers.

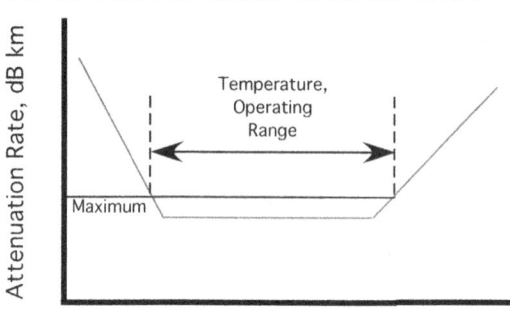

Figure 4-22: Operating Temperature Range

While not an installation characteristic, the operating temperature range of a cable can be a cause of increased attenuation. The temperature operating range is the range within which the cable exhibits an attenuation rate below is maximum (Figure 4-22). A cable operating outside of its operating temperature range will exhibit an attenuation rate that exceeds its specification.

4.7 STANDARDS

4.7.1 TIA/EIA-568-C

TIA/EIA-568-C addresses fiber cabling component specifications and field test instrument requirements. This standard requires that indoor cables comply with ANSI/ICEA S-83-596-2001. TIA/EIA-568-C, B.3 requires that outdoor cables comply with ANSI/ICEA S-87-640-1999.

4.7.1.1 INDOOR CABLES

ANSI/ICEA S-83-596-2001 provides comprehensive dimensional, performance and structural requirements for premises distribution cable. The dimensional requirements (Section 2.3) include tolerances on core and cladding diameters, NA, MFD, and ZDW. This standard requires fiber color-coding in compliance with TIA/EIA-598.

4.7.1.2 OUTDOOR CABLES

ANSI/ICEA S-87-640-1999 provides comprehensive dimensional, performance and structural requirements for outside plant communications cable. The optical and fiber requirements are consistent with those in ANSI/ICEA S-83-596-2001. The default jacket material is high-density polyethylene (HDPE). Ripcords are optional.

4.7.2 COLOR CODING

4.7.2.1 JACKET

Color-coding of cable jackets is optional. TIA/EIA-598-C defines the colors.

Yellow	singlemode
Orange	multimode
Aqua	10 Gbps, 50μ LO

Table 4-2: Cable Jacket Color Codes

Blue
Orange
Green
Brown
Slate
White
Red
Black
Yellow
Violet
Rose
Aqua

Table 4-3: Color Coding Sequence

4.7.2.2 BUFFER TUBES AND FIBERS

TIA/EIA-598 defines color-coding of buffer tubes and fibers. Table 4-3 contains the sequence for the first twelve units. The second twelve units add a dash to the color in this table.

4.8 SUMMARY

In order to install a fiber optic cable without damaging the cable or fibers, the installer needs to know the:

- Short -term, or installation, load
- Long -term, or use, load
- Short term bend radius
- Long term bend radius
- Outer jacket diameter
- Inner jacket diameter
- Buffer tube diameter
- Storage temperature range
- Installation temperature range

With this, and other, information, the installer can handle and install the cable without damage.

4.9 REVIEW QUESTIONS

1. List the structural cable materials that protect the fiber.
2. List five materials that limit the elongation of the fiber.
3. List three descriptive characteristics of the material that provides crush and rodent resistance.
4. List the four most-commonly used cable designs.
5. True or false: the diameter of a tight tube is larger than that of a loose buffer tube.
6. True or false: an installer can apply tension to a premises cable without applying tension to the fibers.
7. True or false: the long-term bend radius of 1 in. diameter cable is 20 in.
8. What are the three possible problems caused by a violation of the bend radius?
9. Why do many outdoor tight tube cables not comply with the requirements of the NEC?
10. Most fiber optic cable manufacturers offer only two NEC compliant ratings. What are they?
11. True or false: OFN series cables need to be grounded.
12. True or false: the latest water blocking technology is a super absorbent polymer.
13. True or false: super absorbent polymer cables cost more to prepare than gel filled, and grease blocked cables.
14. True or false: tight tube cables are always used only indoors.
15. True or false: loose tube cables dominate indoor applications.
16. You are planning a network with a large number of fibers, 432 or 864. Your conduit system already exists, must be used and cannot be replaced. What cable design has the highest chance of meeting your needs?
17. True or false: to minimize splicing cost, use a MFPT or central loose tube design.
18. True or false: a furcation kit is required for use on a premises cable.
19. What two cable designations can be used anywhere inside a building?
20. Define a hybrid cable.
21. Define a composite cable.
22. List the 24 fiber colors in sequence.
23. List the three main cable colors.
24. What does each cable color signify?
25. What is the term that designates a cable with no conductive materials?
26. What are the two bend radii rules?
27. In what situation does each of these rules apply?
28. How many types of loads are important to the installer?
29. List the three temperature ranges.

30. What can happen when each range temperature is exceeded?

31. How many cable diameters are important?

32. List the situations in which each of these diameters becomes important.

33. What types of cables do not need to be grounded?

34. How many jacket colors does TIA/EIA-568-C specify?

35. What are these colors?

36. What type of fiber does each of these colors signify?

37. What cable designs must be terminated in an enclosure?

38. What cable design(s) need not be terminated inside an enclosure?

39. What is a furcation kit?

40. With what cable design is a furcation kit used?

41. In whist location are loose tube cables usually used?

42. In what location is armored cable usually used?

5 CONNECTORS

Chapter Objectives: you learn the language of fiber optic connectors, the types, their advantages, and performance concerns.

5.1 INTRODUCTION

Fiber optic connectors are part of every link. In this chapter, we present

- Function
- Structure
- Performance
- Types
- Installation methods

5.2 FUNCTION

Connectors provide five functions:

- Low power loss
- Low reflectance
- High fiber retention strength
- End protection
- Disconnection

Connectors achieve low power loss through precise alignment of the small fiber cores. Connectors achieve high fiber retention strength by gripping the fiber, either through an adhesive or a mechanical crimp. Connectors achieve end protection by distribution of contact force over an area of increased size, thus reducing the pressure imposed on the end of the fiber. Finally, connectors enable disconnection, or detachment, of a fiber from both optoelectronics and other fibers. In short, connectors are used to create temporary connections and are installed in all locations in which permanent joints are not used. When connectors make physical contact, they can provide low reflectance.

5.3 STRUCTURES

In spite of the apparent complexity and variety of structures, connectors can be described by nine structural characteristics:

- Ferrules
- Latching structure
- Mating structure
- Key
- Back shell
- Strain relief boot
- Crimp sleeve or ring
- Simplex or duplex
- Ferrule cap or cover

5.3.1 FERRULES

5.3.1.1 FERRULE CONTACT

Ferrules in small form factor (5.5.1) and commonly used (5.5.2) connectors make physical contact (Figure 5-1). Most legacy connectors (5.5.3) are non-contact and have loss and reflectance higher than those of contact connectors.

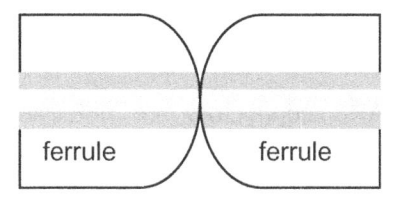

Figure 5-1: Ferrule Contact

This increased loss results from the expansion of the light as it exits a connector. Light exits from a fiber and expands within a cone defined by the NA. In a non-contact connector, the size of the spot on the receiving fiber is larger than the diameter of the receiving core. The light arriving on the cladding of the receiving fiber is lost (Figure 5-2). A contact connector has no such expansion.

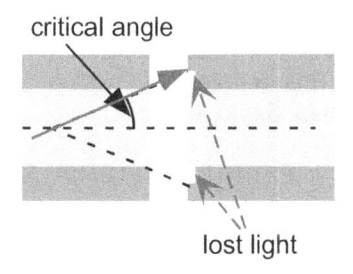

Figure 5-2: Increased Loss From Non-Contact Connectors

5.3.1.2 FERRULE END FACE

Ferrules have three end faces:

- Radius
- Angle physical contact
- Flat

Contact connectors have radiussed end faces (Figure 5-1). Such end faces ensure high performance; that is, low loss and low reflectance (5.4.5). However, this end face cannot provide the lowest possible reflectance. For such reflectance, the angle physical contact (APC) end face is required (Figure 5-3). This singlemode connector has a bevel on the tip of the ferrule at 8° to the perpendicular of the fiber axis. APC products are available for SC, LC, LX.5, and FC connectors.

Figure 5-3: The SC/APC Connector

The APC connectors are not used in some networks because of increased cost. Some FTTH networks use APC for low reflectance.

In current generation connectors, flat end faces are rare and not acceptable. Prior to approximately 1985, flat end faces were dominant.

5.3.1.3 FERRULE DIAMETERS

While there are multiple ferrule diameters, two dominate the industry: 2.5mm and 1.25mm. The 2.5mm ferrule was introduced in the early 1980s.

The 1.25 mm ferrule is used in SFF connector types. It was first used in the late 1990s. This ferrule has loss lower than that of the 2.5 mm ferrule.

5.3.1.4 FERRULE MATERIALS

The ferrule provides low loss through alignment of the fiber cores. With one exception, all connectors have ferrules. The one exception, the Volition (5.5.1.4), has no ferrules; instead, it has precision V-grooves that align the fibers.

Ferrules can be of many different materials. The two most commonly used materials are:

> Ceramic
> Liquid crystal polymer

Occasionally, the ferrule is stainless steel.

As of this writing, ceramic ferrules are required for singlemode connectors, since only ceramic materials can be manufactured at a competitive price and with the precision required for singlemode core alignment.

Unlike the other ferrule materials, ceramic ferrules can be over polished, creating undesirable, excessive undercut (Figure 12-25). Undercutting results increases both power loss and reflectance.

The other materials, which are softer than ceramic, can be re-polished without expensive diamond polishing films. This polishing removes light to moderate damage. Such polishing restores original performance without connector replacement. In addition, soft ferrule materials can be re-polished without risk of undercutting. In an apple-to-apple comparison, the soft ferrule materials provide both reduced initial and life cycle costs.

While not commonly used at the time of this writing, other connector ferrule materials include:

> nickel-plated brass
> aluminum
> zinc
> thermoplastic polymer

Thermoset polymer is used in the MTP/MPO and MT-RJ connectors.

5.3.2 LATCHING STRUCTURES

Connectors have one of two latching mechanisms:

> Push on/pull off (PO/PO) latching mechanism
> Rotating retaining ring

All current generation connectors have a push on/pull off latching mechanism. This mechanism has two advantages. First, it makes the connector easy to remove. Second, it enables the connectors to be closely-spaced in a patch panel. Close spacing enables doubling and quadrupling the number of connectors in a patch panel. The inner and outer housings of the SC connector create this mechanism (Figure 5-4).

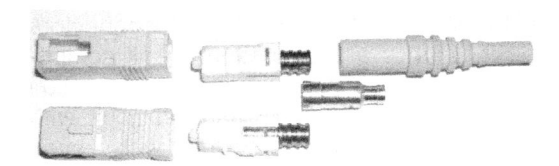

Figure 5-4: SC Connector Design With A Latching Mechanism

Examples of connectors having PO/PO latching are:

- LC
- SC
- MU
- LX.5
- Volition
- Opti-Jack
- MT-RJ
- FDDI
- ESCON

The alternate latching mechanism requires a ring. The ring must be rotated to latch the connector. The need for rotation increases the space required for each connector.

Examples of connectors having this type of latching are:

- ST™-compatible (Figure 5-5)
- FC
- Biconic
- 905 and 906 SMA
- Mini BNC
- D4

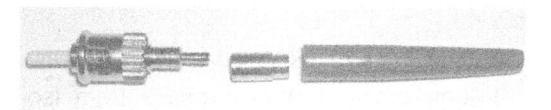

Figure 5-5: Connector With Latching Ring

This ST™-compatible structure includes, from left to right: the ferrule, retaining ring, back shell, the crimp ring, and the strain relief boot.

5.3.3 MATING STRUCTURES

There are two structures for connecting, or 'mating' two connectors:

- Two identical connectors or 'plugs'
- A plug and jack

When the connectors are identical, the connectors require a 'barrel', also known as an adapter, a bulkhead, and a feed through (Figure 5-6). The dominant connector design is of identical plugs.

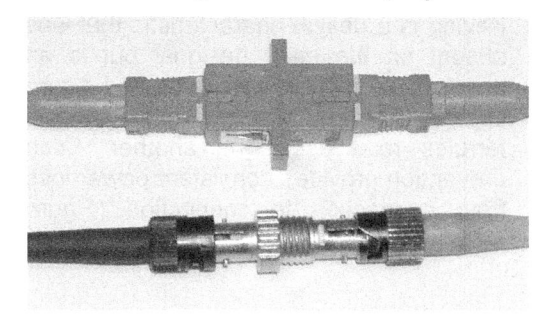

Figure 5-6: Connectors With Plugs And Barrel

When the connectors are not identical, the connectors form a plug and jack pair (Figure 5-7). The jack performs the functions of a plug and barrel in a single structure. In general, jacks can be flush mounted, while barrels cannot. Flush mounting is an advantage in FTTD networks.

Examples of plug and jack connectors are:
- Volition
- Opti-Jack
- MT-RJ

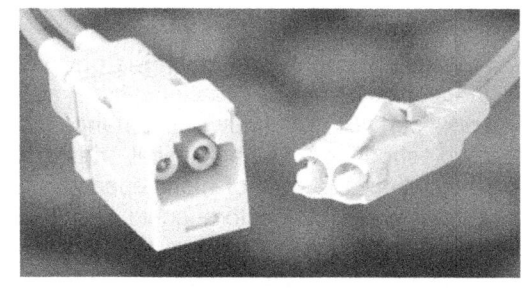

Figure 5-7: Jack (Left) and Connector (Right)[18]

5.3.4 CONNECTOR STRUCTURES

One common structure (Figure 5-4) includes, from left to right:

- outer housing
- connector body, which includes
  - ferrule
  - inner housing

---

[18] Courtesy of Panduit Corporation

- back shell
- crimp ring
- strain relief boot

5.3.4.1 KEY

Keying is a design characteristic that was absent on the early designs, but is an absolute necessity in current and future designs. The key prevents rotation of two ferrules relative to one another. Such prevention provides consistent power loss from connection to connection (Figure 5-8).

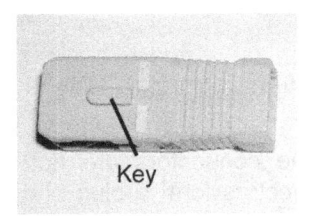

Figure 5-8: Key On Connector

This consistency, referred to as 'repeatability' is specified commonly as a maximum of 0.2 dB.[19]

This author has tested repeatability for both keyed and unkeyed connector styles. For keyed connectors, we find typical repeatability closer to 0.1 dB than to 0.2 dB. For unkeyed connectors, we find typical repeatability to range from 0.5 dB to 1.0 dB.[20]

This improvement from keying is obvious and significant in two situations: initial network certification and maintenance or troubleshooting. During initial network certification, the lack of high repeatability of non-keyed connectors could result in the need to tune connectors to bring link power loss into compliance with requirement for the optoelectronics (OPBA, 8.4.1). Subsequent disconnection and reconnection of any connector could result in loss of the preferred ferrule orientation and in link failure.

During maintenance or troubleshooting, the installer compares current loss measurements to original measurements. There is a possibility that improvement in the alignment of non-keyed connectors could mask a degradation of link components. With this possibility, interpretation of a loss increase is more difficult with non-keyed connectors than with keyed connectors.

5.3.4.2 BACK SHELL

The back shell is the location to which a jacketed fiber is attached (Figure 5-9). The jacketed fiber means a patch cord. The back shell design may, or may not, provide for axial load and lateral load isolations.

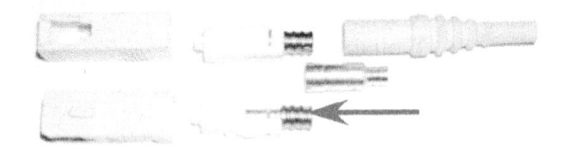

Figure 5-9: Connector Back Shell

5.3.4.2.1 AXIAL LOAD ISOLATION

Axial load isolation, otherwise known as 'pull proof' behavior, results from isolation of the ferrule from motion due to axial tension on the back shell. This isolation can be achieved by inclusion of a spring between the ferrule and the back shell. Such isolation prevents an increase in power loss from tension applied to the cable. SC and LC connectors are pull proof. The ST™-compatible connectors are not.

5.3.4.2.2 LATERAL LOAD ISOLATION

'Wiggle-proof' behavior results from isolation of the ferrule from motion due to lateral pressure on the back shell. Such pressure will result in increased power loss if the ferrule tilts. SC and LC connectors are wiggle proof. The ST™-compatible connectors are not.

- In summary, a pull-proof, wiggle-proof connector provides more reliable operation than one that is neither.

5.3.4.3 STRAIN RELIEF BOOT

The strain relief boot limits the bend radius of the cable or the tight buffer tube as they exit the back shell. The boot

---

[19] We obtained this value from a survey of connector data sheets.

[20] Data are from SMA 906 connectors tested from 1990-1994.

increases the reliability of the connector (Figure 5-10).

Figure 5-10: Strain Relief Boot

5.3.4.4 CRIMP SLEEVE

The crimp ring, or crimp sleeve, grips the strength members of the cable (Figure 5-11). When making patch cords with 1.6 mm-3 mm jacketed fibers, the installer uses this ring. When installing connectors on 900μ tight tubes, the installer does not.

Figure 5-11: Crimp Sleeve

5.3.4.5 SIMPLEX AND DUPLEX

Most connector types are simplex. Some simplex connectors can be converted to duplex through the use of an external clip (Figure 5-12). Such conversion is possible, if the mating mechanism is PO/PO,. Five connector types are duplex by design: Volition, Fiber-Jack, FDDI, ESCON, and MT-RJ.

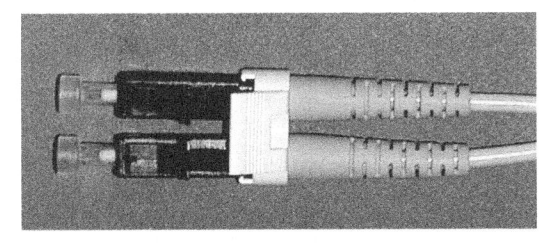

Figure 5-12: LO LC Duplex Connector

If the connector has a retaining ring, such conversion is not possible. Such connector types are: 905 and 906 SMA, Biconic, mini- BNC, ST-™ compatible, FC, and D4.

5.3.4.6 FERRULE CAP

Connector caps or covers keep dust from the end of the fiber. Volition and LX.5 have caps that are integral to the connectors. All other connectors have caps that are separate and forgettable.

5.3.5 COLORS

TIA/EIA-568-C recommends the connector color-coding in Table 5-1. Connector bodies and/or boots can have the color. Barrels should have the same color to indicate the type of fiber behind the patch panel.

Color	Use
Green	APC singlemode
Blue	singlemode
Aqua	50 μ LO
Black	50 μ
Beige	62.5 μ

Table 5-1: Connector Color Codes

5.4 PERFORMANCE

The installer has four connector performance concerns:

- ➢ Maximum insertion loss, dB/pair
- ➢ Typical insertion loss, dB/pair
- ➢ Repeatability, in dB
- ➢ Reflectance, in - dB

5.4.1 DB PER PAIR

The unit of connector power loss is "dB/pair". Such loss is from one fiber to another. Thus, a physical pair is required. The term 'dB/connector' has no meaning.

By this definition, there is no power loss from a transmitter source to a fiber or from a fiber to a receiver. In an absolute sense, power loss occurs at both locations. However such loss is accounted for in the method of measurement of the optical power budget of the optoelectronics (8). In a practical sense, field measurements need not include this loss.

5.4.2 MAXIMUM LOSS

The maximum insertion loss, in dB/pair, is the maximum loss the installer will see when he installs the connectors correctly. However, he should not expect to see this value, any value close to this value, unless he makes errors during installation. Installers use this maximum value in certifying networks.

When the installer installs connectors correctly, the loss is due to intrinsic causes, such as core and cladding diameter variations, core offset, cladding non-circularity, NA mismatch, and offset of the fiber in the ferrule. When the installer makes errors, he introduces extrinsic causes of loss, such as damaged core, bad cleaves, dirt or contamination on the core, and air gaps due to excessive polishing. All data communication connectors are rated at the maximum loss of:

➢ 0.75 dB/pair

This is a 'Magic Value' since it is always true. It applies to singlemode or multimode, keyed and contact connectors.

There is evidence that this value may change in the future. Actual connector loss has dropped since this value was established in the early 1980s. This author will not be surprised if this maximum value drops to 0.5-0.55 dB/pair.

5.4.3 TYPICAL LOSS

The typical insertion loss, in the dB/pair, is the value the installer sees when he installs connectors correctly. Frequently, the installer will see values lower than this value. The installer uses this value in certifying networks (19).

All singlemode or multimode, keyed and contact, connectors have three typical values ('Magic Values'):

➢ 0.15-0.20 dB/pair for connectors which require polishing and have with 1.25 mm ferrules
➢ 0.30 dB/pair for connectors which require polishing and have with 2.5 mm ferrules
➢ 0.40 dB/pair for connectors which require no polishing

5.4.4 REPEATABILITY

Repeatability is the maximum increase in loss, in dB, that a connector will exhibit between successive measurements. The repeatability is used to calculate the range. The range is used to interpret increases in link loss during troubleshooting or maintenance testing (14.8).

5.4.5 REFLECTANCE

Reflectance is a measurement of the relative optical power reflected backwards from a connector. This reflected power can travel back to the light source and be reflected back into the fiber. If this reflected power reaches the receiver with a power level above its sensitivity, the receiver will convert this optical power to a digital 'one'. If the time interval in which the reflected power arrives was a digital 'zero' at the transmitter, the output signal differs from the input signal. Thus, reflectance influences the accuracy with which a fiber optic link will operate. Minimizing reflectance results in maximizing signal accuracy. The objective of a reflectance requirement is to limit the reflected power at the receiver to a value less than the sensitivity of that receiver.

Reflectance occurs at any glass-air interface. This interface, or any interface at which there is a change in speed of light, or RI, can produce a reflection, called a "Fresnel reflection". This reflection occurs in fiber optic connectors because the end faces of mating connectors have surface roughness, which creates microscopic air gaps (Figure 5-13).

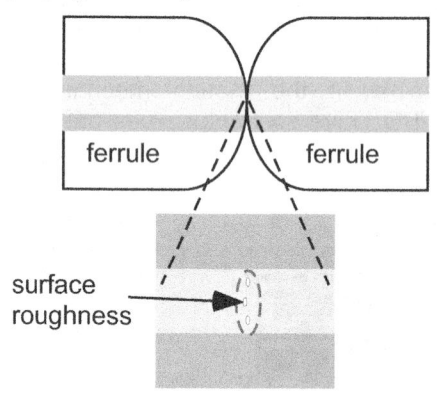

Figure 5-13: Air Gaps Create Reflectance

In connectors, this reflection is called reflectance. Reflectance (Equation 5-1 is stated in units of negative dB, with values from -20 dB to -65 dB. A non-contact connector exhibits a reflectance between -14 dB to -18 dB.

Reflectance =

10 Log (reflected power/incident power)

Equation 5-1

Reflectance is described qualitatively by three terms:

- PC (physical contact)
- UPC (ultra physical contact)
- APC (angled physical contact)

Commonly, but not precisely, PC refers to reflectance of less than –40 dB; UPC, to less than –50 dB; and APC, to less than –55 dB.

Historically, reflectance has been a concern for singlemode connectors only. However, reflectance is now a concern in multimode networks transmitting at and above 1 Gbps. The gigabit Ethernet standard requires multimode connectors with reflectance below -20 dB.

5.5 TYPES

We present connectors in three groups:

- Small form factor (SFF) connectors
- Commonly used connectors
- Legacy connectors

The SFF connectors are expected to become dominant in the future. The commonly used types may be dominant at the time of this writing. The legacy connectors were used in the past but are not used in new installations. Photographs in this section are approximately actual size.

5.5.1 SFF CONNECTORS

While the ST-compatible and SC connector types have had long, and successful lives, both have the same, significant drawback: large size. This size results in increased optoelectronic cost. The large size of the connectors forces optoelectronics manufacturers to space transmit and receive electronics far apart. This large spacing results in half as many fiber ports in a fiber switch as UTP ports in a UTP switch.

The optoelectronics manufacturers requested a reduced size fiber connector that would enable them to reduce the per port cost of switches. Their goal was for cost parity of fiber switches with UTP switches.

Connector manufacturers answered this request with a series of small form factor (SFF) connectors. These SFF connectors obviously meet the need of reduced size (Figure 5-14).

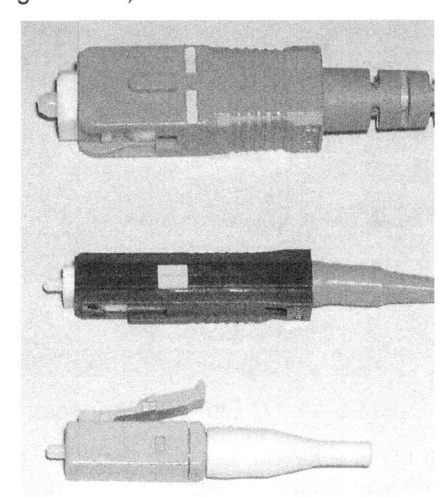

Figure 5-14: Size Comparison of Connectors: SC (top), MU (middle) And LC (bottom)

The request for reduced size fiber connectors resulted in seven SFF connectors:

- LC
- LX.5
- MU
- Volition
- Opti-Jack
- MT-RJ
- MTP/MTO

While the MTP/MTO connector is not usually considered a SFF connector, its characteristics make it one.

5.5.1.1 LC

The LC is a simplex SFF connector that can be converted to a duplex form with a clip. The LC (Figure 5-15 and Figure 5-12) was developed by Lucent Technologies, but is available from other manufacturers.

The LC was developed as a telephone connector to enable telephone companies to increase the density of installed connectors. Dense wavelength division multiplexing (DWDM) is one of the technologies that created this need for increased density. Many optoelectronics manufacturers provide products with the LC interface.

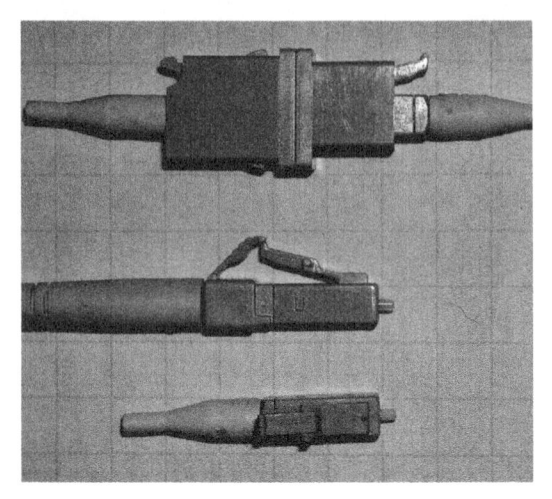

Figure 5-15: LC Simplex Connector

The LC is a keyed, contact, low loss, pull proof and wiggle proof design. The LC has a 1.25 mm ferrule, which is half the diameter of the ferrules used in ST-compatible, FDDI and ESCON connectors. This small ferrule size can cause problems for installers, until they realize that half the diameter means one quarter of the usual polishing pressure.

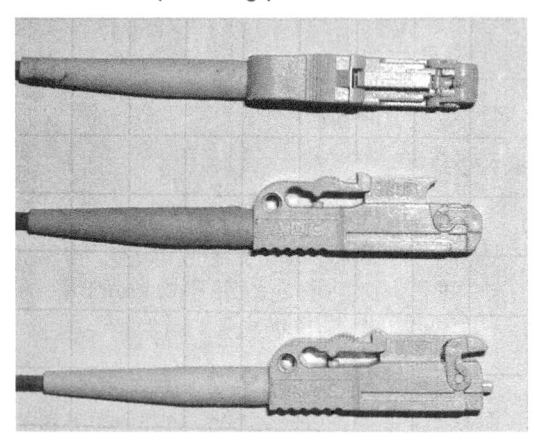

Figure 5-16: LX.5 Connector

5.5.1.2 LX.5

ADC Telecommunications developed the LX.5 (Figure 5-16), similar to the LC, for use in telephone networks. The LX.5 is a keyed, contact, low loss, pull proof and wiggle proof design. The LX.5, available in radius and APC versions, has a unique feature: a built-in dust cover (Figure 5-16, bottom). This cover lifts as the connector is installed.

Similarly, the LX.5 adapter has a built in dust cover (Figure 5-17). The dust covers on the connector and in the adapter increase the eye safety of the product. With multiple wavelengths on singlemode fibers in DWDM networks, the total power level at the connector can be high enough to cause eye damage.

Figure 5-17: SC And LX.5 Barrels

5.5.1.3 MU

The MU is a simplex SFF connector with the appearance of an SC, but with all dimensions reduced in half (Figure 5-18). Designed and developed in Japan, the MU is a keyed, contact, low loss, pull proof and wiggle proof design. Its low loss results from the 1.25mm ferrule. It is unique in that it can be assembled to create duplex, triplex or quadruple connectors.

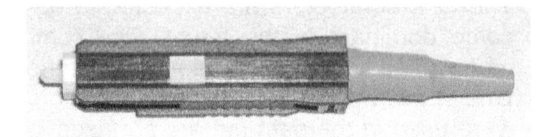

Figure 5-18: MU Connector

5.5.1.4 VOLITION™

The Volition™ duplex connector from 3M[21] is a plug and jack system (Figure 5-19, a jack open with darkened fibers in grooves, and Figure 5-20). The Volition incorporates a design feature that is both evolutionary and revolutionary: 'V'-grooves instead of ferrules.

V-grooves are evolutionary, in that they have been used for alignment of fibers in fusion splicers since the 1970's. V-grooves are revolutionary in that they have not been used in connectors prior to the Volition.

3M used V-grooves for alignment of the fibers: in the jack, fibers rest in precision, molded, plastic V-grooves (Figure 5-19,

---

[21] The generic name of the Volition™ connector is SG. In addition, the Volition is known as the VF-45.

top). In the plug, a custom fiber[22], floats in space. The plug fibers slide into the V-grooves of the jack, with contact of the mating fibers maintained through pressure created by bent fibers.

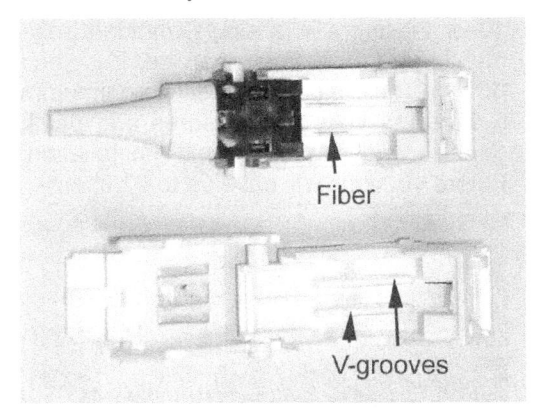

Figure 5-19: Volition™ Jack

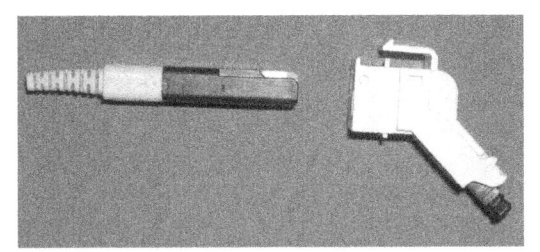

Figure 5-20: Volition™ Plug And Jack

The Volition plugs and patch cords are factory made. The jacks are field installable onto a custom Volition cable. This custom cable is required, since the jack requires fibers with a primary coating of 250µ.

This diameter is smaller than the 900µ tight tubes of commonly used premises cables. Because of this requirement, it is not possible to retrofit existing fiber optic tight tube cable, networks with the Volition connectors.

5.5.1.5 FIBER JACK

In January 1997, Panduit Corporation introduced the first duplex SFF plug and jack connector system, generically known as a Fiber Jack (Figure 5-7). This connector system, trade name 'Opti-Jack™', introduced a new package based on four existing and well proven technologies: 2.5 mm ferrules, quick cure adhesive installation, split alignment sleeves, and an RJ-45 form factor. Panduit's use of well-proven technologies eliminated the risk of using this 'new' product. By its appearance, the Opti-Jack™ is the most rugged of the SFF connectors and is well suited for fiber to the desk (FTTD) applications. Current products require the 'cleave and crimp' installation method.

The 2.5mm ferrules are the same as those in the ST-compatible, SC, FC, FDDI, and ESCON™ connectors. Quick-cure adhesives had been in use since 1992 and were well developed. The split alignment sleeves were commonly used in multimode adapters since 1986. The form was the same as that of the RJ-45, with which network installers and users were familiar.

The Opti-Jack™ is keyed, contact, moderate loss, pull proof and wiggle proof. It is a duplex connector with one ferrule per fiber. This structure allows separate alignment of each fiber for minimum power loss. Installation of the current product is by the cleave and crimp method (5.6.5).

5.5.1.6 MT-RJ

The MT-RJ is available in two versions, the TYCO version (Figure 5-21) and the Corning Cable Systems version (Figure 5-22). The MT-RJ is keyed, contact, and pull proof with two fibers in a single ferrule. Installation is by the 'cleave and crimp' method. Photographs are ~45% actual size.

The MT-RJ is keyed, contact, and pull proof with two fibers in a single ferrule. Installation is by the 'cleave and crimp' method. Manufacturers have reduced their support of this connector type. As a result, this type is rarely used in new installations.

---

[22] This fiber, known as GGP, for glass/glass/polymer, has a glass core, a glass cladding to a diameter of 110 µ and a hard plastic over cladding to a diameter of 125 µ.

Figure 5-21: TYCO/AMP MT-RJ System

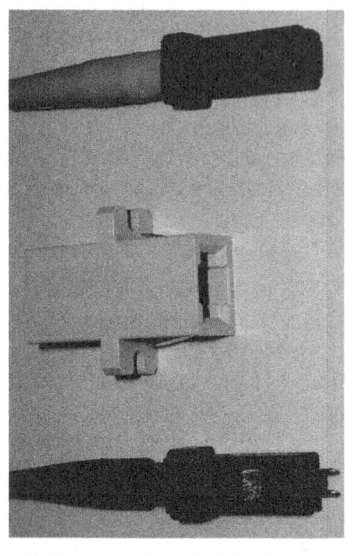

Figure 5-22: Corning Cable Systems MT-RJ System

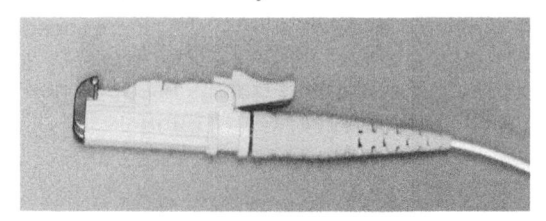

Figure 5-23: E2000 Connector

5.5.1.7 E2000

The SI2000 is a European connector. It has an appearance similar to that of the LX.5. has an integral dust cover (Figure 5-23). It differs from the LX.5 in the structure of the latch.

5.5.1.8 MTP/MPO

MTP/MPO connectors (Figure 5-24) have 12 or 24 fibers in a single, molded polymer ferrule. The fibers can be, but need not be, in ribbons. These connectors reduce the size of enclosures and patch panels and enable rapid installation. Future versions will have up to 72 fibers.

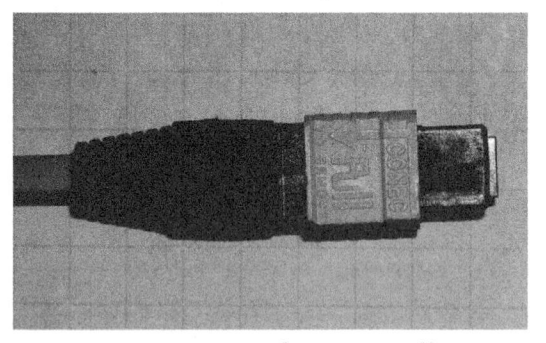

Figure 5-24: MTP® Connector[23]

This MTP is used with pre-terminated cable systems. Such systems are used in data centers. Such centers use pre-terminated cables, with modules on the ends. The modules break out the fibers to single fiber connectors, such as SC or LC. This cable and module system reduces installation time significantly. The costs of pre-terminated systems vary, but are close to the those of field installed systems.

Some 12-fiber MTP/MTO connectors have the same specifications as keyed and contact connectors: 0.75 dB/pair maximum and 0.3 dB/pair typical. 24-fiber MTP/MTO connectors have increased loss: 1.0 dB/pair maximum and 0.4 dB/pair typical.

5.5.2 COMMON TYPES

The two most popular, or commonly used, connector types are the ST™-compatible[24] and the SC.

---

[23] Registered trademark of the US Conec Corp.

[24] ST is a trademark of Lucent Technologies.

5.5.2.1 ST-Compatible

Introduced in 1986, the ST-compatible (Figure 5-25 and Figure 5-5) is keyed, contact, moderate loss connector that is neither pull proof nor wiggle proof. Its installation is significantly easier and faster than that of predecessor connector types. The ST-™ compatible connector has a cost lower than that of the SC. However, the ST™-compatible connector requires more space on a patch panel than does the SC. This space is required for rotation of the retaining ring. A consequence of this space requirement is increased patch panel and enclosure cost. In locations requiring more than 12 connectors, the ST-™ compatible connectors can have a total installed cost higher than that of the SC connectors.

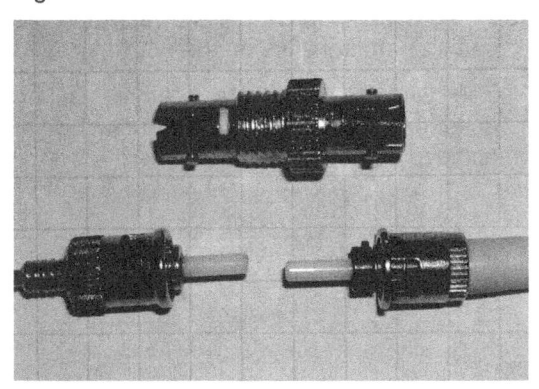

Figure 5-25: ST-™ Compatible Connector

5.5.2.2 SC

First available in the US in 1988, the SC (Figure 5-26 and Figure 5-27) was developed by Nippon Telephone and Telegraph (NTT) in Japan. The SC became commonly used after it became the connector type required for compliance with the Building Wiring Standard, TIA/EIA-568. Because of its exclusive recommendation in TIA/EIA-568 and TIA/EIA-568-A, the SC dominated multimode data installations during 1995-2000.

The SC is keyed, contact, moderate loss, pull proof and wiggle proof. Because of these latter two characteristics, the SC is more reliable than the ST-compatible.

The pull and wiggle proof behaviors result in a price higher than that of the ST-compatible connector. This behavior makes damage of the fiber difficult by controlling the force with which ferrules make contact. In contrast, the installer can apply excess contact force during the insertion of an ST-compatible connector.

A second benefit of this behavior is immunity to damage from 'pull and snap'. An SC connector cannot be pulled and allowed to snap back into an adapter. An ST-compatible connector can. Usually, such action results in damage to the fiber in an ST-™ compatible connector.

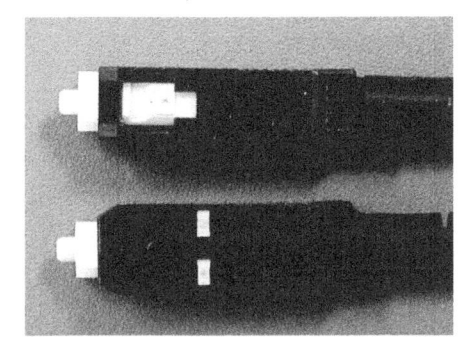

Figure 5-26: SC Simplex Connector

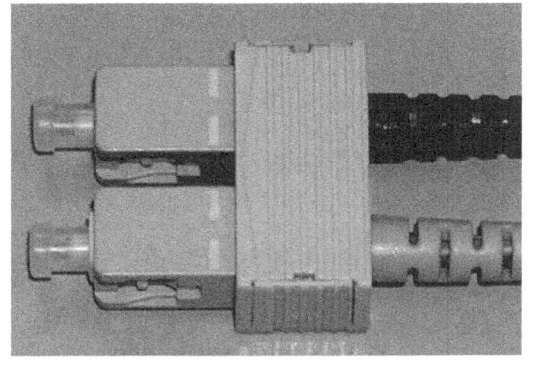

Figure 5-27: SC Duplex Connector

This author has observed this improved reliability: during fiber optic connector installation training, we experienced a damage frequency for ST-compatible reference leads that is 3-6 times that of SC reference leads. Thus, in spite of the increased price, the life cycle cost of the SC connector will be lower than that of the ST-compatible connector.

The SC differs from the ST-compatible in two additional aspects: insertion method and ability to be duplexed. The SC insertion method is 'push on/pull off'. Because of this method, the SC connectors can be more closely spaced in a patch panel than can the ST-compatible, which requires space for rotation. In practice, this reduced spacing results in a connector density of two to four times that possible with ST-compatible connectors. This increased density results in a reduction in the total installed cost of SC.

The SC can be duplexed, either through use of left and right outer housings or an external clip (Figure 5-27). This capability results in increased network reliability because it is difficult to reverse the fibers in a duplexed SC connector. Thus, use of SC connectors results in three cost reductions:

- Reduced life cycle cost
- Reduced enclosure cost
- Increased link reliability

5.5.3 LEGACY CONNECTORS

In this section, we present the legacy connectors, those that are not commonly used. These connectors are:

- SMA 905
- SMA 906
- Biconic
- Mini-BNC
- ESCON
- FDDI, MIC
- FC

5.5.3.1 SMA, BICONIC AND MINI-BNC

Early connector types, such as the SMA 905 (Figure 5-28), SMA 906 (Figure 5-29), Biconic (Figure 5-30), and mini-BNC (Figure 5-31) had retaining rings and flat end faces with deliberate air gaps. The gaps resulted in high reflectance and increased power loss. These four types are not keyed.

Figure 5-28: 905 SMA Connector

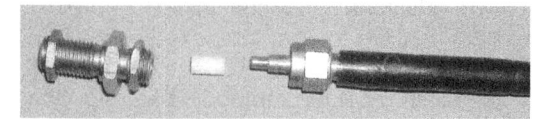

Figure 5-29: 906 SMA Connector

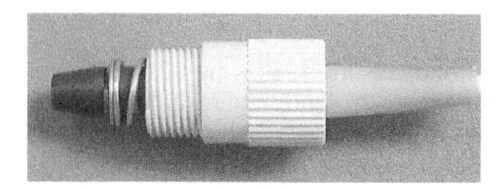

Figure 5-30: Biconic Connector

5.5.3.2 FC AND D4

Originally developed in Japan, the FC (Figure 5-32) and D4 (Figure 5-33) connectors have some of the characteristics of both the ST-compatible and the SC connector types. Like the ST-compatible, the both have a rotating retaining rings, which requires a relatively large spacing in a patch panel. Like the SC, both are keyed, contact, moderate loss, pull proof and wiggle proof. The FC/APC has a beveled end face and the same characteristics as the FC.

The D4 has a unique feature. During assembly, the key can be adjusted to achieve the lowest possible loss.

Figure 5-31: Mini BNC Connector

Figure 5-32: FC Connector

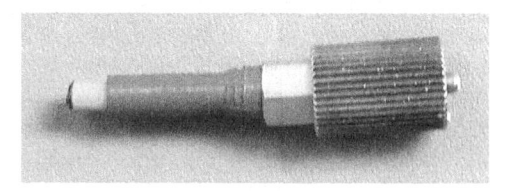

Figure 5-33: D4 Connector

5.5.3.3 FDDI AND ESCON

All the previous connectors were simplex. The last two, FDDI (Figure 5-34) and ESCON (Figure 5-35 and Figure 5-36) are duplex connectors. Both use the 2.5 mm ferrule. Both are large, keyed, contact, pull proof and wiggle proof. The FDDI connector is has a fixed shroud. In contrast, the ESCON connector has a movable shroud. Cleaning the ferrules on an ESCON connector is easy. In Figure 5-36, the shroud has been pushed back to reveal the ferrules. Photographs are reduced in size.

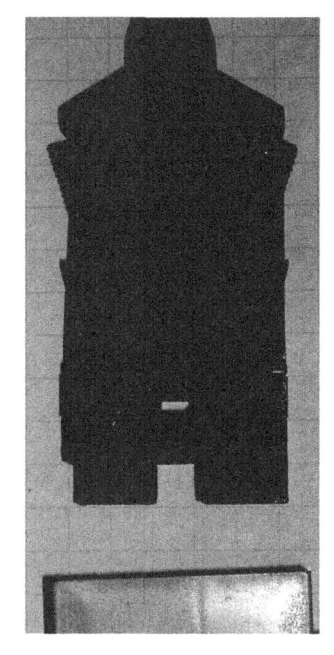

Figure 5-34: FDDI Connector

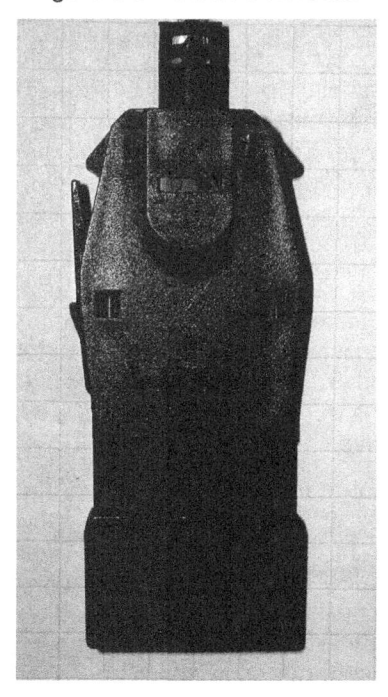

Figure 5-35: ESCON Connector

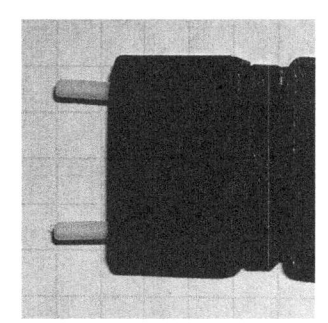

Figure 5-36: ESCON Ferrules Exposed

5.6 INSTALLATION METHODS

Fiber optic connectors can be installed by many methods. These methods were developed to address different concerns, such as reliability, convenience, time to install, and cost to install.

Most of the development of methods has been to reduce one or more of the cost factors. In most cases, the effort to reduce the total installed cost resulted in increased connector cost. These methods have become reasonably well received:

- Epoxy
- Hot melt adhesive
- Quick cure adhesive
- Crimp and polish
- No polish, no adhesive
- Fuse On or Splice On

5.6.1 EPOXY

The epoxy, cure, and polish method was the first method used for installation of connectors. These epoxies can require a slow cure at room temperature, a fast cure at room temperature, or a slow or fast cure at an elevated temperature.

This method has four advantages:

- High resistance to degradation
- Loss stability over a wide temperature range
- High installation process yield
- Ability to be used with the lowest cost connectors

Epoxies are considered to have the high resistance to a wide range of conditions. As such, epoxies can provide the highest reliability connectors. Other adhesive systems have reduced resistance. As an example, connector epoxies can resist degradation to temperature of 105°C., while the original Hot Melt™ adhesive system was specified to 85°C.

Epoxies provide a good match of thermal expansion coefficients of epoxy, fiber and ceramic ferrules. This matching results in minimal movement of the fiber in the ferrule over a wide temperature range. Such limited movement results in stable power loss. The temperature stability test data this author has reviewed indicate loss stability of ±0.1-0.2 dB over a wide temperature range.

Use of epoxies tends to result in high process yield.[25] This high yield is a result of the bead on the ferrule (Figure 5-37). This bead supports the fiber during polishing. This support nearly eliminates damaged ends, which represent 95% of connector losses during installation.

Figure 5-37: High Yield From Large Bead

Labor cost can be low for factory-installed connectors. The combination of low labor cost, low epoxy connector cost, and high process yield favors the use of epoxies as a factory installation method.

This method has three disadvantages.

- Inconvenient use
- Low installation rate
- Need for power

The first two disadvantages are related as they translate to high installation man-hours per connector. The third disadvantage is the need for power to heat curing ovens.

The use of epoxy requires mixing of two parts, transfer of the mixture into a syringe or automated injection mechanism, and clean up of excess epoxy. Because of these steps, the rate of installation of epoxy connectors is low, typically 6-8 per man-hour. This low rate results in a high total installed cost when field installation labor rates apply. Such rates exceed $30/hour. In comparison, factory installation labor rates can be $10/hour and below. In developing countries, this rate can be significantly less.

In addition, the fraction of the total time spent in installation by field installers (or percent utilization) is much lower than that for factory installers. Field utilizations can be 80% and below, while factory utilizations are above 90%.

Reduced labor utilization increases the cost impact of the low installation rate with epoxy. Because of the potentially high labor cost of this method, connector manufacturers developed other methods.

5.6.2 HOT MELT™

Because of the inconvenience and time impact of the installation with epoxy, 3M developed the Hot Melt installation method. In this method, a hot melt adhesive is pre-installed into the connector. The installer preheats the connector to soften the adhesive so that he can install the fiber. The installer installs the fiber and/or cable into the connector and allows the connector to cool in a heat sink (the cooling stand).

Once cooled, the installer removes the excess fiber and polishes the end in a two step polishing procedure. This installation method requires a 3M polishing film, which is not easily clogged by the hot melt adhesive.

The hot melt process eliminates the time factors of preparing and injecting the epoxy. In addition, this method eliminates the mess and inconvenience of epoxy. These eliminations allow an increased installation rate, 12-16 per hour. This increased rate can result in a reduced installation cost.

However, the Hot Melt™ connectors are more expensive than are epoxy connectors. In addition, the Hot Melt method requires power for the heating oven and

---

[25] Yield is the ratio of the number of acceptable connectors to the total number of connectors installed.

proprietary oven, holders and polishing film. In spite of these factors, many installation situations can achieve a total installed cost with the hot melt method that is lower than that with the epoxy method. This is due to the favorable pricing of the 3M hot melt connectors.

One potential disadvantage is reduced upper temperature operation. The original hot melt connectors were rated for use up to 85°C. However, one version of the singlemode connector is rated for use up to 100°C.

5.6.3 QUICK CURE ADHESIVE

Two manufacturers, Lucent Technologies and Automatic Tool and Connector, addressed the disadvantages of the proprietary parts of the Hot Melt method and the inconvenience of the epoxy method. They developed 'quick cure' adhesive methods.

The first advantage is elimination of the requirement for power. These methods, the two part adhesive (from Lucent Technologies and Automatic Tool and Connector), enabled an increase in the rate of installation. This increased rate, typically 15/hour, can result in reduced installation cost.

A second advantage of quick cure adhesives is their compatibility with two ferrule materials: ceramic ferrules and some liquid crystal polymer (LCP) ferrules. This combination of low cost connectors and high installation rate can result in a low total installed cost. Installers can achieve this low installed cost if they achieve a high process yield.

There are three disadvantages of quick cure adhesives. The first two disadvantages are premature hardening of the adhesive and minimal support of the fiber during polishing. These two disadvantages contribute to a reduced process yield and increased cost.

Premature hardening occurs when the fiber is coated with the hardener or accelerator and inserted into a connector loaded with adhesive. If the installer inserts the fiber slowly, the adhesive cures and the fiber locks up before it is fully inserted. In this situation, bare fiber exists inside the connector. Such bare fiber causes a reduction in the reliability of the connector.

For example, SC connectors allow the fiber inside the back shell to flex slightly during insertion into a patch panel or a receptacle. Repeated flexing of such bare fiber can result in breakage of the fiber. This author expects this failure to occur in any connector that is pull proof and wiggle proof, such as the FC, LC, and Opti-Jack. This author has observed such failure in as few as 5 cycles.

The minimal support of the fiber during polishing occurs because the adhesive cannot produce a large bead on the tip of the ferrule. This inability is due to the nature of the adhesive: quick cure adhesives are edge-filling adhesives. Unlike the bead on epoxy or hot melt connectors, the small bead on the tip of the ferrule can allow the fiber to break below the surface of the ferrule during end finishing. Without extreme care during end finishing, the installation yield can be low.

For example, typical yield during training of novices installing epoxy and Hot Melt™ connectors is 90-95%. Typical yield for these same novices with quick cure adhesives is 50-60%. However, typical yield of quick cure adhesive connectors for experienced professionals is 90-95 %.

The third, and final, disadvantage of the use of quick cure adhesives is reduced reliability. Some of the quick cure adhesives have exhibited degradation when exposed to a wide temperature range, a wide humidity range, rapidly changing temperature, or rapidly changing humidity. While there are exceptions, quick cure adhesives are usually used for networks within office buildings.

5.6.4 CRIMP AND POLISH

Several manufacturers, AMP, 3M and Automatic Tool and Connector, have offered connectors that have a mechanical method of gripping the fiber. These connectors require polishing. During polishing, these products are more susceptible to damage than are connectors installed with quick cure adhesives. This susceptibility to damage occurs because there is no bead supporting the fiber. Of the five methods presented thus far, this method has the lowest acceptance.

5.6.5 CLEAVE-AND-CRIMP

For all the installation methods presented, the largest cause of reduced yield and increased cost is damaged or shattered fiber ends. This damage occurs during polishing. In addition, a significant amount of time is consumed by preparation, injection, and use of adhesives.

The 'cleave and crimp' installation method addresses both yield loss and time consumption by elimination of both adhesives and polishing. Such elimination offers the potential of increased installation rate and reduced installation cost. This increase in installation rate may be only a potential cost reduction, since the cost of these connectors is higher than that of connectors installed by other methods.

'Cleave and crimp' is a generic term for no epoxy, no adhesive, no polish connector products with the trade names LightCrimp™ Plus (TYCO/AMP), Opti-Crimp® (Panduit Corporation), Unicam™ (Corning Cable Systems), and others.

All these products require the same steps:

- Cleave the fiber
- Insert the fiber
- Crimp the connector to the fiber

The connector contains a pre-installed fiber stub that has been polished by the manufacturer. In essence, this connector has a mechanical splice in its back shell.

The primary advantage of this installation method is reduced installation time, with the potential for reduced installation cost. Field installation rates can be 30-40/hour. One web site claims 60/hour. A second, potential, advantage is reduced training cost.

The realization of reduced installation cost depends on five factors. Since the cost of connectors with this method is high and these five factors vary widely from installation to installation, the increased installation rate of this method may or may not result in reduced total installation cost.

These five factors are:

- Total loaded hourly labor rate
- Labor time utilization
- Installation rate
- Installation yield
- Connector cost

These factors do not always combine to produce the lowest total installed cost. In spite of the complexity of the cost analysis, we can develop some rough guidelines. The higher the total loaded labor rate, the more likely the cleave-and-crimp method will result in the lowest total installed cost.

As the number of connectors per location drops, the labor time utilization drops. As this utilization drops, the probability of this method for lowest total installed cost increases. Such situations include the desk locations in a fiber to the desk network (FTTD) but not necessarily a central equipment room in a vertical back bone or in an FTTD network.

This author's testing and use of more than 10,000 cleave and crimp connectors in training indicate that the cleave and crimp method yields and power losses are close to those of epoxy, Hot Melt and quick cure methods.

5.6.6 FUSE ON CONNECTORS

The final installation method is that of 'fuse on' or 'splice on' connectors (SOC). These connectors have a pre-installed fiber with a fiber stub protruding from the back shell. The fiber in the connector is fusion spliced to the fiber in the cable. Such products are available from Corning Cable Systems, Alcoa-Fujikura, Sumitomo, Fitel, Clearfield, and Diamond.

Some of these manufacturers require use of a custom splicing machine. Others require use of a special holder with a stan-dard splicing machine. These latter machines have removable holders.

Such connectors are useful in situations of replacement or limited space. In replacement, the existing enclosure does not allow space for splice trays. In some situations, such as military ships, the space available does not allow for the space required by enclosures or splice trays. Such connectors have a cost higher than that of those of other installation methods.

All such products are one use. In other words, if the connector is high loss, either because of contamination of either cleave surface or because of a bad cleave, the connector is lost.

Recent information from users suggests that the power loss and process yield for this type of product are both acceptable.

5.7 PIGTAIL SPLICING: THE LOWEST COST METHOD

In the past, fusion splicer cost was much higher than at present. However, the availability of splicing machines in the $5000-$8000 range changes the total installed cost analysis of fusion spliced pigtails. If the total number of is between 725 and 3000, the fusion splicing machine will pay for itself in reduced total installation cost. After paying for itself, the splicing machine becomes an investment that continues to pay dividends. This analysis is available at:
http://www.ptnowire.com /tpp-V3-I2.htm.

The advantages of this method are low cost pigtails, high process yield (nearly 100%), fast installation at rates from 12=24/hour, high reliability, and factory low loss and low reflectance.

The main disadvantages are increased enclosure space and cost for splice trays. In summary, the best, meaning lowest cost, method of installing connectors can be to avoid installing them!

5.8 SUMMARY

In summary, the installer needs to know the following information about the connectors he is to install.

- Connector type
- Core diameter
- NA
- Maximum loss, dB/pair
- Average loss, dB/pair
- Repeatability, or range
- Reflectance

The installer needs to know the connector type, the core diameter and the NA to choose reference leads for testing. He needs to know the maximum and typical connector loss in order to calculate the acceptance value for certification of network (19).

He needs to know the range, or the repeatability, for troubleshooting. The range will indicate the maximum increase in loss that he can accept as indicating normal behavior of the connectors. Finally, He needs to know the reflectance requirement in order to certify the connectors.

5.9 REVIEW QUESTIONS

1. True or false: contact connectors have higher loss than non-contact connectors.

2. True or false: contact connectors have lower reflectance than non-contact connectors.

3. True or false: contact connectors dominate today's market.

4. True or false: non-contact connectors are common today.

5. True or false: connector loss is a maximum of 0.75-dB/ connector.

6. True or false: more reflectance is better than less reflectance.

7. True or false: more reflectance provides lower BER than less reflectance.

8. True or false: small form factor connectors are legacy connectors.

9. True or false: SFF connectors drive network costs up.

10. True or false: a D4 is a SFF connector.

11. True or false: an MT-RJ is a simplex connector.

12. What characteristic of the fiber contributes to the loss of non-

contact connectors but not to that of contact connectors.

13. List 9 characteristics of the connector structure.

14. How many types of end faces are there?

15. What are the types in Question 14?

16. List 5 connector colors.

17. What does each color indicate?

18. What does SFF mean?

19. The diameter of an SC ferrule is:

20. The diameter of an LC ferrule is:

21. Is an LC connector a simplex or duplex?

22. Is the MTP/MPO a simplex or duplex connector?

23. True or false: There are two ferrules in a Volition connector.

24. What type of connector is this?

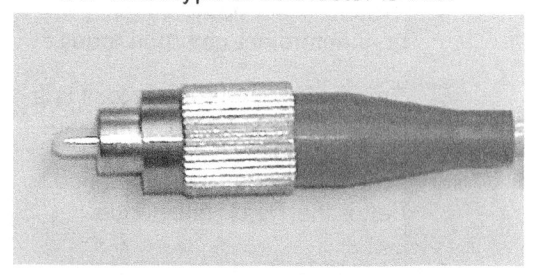

25. What type of connector is this?

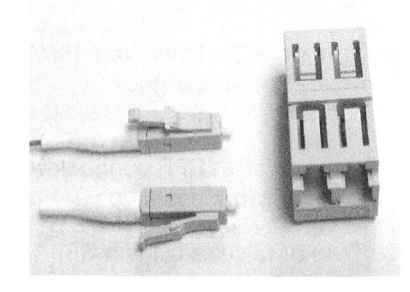

26. What type of connector is this?

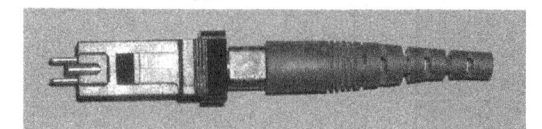

27. What type of connector is this?

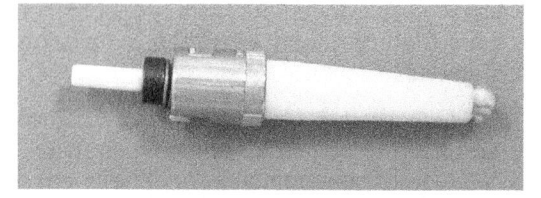

28. Why is the range of a connector important?

29. You observe a network technical start to install a green connector into a blue barrel. What should your response be?

30. True or false: when an installer installs connectors, he will often see the maximum loss value.

31. True or false: a normal experience is a measured value of 0.75 dB/connector.

32. True or false: a bead of the tip of the ferrule is undesirable for polishing.

33. Make a chart of all connector types presented in the text. On that chart, indicate the following: type; contact or not; latching mechanism, (PO/PO or ring); mating type, (plugs or plug and jack); end face type; simplex, duplex, or both; keyed or not; pull proof or not; wiggle proof or not.

34. What is the reason that the angle on an APC connector is 8°?

35. What is the reason that an MTP/MTO connector can be considered a SFF?

36. What connector installation method is most often used in cable assembly factories?

37. Define PC.

38. Define UPC.

39. Define APC.

40. What characteristic is differentiated by the designations in Questions 38-40?

41. Which of the designations in Questions 38-40 and identified in Question 41 has the highest/best performance?

6 SPLICES

Chapter Objectives: you learn the process, types, advantages hardware, performance, and language of fusion and mechanical splicing.

6.1 FUNCTIONS & LOCATIONS

Splices are permanent joints between two fibers. Splices are the preferred method of connection whenever the fibers need not be disconnected or the signal rerouted at the location of the connection. A splice has two functions:

- Low loss
- High strength

Installers splice in two locations: mid span and pigtail. Installers perform mid span splicing whenever the following conditions exist:

- The cable cannot be obtained in a single length
- The cable cannot be installed in a single length
- The cable has experienced backhoe fade, posthole driller fade, rodent fade, shark fade or any other cause of broken fiber(s)

Pigtails are short length of cables or tight buffer tubes with a connector installed on one end. Installers perform pigtail splicing in order to reduce installation cost or installation time. Instead of field installing singlemode connectors, installers perform singlemode pigtail splicing.

6.2 PROCESS

The process of splicing requires the following five steps:

- Cable end preparation
- Enclosure preparation
- Splicing
- Placing fiber and buffer tubes in enclosure
- Closing the enclosure

The last two steps are known as 'dressing the enclosure'.

6.3 TYPES

6.3.1 INTRODUCTION

The two methods of splicing are:

- Fusion
- Mechanical

In both methods, two prepared fiber ends are core-aligned or cladding aligned to provide low power loss. Each splicing method has different characteristics, requirements and advantages.

6.3.2 TOOLS

The installer requires the same basic tool kit for both methods of splicing. This kit consists of cable end preparation tools and a precision cleaver. This basic kit costs approximately $2000 with a precision cleaver. With a low cost cleaver, the kit can cost $500-$1000. This author does not recommend use of low cost cleavers.

Both methods require cleaving of the fiber ends to produce nearly perfect end faces. Such end faces are flat and perpendicular. With a precision cleaver, typical end face angles are less than 1°.

Fusion splicing requires one additional tool, the fusion splicer. Mechanical splicing may require a tool to hold the splice during installation.

6.3.3 FUSION SPLICING

Fusion splicing is the process of fusing, or welding together, two fibers. When fiber ribbons are spliced together, the method is 'ribbon splicing' or 'mass splicing' (4.3.3).

6.3.3.1 PROCESS

This method requires a fusion splicer. The fusion splicer provides two functions: the precise alignment of the fibers to each other prior to splicing and control of the splicing operation.

The fusion splicer provides precise fiber alignment in two ways:

- Passive alignment
- Active alignment

The splicer provides passive alignment through the use of a precision 'V' groove (Figure 6-1). With such a groove, the splicer design operates with two implicit assumptions;

1) the fiber diameters and
2) the core-cladding concentricity are precise enough to achieve low power loss.

These assumptions are valid for both the fiber made in North America and for much, but not all, of the fiber made overseas.

Figure 6-1: 'V-Groove'

Active alignment enables the splice to have the lowest loss, in spite of conditions of imperfect fiber. In other words, active alignment compensates for the lack of perfection in one or both of the fibers. Such lack of perfection includes the following conditions:

- different core diameters
- different MF diameters
- different cladding diameters
- offset cores
- non-circular cladding

The splicing machine provides active alignment with one of two mechanisms:

- Profile alignment (PAL or PAS)
- Local injection and detection (LID)

In order to provide alignment by the profile method (Figure 6-2), the splicing machine incorporates a collimated light source, a microscope, image digitization software and image analysis software. The collimated light makes visible the core-cladding boundaries of both fibers. The digitized images are analyzed by software to define this boundary in perpendicular views. Control software adjusts one fiber so that the core offset is minimized in both axes.

Such alignment occurs for singlemode fibers. For multimode fibers, the splicer may align the cladding to minimize loss.

After the splice has been made, this same software analyzes the image to provide a value of the splice loss. This value is an estimate. The installer can measure the accurate splice loss only with an OTDR. In addition, accurate splice loss is an average of the loss in both directions (15.7.4.3).

Figure 6-2: Profile Alignment Splicer

In order to provide alignment by the local injection and detection method, the splicing machine includes a laser light source and a detector. Both source and detector are incorporated into structures that bend the fibers being spliced. The bend at the source launches light into one fiber; the bend at the detector taps light from the second fiber. Control software adjusts one fiber so that the power at the detector is maximized.

After the splicing machine aligns the fibers, it creates an RF electrical arc across the fibers. The machine controls the arc current, arc time and overrun. These parameters determine the power loss and strength of the splice. As these parameters are different for different fibers, the installer must set them. This setting is by choosing the type of fiber from a menu. The menu defines these parameters.

The arc current and time control the energy provided to the fibers. This energy controls the strength of the splice. The correct overrun provides results in a splice with the same diameter as the fibers.

6.3.3.2 USE

Installers use fusion splicing in most initial installations and in restoration. However, for rapid restoration and for splicing in manholes, installers use mechanical splicing.[26]

6.3.3.3 ADVANTAGES

The four advantages of fusion splicing are:

- Low power loss
- Low to no reflectance
- High strength
- Low cost

Often, fusion splices result in 0 dB loss. In contrast, mechanical splices rarely exhibit 0 dB loss. Because of this low loss, applications that require lowest possible loss require fusion splicing with active alignment.

Theoretically, there could be low reflectance: if the RIs of the two fibers are slightly different, some reflection could occur. However, even with 36 years of experience in fiber optics, this author has never seen reflectance from a properly made fusion splice Evidently, the difference between the RIs is so low that no reflectance is the rule.

The only unique consumable required for fusion splice is the splice cover. Low cost per splice results from the low cost of the splice cover ($0.40-$1.00).

High strength results from the welding process, since a properly made splice is stronger than both fibers. The exception to this strength advantage occurs in mechanical splices that use an adhesive. Such splices may have strength equal to that of fusion splices.

6.3.3.4 DISADVANTAGES

Fusion splicing has two disadvantages:

- High cost of the fusion splicer
- Potential for disruption of the profile of a multimode fiber (3.4.2)

At the time of this writing, fusion splicer cost ranges from $5000 to $30,000. The cost of used fusion splicers can be as low as $3000. The more expensive the fusion splicer, the faster it operates. In addition, the ribbon fusion splicer has a cost higher than that of single fiber splicer. For a small number of splices, fusion splicing may be cost prohibitive, even with a rented splicer.

The disruption of the multimode core profile is a possibility with mixed data support. In December of 2001, FOTEC and Pearson Technologies Inc. performed such testing with a prototype FOTEC multimode bandwidth tester. This testing demonstrated a reduction in bandwidth through mid-span fusion splices of standard multimode fibers. The reduction was approximately 30 % on multimode links of 200-220 m. The same links exhibited high bandwidth when the mid-span location contained mechanical splices or connectors. This testing suggested the possibility of failure of gigabit links with mid span splices when the links were near the maximum distance of the 803.2ae standard. However, testing by fiber manufacturers indicates no significant bandwidth reduction on fusion spliced, laser-optimized fibers.[27]

6.3.4 MECHANICAL SPLICING

Mechanical splicing is the process of inserting two prepared fiber ends into a mechanical splice (Figure 6-3). The mechanical splice provides two functions:

- Precise alignment of fibers
- The function of a splice cover

---

[26] Manholes can collect methane, which could be ignited by the arc of a fusion splicer.

[27] Private communication, Phillip Bell (Corning Optical Fiber) to the author, 3/3/05.

Most mechanical splices are either multimode or singlemode. One, the 3M FibrLok™ can be used on either multimode or singlemode fibers. This author recommends use of this product, part number 2529. Of course, it may be possible to use a singlemode mechanical splice on multimode fiber.

6.3.4.1 ALIGNMENT

Mechanical splices create precise alignment by at least three mechanisms. Some splices have a precision capillary tube for alignment. Other splices use a precision etched silicon substrate. Finally, one splice, the 3M Fibrlok II®, has a precision 'V' groove that is unique.

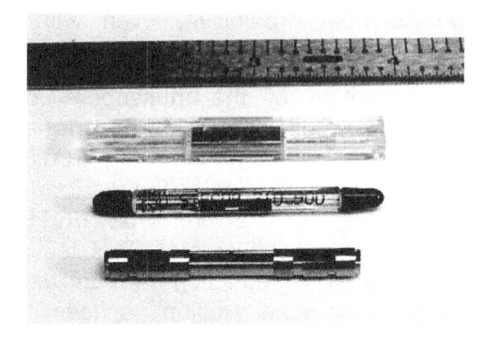

Figure 6-3: Mechanical Splices

6.3.4.2 GEL

At their centers, all mechanical splices have index matching gel. This gel fills any air gap that may result from the end faces of either fiber being less than perfectly perpendicular. Such filling reduces loss and reflectance. In some singlemode fibers, the index matching gel eliminates all reflectance.

6.3.4.3 ADVANTAGES

Mechanical splicing is used in three situations:

- Fast restoration
- By organizations with a large number of splice teams
- On old multimode fibers

In fast restoration, both ends of a length of cable are prepared and inserted into mechanical splices. The splices are placed in trays. The trays are placed in enclosures. This cable and enclosures form a fast restoration kit.

When a cable is broken, the installer brings the kit to the broken cable, prepares two ends, and inserts these ends into the fast restoration cable. Through advance preparation, the installer reduces restoration time by approximately 50 %.

Organizations with a large number of splicing teams may choose mechanical splicing because of low tool cost: the installer requires no fusion splicer. Purchase or rental of a fusion splicer can increase the total cost above that of mechanical splicing.

A third situation is splicing of old multimode fibers: the melting of the fibers can disrupt the core profile, resulting in reduced bit rate capacity. This reduction may not be significant for bit rates below 200 Mbps, but can be significant for Gigabit Ethernet rates and above on old multimode fibers.

6.3.4.4 DISADVANTAGES

Mechanical splicing has one major disadvantage and one minor disadvantage. The major disadvantage is high cost in large installations: at a price of $8-$26, a 1000 splice installation will cost $8000-$26,000. At $8,000, a fusion splicer becomes an cost favorable alternative.

A minor disadvantage is increased installation time: the fibers may need to be tuned to achieve low loss. Tuning is the process of rotating one or both fibers to obtain the desired loss. Some singlemode mechanical splices allow tuning. Some singlemode mechanical splices require tuning.

6.4 SPLICE HARDWARE

Splice hardware consists of four primary components and secondary components. The primary components are the:

- Fusion splice cover
- Splice (Figure 6-4)
- Splice tray (Figure 6-5)
- Splice enclosure (Figure 6-6 and Figure 6-7)

The secondary components include: gaskets, internal, patch panels, strength member gripping mechanisms, moisture seals, grounding strips, pressure valves, locking mechanisms and plugs.

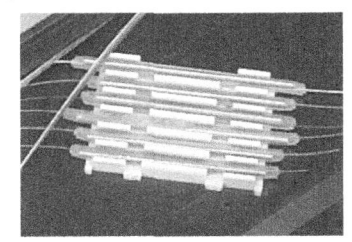

Figure 6-4: Fusion Splices in Splice Covers[28]

Figure 6-5: Half Length Splice Tray

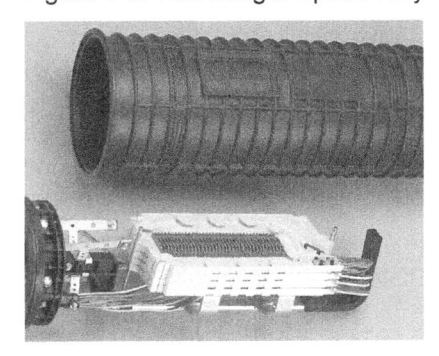

Figure 6-6: Outdoor Splice Enclosure

Figure 6-7: Indoor Splice Enclosure

---

[28] Figures 6-4 to 6-6 are courtesy of Preformed Line Products.

6.4.1 SPLICE COVER

The installer places a splice cover over the bare glass of each fusion splice. This cover isolates the splice from the environment, and supports and protects the splice.

Splice covers can be heat shrinkable (Figure 6-8) or adhesive (Figure 6-9). The shrinkable splice cover consists of two layers of plastic separated by a steel rod. The rod prevents the fiber from bending when the cover shrinks. The installer places the heat shrinkable cover on the fiber prior to splicing.

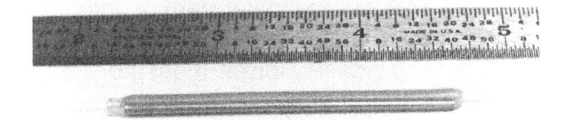

Figure 6-8: Shrinkable Fusion Splice Cover

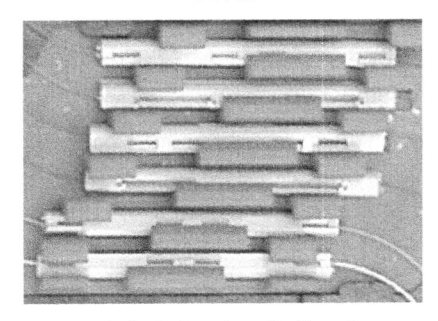

Figure 6-9: Adhesive Splice Covers

In contrast, the installer places the adhesive splice cover on the fiber after splicing, or, when he realizes he forgot to place the heat shrinkable cover on the fiber prior to splicing!

6.4.2 SPLICE TRAY

The splice tray houses and protects the splice and excess fiber (Figure 6-5 and Figure 6-10). Excess fiber is required to allow making the splice outside of the tray. While not a rule of thumb, outdoor splices require 3-4' of fiber for each end of each splice. In contrast, indoor splices may require only 1-2.5' of fiber for each end of each splice.

Trays are of two sizes, half (Figure 6-5) and full (Figure 6-10). Full trays are used in outdoor enclosures. Half trays are used in indoor enclosures and in some FTTH enclosures.

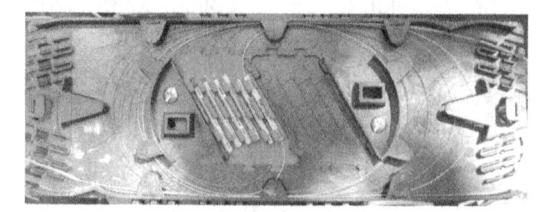

Figure 6-10: Full Length Splice Tray

Trays are designed for the number and type of splice. Most trays will allow 6-36 splices. In addition, the trays will contain a splice holder designed for either fusion splice covers or a specific mechanical splice. This splice holder can be integral to the tray (Figure 6-5) or an insert (Figure 6-10). The splice holder must be compatible with the splice cover or the mechanical splice.

6.4.3 SPLICE ENCLOSURE

The enclosure houses and protects the splice tray(s) and excess buffer tube. Excess buffer tube allows the splice tray to be outside of the splice enclosure while the installer makes the splice. While not a rule of thumb, outdoor splice enclosures can require four to five feet of buffer tube per cable per end. The enclosure can include an internal patch panel, complete with barrels.

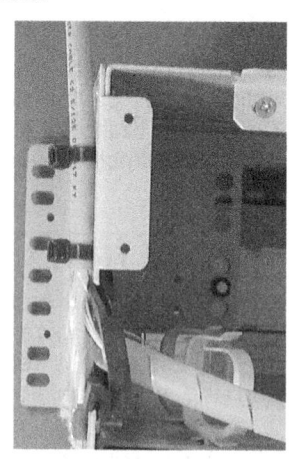

Figure 6-11: Gripping Mechanism For Indoor Splice Enclosure

Indoor (Figure 6-7) and outdoor (Figure 6-6) enclosures have similarities and differences. One similarity is the provision for mechanism(s) to grip the cable strength members, so the cable cannot be pulled from the enclosure (Figure 6-11 and Figure 6-12). Another similarity is incorporation of an integral patch panel (Figure 6-7). A difference is that not all indoor enclosures provide space for splice trays. Another difference is that not all outdoor enclosures include patch panels.

Figure 6-12: Gripping Mechanism For Outdoor Splice Enclosure

Other differences include isolation of the interior from the environment, including moisture and dust. Some outdoor, aerial enclosures do not provide moisture isolation. Instead, they provide weep holes so that condensed moisture can drain. Unlike outdoor enclosures, indoor enclosures rarely provide such isolation.

Some below ground level outdoor enclosures provide for internal pressure to prevent moisture ingress (Figure 6-13). Such enclosures have gaskets to prevent such ingress and release of pressure.

Figure 6-13: Valve For Internal Pressure

6.5 PERFORMANCE

Splice loss is measured in dB. Splice loss of both fusion and mechanical splices is less than 0.15 dB, though some organizations allow higher or require lower values. Properly installed fusion and mechanical splices achieve this value, although fusion splices tend to have lower loss than do mechanical splices.

While maximum loss is less than 0.15 dB, typical loss is less than 0.10 dB for matched fibers. This value is based on the author's experience with more than 10,000 fusion and mechanical splices.

Mismatched fibers will have losses higher than these values. This loss increase is due to mismatched core diameters, MFDs, and mismatched NAs.

As you will learn (15.7.4.3), true splice loss is the average of losses in both directions. Thus, a splice can be acceptable even if the loss in one direction is higher than the maximum acceptance value.

The Fiber Optic Association advanced certification process (CFOS) requires all splices to be a maximum of 0.15 dB. The Building Wiring Standard, TIA/EIA-568 B allows splices to be as high as 0.3 dB. Some organizations have allowed splices to be as high as 0.5 dB, probably for cost reasons. Finally, some organizations, have required maximum values as low as 0.08 dB. In summary, the choice of a maximum splice loss may be more of a political or cost decision rather than a technical decision.

6.6 SUMMARY

The installer needs to know the following characteristics of splices:

- Splice type
- Cladding diameter
- Mode type
- Maximum loss, in dB
- Instructions for installing mechanical splices
- Instructions for installing splices into the enclosure

6.7 REVIEW QUESTIONS

1. From the figures below, identify a fusion splice.
2. From the figures below, identify a mechanical splice.

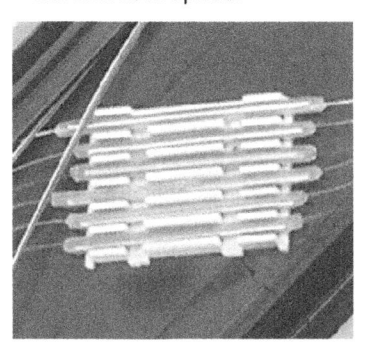

Figure 6-14: Review Figure 1

Figure 6-15: Review Figure 2

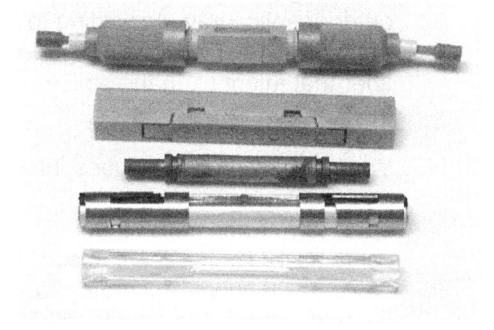

Figure 6-16: Review Figure 3

3. True of False: a properly installed splice has a maximum loss higher than that of a properly installed connector pair.
4. True of False: a splice has a typical loss lower than that of a connector pair.
5. What is the function of a splice cover?
6. In what situation is splicing preferred?

7. In what situation is a splice undesirable?

8. Ribbon splicing is also known as _____.

9. What characteristic does a fusion splice have that a mechanical splice may not have and most connectors do not have?

10. Under what conditions will fusion splicing be less expensive than mechanical splicing?

11. Under what conditions might mechanical splicing be preferred?

12. Identify one situation in which mechanical splicing can be less expensive than fusion splicing.

13. Provide two reasons that mechanical splicing can be more expensive per splice than fusion splicing.

14. Why is there excess fiber in the splice tray?

15. Why is there excess buffer tube in the splice enclosure?

16. You are placing 12 fibers in an outdoor splice tray. According to the text, what is the total length of fiber that might be in the splice tray?

17. You are placing buffer tubes into a mid-span, outdoor splice enclosure. Each buffer tube contains 12 fibers. The cable contains 96 fibers. According to the text, what is the total length of buffer tube that might be in the enclosure?

18. What is the total length of jacket that you might remove from a cable end in order to splice one outdoor cable to another?

19. Provide one reason that fusion splicing can be more expensive than mechanical splicing.

20. Provide a reason to have both heat shrinkable and adhesive splice covers in a splicing kit.

21. What three factors can cause a splice of mismatched fibers to have high loss?

22. Provide a reason that active alignment can produce splice loss lower than that of passive alignment.

7 PASSIVE DEVICES

Chapter Objectives: in this chapter, you learn the installation concerns for, types of, and functions of passive devices.

7.1 INTRODUCTION

A passive device is one that manipulates light as light. No optical to electrical to optical conversion is required.

The use of passive devices is increasing. Such devices are used in two applications: metropolitan telephone networks, long haul telephone networks, and FTTH networks. In these last two types of networks, the most cost-effective method of increasing capacity is addition of wavelengths.

7.2 TYPES

Multiplexing and de-multiplexing wavelengths require passive components. Such components are:

- Couplers
- Splitters
- Wavelength division multiplexors and de-multiplexors
- Optical amplifiers
- Dispersion compensators
- Switch
- Rotary joint

7.2.1 COUPLERS

Couplers (Figure 7-1) combine different optical signals, each with a unique wavelength, onto the same fiber. Coupling can be in one direction (Unidirectional, Figure 7-2) or in opposite directions (bi-directional, Figure 7-3).

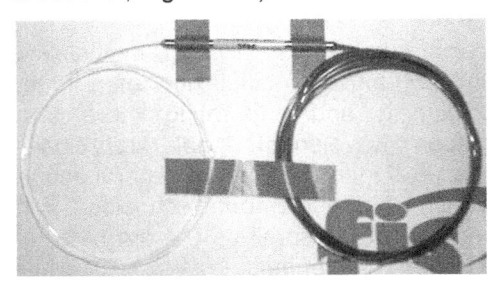

Figure 7-1: Coupler

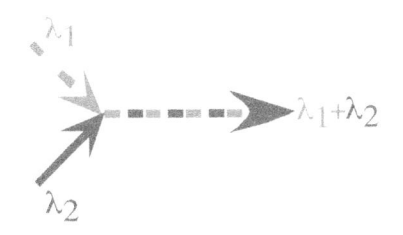

Figure 7-2: Unidirectional Coupler Function

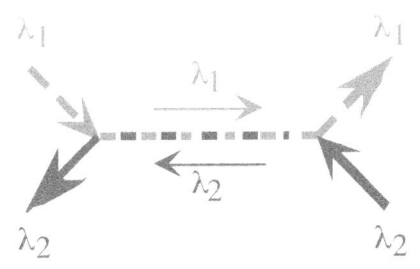

Figure 7-3: Bi-Directional Coupler Function

Two types of optical couplers are:

- Fused, Biconic taper fiber
- Arrayed waveguide grating (AWG)

Fused couplers are singlemode fibers from which the cladding has been removed. The remaining cores are twisted and melted together. This core region is stretched to result in a final diameter equal to the diameter of the original core. Cladding is restored to the core region.

AWGs use the principle of interference to separate wavelengths in one direction. In the opposite direction, the AWG multiplexes wavelengths.

7.2.2 SPLITTERS

Splitters perform the function of splitting a single input signal to multiple outputs (Figure 7-5). If the outputs contain all the wavelengths from the input (Figure 7-4), the device is a simple splitter. Splitters can operate uni-directionally (Figure 7-4 and Figure 7-5) or bi-directionally (Figure 7-3).

Figure 7-4: All Wavelengths In Output Of Splitter

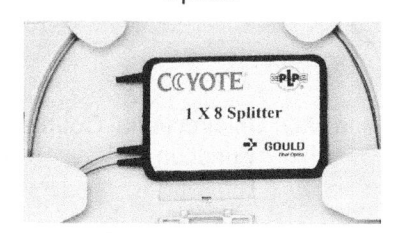

Figure 7-5: A Splitter[29]

If the outputs contain separate wavelengths, the device is a wavelength division de-multiplexor (Figure 7-6).

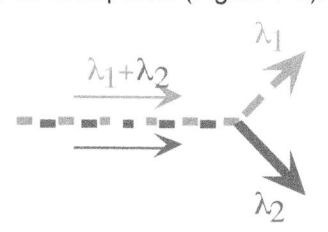

Figure 7-6: Wavelength Division De-Multiplexor Function

7.2.3 WAVELENGTH MULTIPLEXING

7.2.3.1 WDM

WDM is the transmission of two widely separated wavelengths (Figure 7-7). For multimode, these wavelengths are 850 nm and 1300 nm. For singlemode, these wavelengths are 1310 nm and 1550 nm. While possible in custom transmission systems, use of multimode WDM is rare.

7.2.3.2 CWDM

A singlemode technology, coarse wavelength division multiplexing and de-multiplexing (CWDM) refers to the joining and separating of more than two wavelengths that are moderately separated.

CWDM is defined in ITU-T G.694.2. The wavelength range is 1271 to 1611 nm with separations of at least 20 nm between adjacent wavelengths. The range and spacing limit CWDM to 16 wavelengths.

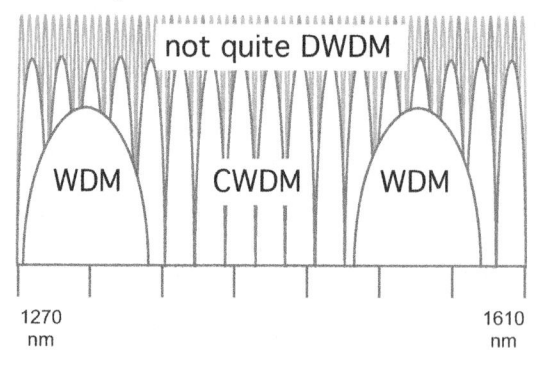

Figure 7-7: CDWM and DWDM

Because of use of wavelengths near 1310 nm, the transmission is limited to ~ 60 km and 2.5 Gbps (OC-48). These limitations make CWDM well suited for metropolitan telephone networks.

FTTH networks are CWDM networks. Transmission is bidirectional on a single fiber. Transmission to the customer is with 1550 nm for CATV signals and 1490 nm for telephone service and Internet signals. Transmission from the customer is 1310 nm for telephone service and Internet signals.

The primary advantage of CWDM is reduced optoelectronics cost. The moderate wavelength separation allows use of uncooled lasers. Such lasers are lower in cost than cooled lasers, which are required for the closely spaced wavelengths of DWDM.

7.2.3.3 DWDM

Dense wavelength division multiplexing (DWDM) and de-multiplexing refers to the combining and separating more than sixteen wavelengths that are closely spaced (Figure 7-7). In its general usage, DWDM is a singlemode technology, with separations as small as 0.4 nm between adjacent wavelengths.

---

[29] Courtesy of Preformed Line Products

.According to the ITU standard G.692[30], as many as 200 wavelengths can be launched into a single fiber (Figure 7-7). These wavelengths can be unidirectional (Figure 7-2) or bi-directional (Figure 7-3). In 1999, Lucent announced a DWDM demonstration of 1000 wavelengths launched into a fiber.

7.2.3.4 OADM

Optical add-drop multiplexors (OADM) and re-configurable optical add-drop multiplexors (ROADM) are devices that add and drop wavelengths (Figure 7-8). Both enable mid-span dropping and adding wavelengths in a link. These two products incorporate circulators, Bragg gratings and array waveguide gratings.

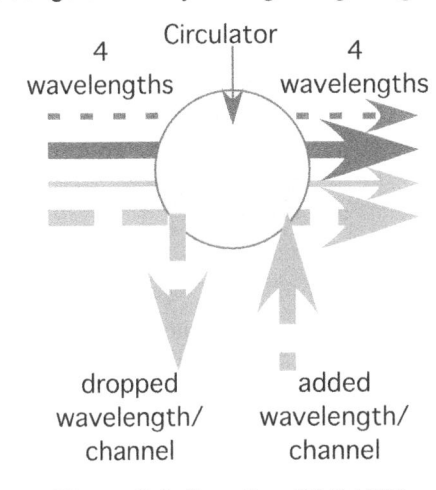

Figure 7-8: Function Of OADM

7.2.4 OPTICAL AMPLIFIERS

Optical amplifiers provide amplification of an optical signal. They require power for operation, but are considered passive devices. Optical amplifiers are so considered because they manipulate light in the optical regime without an optical-electrical-optical conversion (Figure 7-9).

Amplifiers are of two types:

➤ EDFA
➤ Raman

---

[30] The title for this standard is "Optical Interfaces for Multichannel Systems with Optical Amplifiers."

Figure 7-9: EDFA Demonstration System

Such amplification is most commonly performed with an erbium-doped fiber amplifier (EDFA) or a Raman amplifier. The EDFA amplifies in the direction of the transmission. The atoms in the erbium-doped fiber are excited by laser power at a wavelength of 980 nm or 1480 nm. Incoming photons from the signal fiber strike these excited atoms. These excited atoms respond to the disturbance by emitting a photon at the same wavelength as that of the incoming photon. This process repeats itself as the original photons and the emitted photons travel through the amplifier fiber (Figure 7-10). This process can result in a gain of 20 dB.

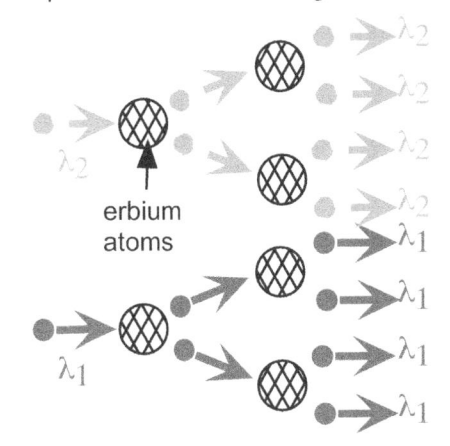

Figure 7-10: The EDFA Amplification Process

Most amazing and useful of the EDFA properties is its ability to amplify a wide range of wavelengths simultaneously. This ability simplifies the design of DWDM networks by allowing a single amplifier for amplifying multiple wavelengths. The alternative would be multiple amplifiers. While a wide range is possible, the gain throughout the wavelength range is not uniform and is adjusted with attenuating filters.

The alternate type of optical amplifier is a Raman amplifier. In a Raman amplification system, the amplification occurs in the transmission fiber. The Raman amplifier is at the end of the link. It launches 1426-1454 nm light towards the input end. The signals are amplified by a method stimulated Raman scattering. Raman amplification has the advantages of reduced noise generation, and a relatively flat gain over a wide wavelength range. Many transmission systems incorporate both EDFA and Raman amplification.

7.2.5 DISPERSION COMPENSATORS

Dispersion compensators reverse chromatic dispersion that occurs in a fiber. By such reversal, optical signals can travel long distances before they need to return to the electrical regime to regeneration.

Review of the IRs (26.1) reveals that many fibers experience an increase in IR with increasing wavelength. Increased IR means reduced speed. Compensation is performed by causing wavelengths that travel fast in a fiber to travel slow in the compensator; and vice versa.

There are two types of compensators: compensating fibers and Bragg grating compensators. Compensation fibers have a dispersion slope that is opposite to and higher than that in the transmission fiber (Figure 3-13). Since the slope is higher, the compensating fiber is shorter than the transmission fiber. Since such fibers have high attenuation, they are either preceded or followed by an EDFA.

Bragg grating compensators delay wavelengths within the spectral width to reduce chromatic dispersion. Such compensators do not exhibit the high power loss exhibited by compensating fibers.

7.2.6 SWITCHES

Optical switches move optical power from one input path to one or more output paths. Such switches can reroute all wavelengths or individual wavelengths. Some switches have MEMs, micro-electro-mechanical systems, in which microscopic mirrors (Figure 7-11 and Figure 7-12), with diameters of approximately 20 μ, reflect light from one input port to any of multiple output ports.

Movement of the mirrors is by piezo-electric crystals.

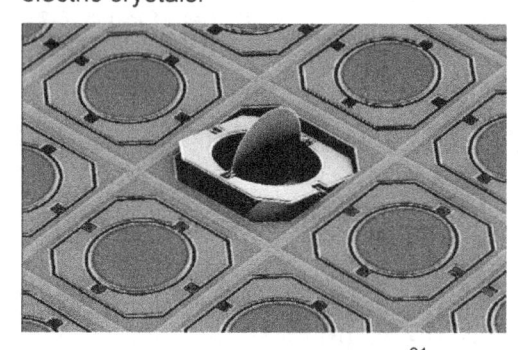

Figure 7-11: MEMs Mirrors[31]

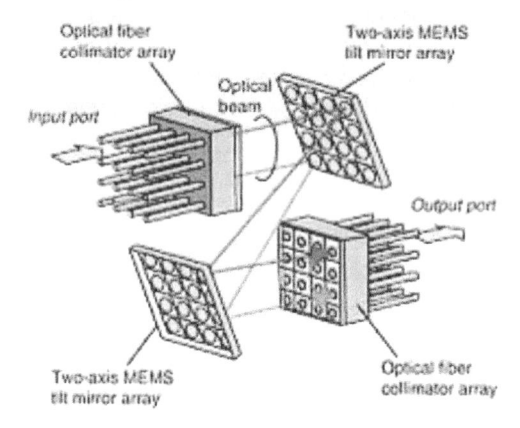

Figure 7-12: Optical Switch Mechanics[32]

7.2.7 ROTARY JOINTS

The rotary joint (Figure 7-13) is a passive optical device that allows transmission of an optical signal from a rotating structure to a stationary structure. Such a function is needed for undersea, remotely operated vehicles (ROVs) and for sensor arrays. Such ROVs are used for inspection of the underside of ships, for repair and maintenance of undersea fiber optic cables, and for undersea exploration at depths unattainable by a manned vehicle. Finally, such a function is needed for sensor arrays dragged behind ships for submarine detection.

---

[31] Courtesy www.tf.uni-kiel.de
[32] Courtesy opticalengineering.spiedigitallibrary.org

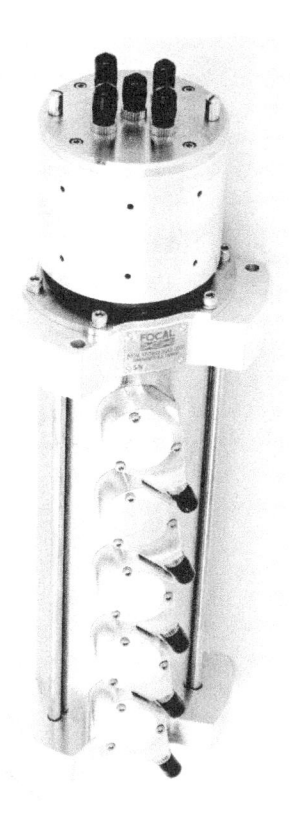

Figure 7-13: Five Fiber Rotary Joint[33]

7.3 TWO INSTALLATION CONCERNS

For the installer, passive devices create two concerns:

- Power loss
- Reflectance (5.4.5)

Any component in the optical signal path has a power loss. As a result, all passive devices reduce the power delivered to the receiver. Excessive power loss in a passive device can result in link failure.

It is not possible to test through some passive devices. For example, OTDR testing through optical amplifiers cannot be done, as the amplifiers contain filters that block light traveling back towards the input end of the fiber. In addition, some tests may be meaningless. For example, a power loss test through an optical amplifier is meaning less. In addition, such a test can be misleading, because optical amplifiers create optical power at wavelengths other than the transmission wavelengths. This power, called amplified spontaneous emission (ASE) would be measured as signal power, which it is not.

In addition, all passive devices are connected to other link components. Such connections create the potential for reflectance. Excessive reflectance can result in inaccurate signal transmission (5.4.5).

Passive devices have three types of power loss:

- Connection loss
- Intrinsic loss
- Extrinsic loss

Connection loss occurs in all passive devices, as they have connectors or splices on the inputs and outputs. Intrinsic loss occurs due to the function of the device. For example, a 1x2 splitter must have a power drop of at least 50 %, or 3.01 dB, between input and output ports (Equation 2-2). For a 1x4 splitter, the power drop is at least 75%, or 6.02 dB.

Extrinsic power loss is in excess of the intrinsic loss. The extrinsic loss is a result of our inability to make perfect passive devices. For example, the typical power loss through a 1x2 splitter is 3.5-4.0 dB. Since the intrinsic loss is 3.01 dB, the extrinsic loss is the difference, or 0.49-0.99 dB.

7.4 SUMMARY

The installer may install passive devices. As such devices introduce power loss and reflectance to the link, the installer will need to know maximum and typical power losses of such devices. With this knowledge, the installer will be able to interpret link power loss measurements. In addition, the installer will need to ensure that the connectors on all passive devices remain low loss and low reflectance.

7.5 REVIEW QUESTIONS

1. What are the two most important installation concerns for passive devices?

---

[33] This is a Focal Model 242 from Kaydon Power & Data Technologies. Photograph is courtesy of Kaydon Power & Data Technologies.

2. What is the proper name for a splitter that routes each wavelength to a separate output port?

3. Define WDM.

4. Define CWDM.

5. Define DWDM.

6. Why is an optical amplifier considered a passive device?

7. What is the acronym for an optical amplifier?

8. What element is used to create the amplification in an optical amplifier?

9. Calculate the intrinsic power drop for a 1x16 splitter.

10. If the actual power loss in Question 9 is 14 dB, what is the extrinsic power loss?

11. Calculate the intrinsic power loss through the combination of a 1x4 splitter and a 1x8 (Figure 7-14).

1x8 splitter

1x4 splitter

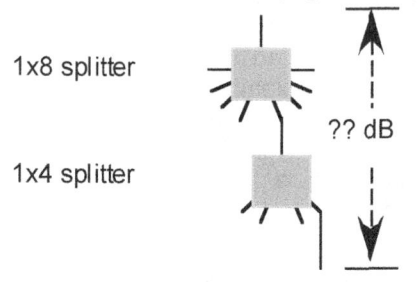

?? dB

Figure 7-14: Splitter Network For Question 11

8 OPTOELECTRONICS

Chapter Objectives: you learn the language, types and numbers for optoelectronics. With this knowledge, you will recognize power loss values that represent conditions under which the optoelectronics may not function.

8.1 INSTALLATION CONCERN

The term optoelectronics describes both the transmitter and receiver, since both function with electrical and optical signals. The transmitter converts an electrical signal to an optical signal; the receiver performs the reverse conversion. The key functional concern is accuracy of signal transmission. The optoelectronics achieve high accuracy when:

- The power level at the receiver is proper
- The optical signal to noise ratio (OSNR) is above a minimum value
- The pulse dispersion is sufficiently low

Proper power level means enough power and not excessive power. The only impact an installer can have on the optoelectronics is reduced power at the receiver. Such power reduction will occur if the installer allows dirt into the ports of the transmitter or receiver.

The optical signal to noise ratio (OSNR) is important for links on which an optical amplifier operates. This amplifier adds optical noise to the signal due to amplified spontaneous emission (ASE). This optical power is amplified when multiple EDFAs operate on a link.

Of course, the installer cannot increase the dispersion, as it is controlled by factors that he cannot degrade. However, installers need to be aware that excessive dispersion due to excessive transmission distance can result in link failure.

8.2 TYPES

8.2.1 TRANSCEIVERS

Transceivers are devices that transmit and receive simultaneously. They have multiple designations, depending on the signals they convert:

- GBICs
- SFP
- SFP+
- XFP
- XENPAK
- QSFP
- QSFP+
- CFP
- SFP
- Media Converters

Each of these designations can have multiple versions, with multiple wavelengths and transmission distances. For example, the SFP transceiver comes in seven versions;

- -SX, 850 nm, multimode
- -LX, 1310 nm, 10 km
- -EX, 1310 nm, 40 km
- -ZX, 1550 nm 80 km
- -EZX, 1550 nm, 120 km
- -BX, 1490/1310 nm 10 km
- -SFSW, bidirectional on one fiber

8.2.2 TRANSPONDERS

Transponders are transceivers that convert a signal at one wavelength to the same signal at a different wavelength. Such devices are used in CWDM and DWDM networks. Some transponders regenerate and reshape the optical pulse. Some merely repeat the pulse at the new wavelength.

8.2.3 LIGHT EMITTING DIODES

LED sources have the characteristics of:

- Large spot size (150-250 μ)
- Circular spot shape
- Relatively large angle of divergence (0.20-0.25 NA)
- Relatively low bit rate/ bandwidth capability (< 200 Mbps)

- Relatively low launch power (1-10 µW)
- Wide spectral width (50-200 nm)
- Relatively low cost
- Multimode operation at 850 nm and 1300 nm

The large spot size, circular spot shape and large angle of divergence result in light filling most of the cross section of the core and most of the angle of acceptance defined by the NA. Under such launch conditions, the light experiences high modal dispersion, the maximum fiber attenuation, and maximum connector loss. These characteristics make the LED best suited for use with multimode fiber in short distance data links.

8.2.4 LASER DIODES

Laser diode (LD) sources have the following characteristics:

- Small spot size
- Rectangular spot shape (2µ x 10µ)
- Small angle of divergence
- Relatively high bit rate/ bandwidth capability (≤ 40 Gbps)
- Relatively high launch power (>1 mW)
- Narrow spectral width (> 5 nm)
- Relatively high cost

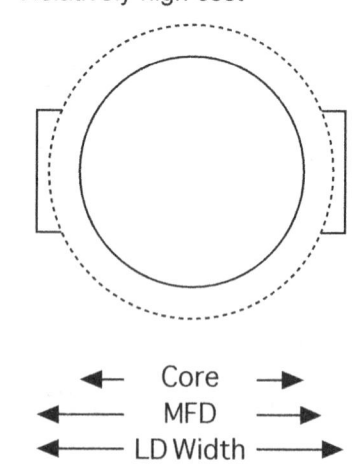

Figure 8-1: Relative Sizes Of LD And Singlemode Core

The width of the singlemode LD results in power launched into the cladding (Figure 8-1). Such power must be removed to provide accurate insertion loss measurements (14.3, Figure 14-28).

Singlemode LDs are common at 1310 nm and at wavelengths between 1310 nm and 1550 nm. Multimode LDs are uncommon in transceivers

The small spot size, small angle of divergence, high bandwidth, high launch power and high cost make the LD well suited for use with singlemode fiber in long distance links. However, the rectangular spot shape makes alignment to the singlemode core relatively expensive.

The power level of LDs can be misleading. In the specific case of FTTH, the 1550 nm wavelength is often amplified with an EDFA (7.2.4) to a power level as high as +20 dBm (0.1 W). At this high power level, eye damage is possible.

8.2.5 VERTICAL CAVITY SURFACE EMITTING LASERS

VCSELs have the characteristics of:

- Medium spot size (30µ)
- Annular spot shape (Figure 8-2)
- Small angle of divergence
- Relatively high bit rate/ bandwidth capability
- Narrow spectral width (≤ 1 nm)
- Relatively low cost
- 850 nm multimode operation

The annular spot shape (Figure 8-2) of VCSELs avoids a potential splitting of pulses. This splitting can occur when the highly collimated light of a laser diode enters the center of the core of some 'old' multimode fibers (Figure 8-3).

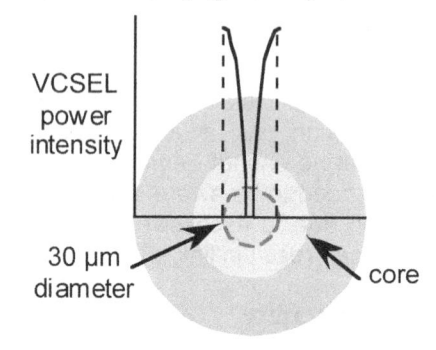

Figure 8-2: VCSEL Annular Launch Area

Index of refraction

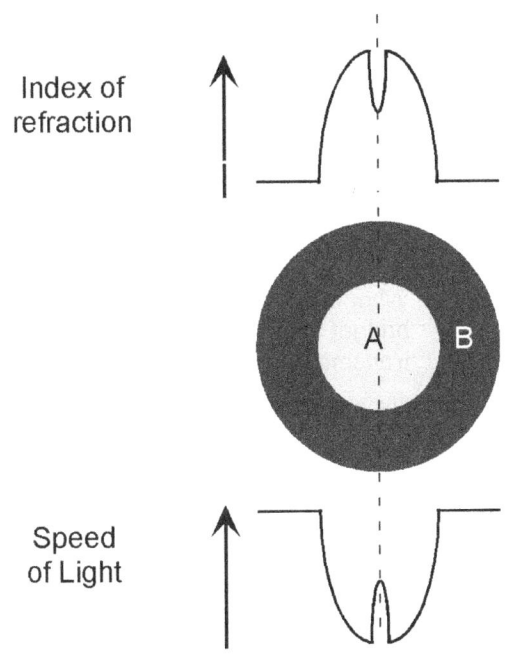

Speed of Light

Figure 8-3: Profile Dip In Old Multimode Fibers

A VCSEL emits most of the optical power in an annular ring with an inner diameter of 9 μ and an outer diameter of 30 μ. This annular launch condition and the small angle of divergence result in low dispersion in multimode fibers. This low dispersion enables 1 Gbps and 10 Gbps transmission to significant distances on multimode fiber. At the time of this writing, 40 Gbps VCSELs have become available, but are not included in the multimode data standards. In summary, the VCSEL combines the performance advantages of a LD with the cost advantage of an LED.

8.3 RECEIVER TYPES

There are two types of receivers: photodiodes and avalanche photodiodes. Photodiodes are used for relatively high receiver power. Avalanche photodiodes are used for very low receiver power.

Rarely does the installer need to know details of the receiver. However, in the event of reduced transmitter output power, the installer may need to know the sensitivity of the receiver. The sensitivity of the receiver is the minimum power level, in dBm, at which the receiver will function at the specified level of accuracy. Below this power, high bit error rates (BERs) are possible.

8.4 PERFORMANCE

8.4.1 POWER BUDGET

The installer needs to know the wavelength and the optical power budget of the optoelectronics, in dB. We call this characterristic the optical power budget available (OPBA). The OPBA is the maximum power loss that can occur between a transmitter and receiver while still allowing the pair to function at the specified level of accuracy. If the link loss exceeds the OPBA, the optoelectronics and the link loss are mis-matched. Table 8-1 presents OPBA for fiber data standards.

Standard	Core, μ	l, nm	OPBA, dB
10BASE-FB	62.5	850	12
10BASE-FL	62.5	850	12
10BASE-FP	62.5	850	16-26
100BASE-F	62.5	1300	11
100BASE-SX	62.5	850	4.0
1000BASE-S	62.5	850	2.33
1000BASE-S	62.5	850	2.53
1000BASE-S	50	850	3.25
1000BASE-S	50	850	3.43
1000BASE-L	62.5	1300	2.32
1000BASE-L	50	1300	2.32

Table 8-1: OPBA For Fiber Standards

The actual loss of the link, known as the insertion loss and optical power budget requirement, must be less than or equal to the optoelectronics OPBA. The installer can calculate the expected loss. This calculation appears in Table 19-3.

No. Pairs	Distance, m	Loss, dB
2	300	2.77
3	280	3.28
4	250	3.92
5	220	4.45
6	180	5.02

| 7 | 120 | 5.62 |

Table 8-2: 850 nm 10 Gigabit Distances Vs. Number of Connector Pairs[34]

In addition, the installer may need to know the sensitivity of the receiver, in dBm, in order to troubleshoot receiver problems.

8.4.2 OPTIONAL OPTICAL POWER BUDGETS

Under certain conditions, some standards, allow for OPBAs that are higher than the values stated in the standard. For example, 10GBASE-SX allows increased OPBAs with reduced link length (Table 8-2).

8.5 DIRECT VS. COHERENT DETECTION

Most of this book is focused on power loss. The assumption that sufficiently low power loss enables transceivers to function properly. In most situations, this is true.

However, there is another factor, which is a concern of the designer, but not the installer. That concern is sufficient optical signal to noise ratio (OSNR). The OSNR must be above a minimum value. Links that include optical amplifiers have optical noise. These amplifiers create this wide spectral width noise. This noise is included in any measurement of power level at the receiver. Thus, a receiver power measurement does not necessarily indicate a power level at the receiver that results in accurate transmission.

8.6 SUMMARY

The installer is concerned with two optoelectronic characteristics:

- The wavelength or wavelengths
- The optical power budget available (OPBA), in dB.

8.7 REVIEW QUESTIONS

1. What are common testing wavelengths?
2. For the installer, what are the two most important optoelectronics characteristics?
3. What is a typical optical power budget available from a transmitter-receiver pair?
4. Why is the installer concerned with the optical power budget?
5. Why is the installer concerned with the wavelength?
6. A 62.5 μ link has a power loss of 2.54 dB. According to Table 8-1, on which protocols will not run on this link?
7. By comparing the OPBA to the measured link loss, what statement can the installer make?
8. A 50 μ link has a power loss of 3.44 dB. According to Table 8-1, which protocols will not run on this link?
9. Your installer tells you that the link should have a loss less than 3.6 dB. It has a loss of 3.8 dB. You plan to transmit 100BASE-SX. Will the link work?
10. When is the OPBR calculated?
11. After comparing the link loss to the OPBA, what statement can the installer make?

---

[34] '10 Gigabit Ethernet over Fiber Operation: More Range Than You Thought', Marty Anderson, Corning Inc. document WPEMEA1002_EN-1.pdf

9 HARDWARE

Chapter Objectives: you learn of the types, locations and functions of hardware commonly used in fiber optic networks.

9.1 FUNCTIONS

Fiber optic hardware increases the reliability or convenience of use of the network. Hardware increases the reliability by protecting network components or limiting what can be done to such components during use. For example, many hardware components limit bend radius of cable to prevent fiber breakage, excess power loss, and delayed failure.

9.2 TYPES

9.2.1 ENCLOSURES

Enclosures (Figure 9-1 and Figure 9-2) protect cable ends, buffer tubes and fibers. The break out cable requires no enclosure, as this design provides protection for each fiber.

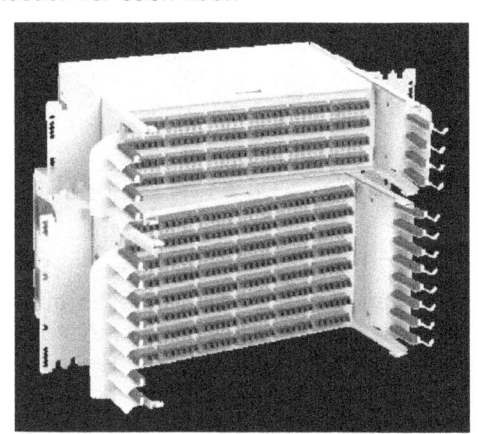

Figure 9-1: Indoor Enclosures[35]

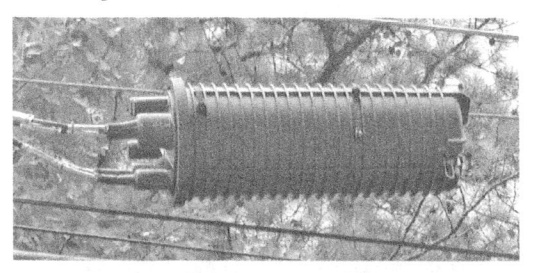

Figure 9-2: Outdoor Splice Enclosure

Both types of enclosures can accept multiple cables, although outdoor enclosures require a minimum of two cables. Both types include mechanisms for gripping the cable strength members.

Indoor enclosures can be rack mounted (Figure 9-1) or wall mounted (Figure 9-3). They can provide front and/or rear access to the interior. The front of the enclosure can be flush with the front of the rack or recessed to create an integral patch cord shelf. Enclosures can allow side, top or bottom cable entrances. Some indoor enclosures allow space for splice trays. Some include mechanisms to restrict bend radius (Figure 9-3).

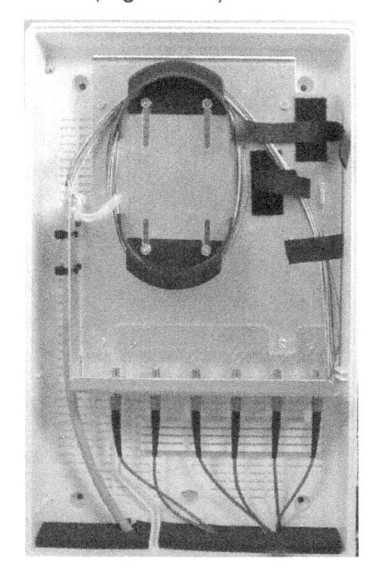

Figure 9-3: Wall Mounted Enclosure

Enclosures are required because termination and splicing require removal of cable structural materials. Such removal eliminates the protection provided by these materials.

Indoor enclosures are sized to the number of fiber ends. The use of SFF connectors doubles the number of connectors possible in an enclosure. With

---
[35] Courtesy of Panduit Corporation

SFF connectors, the number of fiber ends in a 1U enclosure can be as high as 48.[36]

This increase in count increases the total length of fiber and/or buffer tube in an enclosure. With 48 ends in a 1U enclosure and 3' of fiber per end, the total fiber length would be 144'. Such a large amount of fiber can result in increased maintenance cost.[37] In short, packing enclosures with as much fiber as possible will reduce hardware cost. But such packing may increase maintenance cost, and therefore, be counterproductive.

Outdoor enclosures can be single shell or multiple shells. The shell can be a single piece (Figure 9-4) or multiple pieces (Figure 9-5). With a single piece shell, moisture seals are provided at the ends. With a multiple piece shell, moisture seals are provided at the ends and along the perimeter.

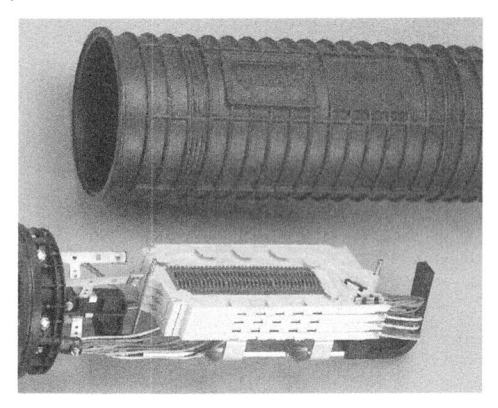

Figure 9-4: Single Piece Enclosure Shell

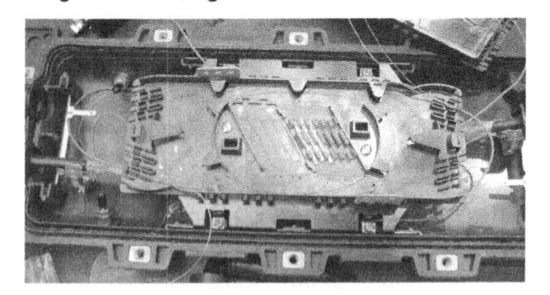

Figure 9-5: Multiple Piece Enclosure Shell

While uncommon at time of this writing, double shell enclosures have space between the shells that is filled with a re-enterable moisture blocking gel. Outdoor enclosures are sized to the number of splices and the number of cables.

Most enclosures meet the requirement for connection of a small number of cables, either for mid span connection or for mid span drops. FTTH/PON enclosures may need may require up to 33 cables, one feed cable and 32 small, single fiber cables. This require-ment can result in increased enclosure size and increased number of cable ports. Some FTTH enclosures achieve size reduction by reducing the number of ports (Figure 9-6).

Figure 9-6: Small FTTH Enclosure.

9.2.2 PATCH PANELS

Patch panels are metal or plastic plates that accept connector barrels. Although these panels can be independent of an enclosure, they are more commonly inserts, or modules (Figure 9-7), that are installed into enclosures. The module form allows for convenient access of an individual connector or a group of connectors from the front of the enclosure.

Figure 9-7: SC Patch Panel Modules

---

[36] 1U is 1.75" high; 2U, 3.5" high, and so forth.

[37] According to Murphy's Law, the fiber in need of repair will be the fiber on the bottom. In addition, repair of one fiber may result in damage to another.

9.2.3 INNER DUCT/SUB DUCT

Installers use inner duct (Figure 9-8) in two situations: for underground cable systems and for indoor cable systems. In underground systems, the inner duct, or flexible conduit, is used to create a crush resistant, rodent resistant, dielectric cable system. The inner duct provides a crush resistance that armor would provide. In addition, the inner duct is large enough that rodents cannot chew through the duct to damage the cable. Finally, elimination of armor reduces the cable cost and the installation cost.

Figure 9-8: Inner Duct

In indoor cable systems, inner duct segregates and protects fiber cable in cable trays, cable troughs and cable raceways. Such segregation increases network reliability in two ways. First, this segregation reduces the possibility that the fiber cable is cut for rerouting. Second, such segregation eliminates the placement of heavy copper cables on top of fiber cables.

When heavy copper cables on top of the fiber cables in a ladder tray, the fiber cables can experience bend radius violation, fiber breakage and excess power loss. Finally, installers cannot attach cable ties to fiber cables inside an inner duct. Excessively tight cable ties cause increased loss.

Installers use sub duct to utilize conduit volume efficiently and prevent installation problems. Unless installed at the same time, fiber cables are restricted to one per sub duct.

This restriction avoids two problems. The first problem occurs when the pull rope snakes around the first cable installed. Such snaking creates two problems. First, the second cable will experience high friction that results in high installation load. Such a load may violate the installation load rating of the cable.

Second, the second cable is forced to violate its bend radius by snaking around the first. Both violations can result in fiber breakage, increased power loss and reduced reliability.

9.2.4 STORAGE LOOP HOLDERS

The storage, or service, loop allows an aerial cable to be rerouted to a location without breaking and splicing. Storage loop holders (Figure 9-9) control the bend radius of the cable while simplifying storage of service loops. Placed in pairs at a fifty-foot spacing, the storage loop holder creates a 100-foot service loop.

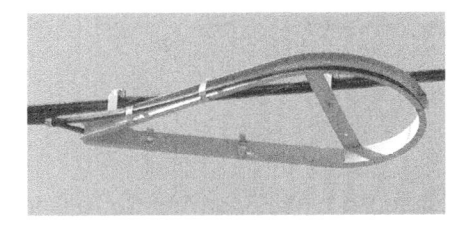

Figure 9-9: Storage Loop Holder

9.2.5 HANGING HARDWARE

Cable hanging hardware includes all products that enable an aerial cable system to be installed in a reliable manner. At its simplest, cable hanging hardware consists of a stranded steel wire, called a messenger or strand cable. The installer attaches the fiber optic cable to the messenger wire by wrapping a smaller wire around both the fiber cable and the messenger.

At its most complex, cable-hanging hardware includes the products in Figure 9-10 to Figure 9-12. Each of these products serves a function based on its location.

9.2.6 ROUTING HARDWARE

Cable routing hardware consists of the hardware used indoors to control and route the cables.[38] Figure 9-13 demonstrates the bend radius control of cables from an overhead trough system into racks. Figure 9-14 demonstrates the bend radius control in the trough. Figure 9-15 demonstrates bend radius control within the rack.

Figure 9-10: ADSS Pole Cable Grip

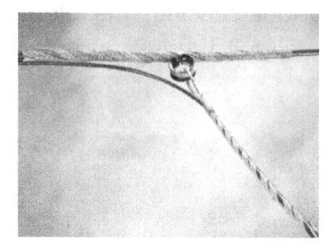

Figure 9-11: Mid Span Tap Cable Grip

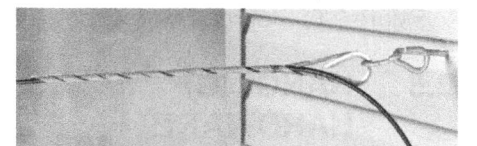

Figure 9-12: End Cable Support[39]

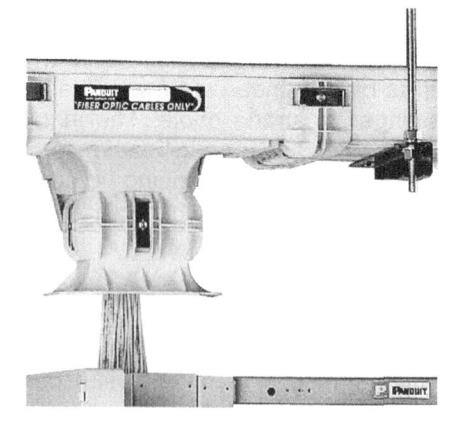

Figure 9-13: Bend Radius Control From Trough To Rack

---

[38] Figure 9-15 to Figure 9-17 are courtesy of Panduit Corporation.

[39] Figure 9-12 to Figure 9-14 are courtesy of Preformed Line Products.

Figure 9-14: Bend Radius Control In Trough

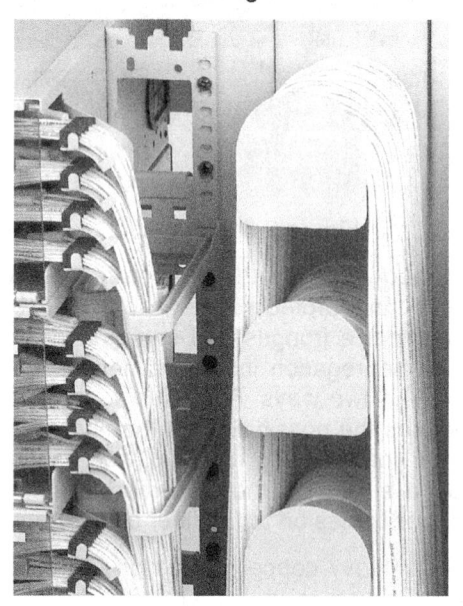

Figure 9-15: Bend Radius Control In Rack

9.3 SUMMARY

The prime function of hardware is provision of reliability. Hardware provides this protection by confining cables and connectors to a restricted access volume and by limiting the bend radius of cables placed in the hardware. While these functions determine the some of the design features of such hardware, ease of use determines other, equally important, design features.

9.4 REVIEW QUESTIONS

1. What is the prime function of all fiber optic hardware?

2. What are the reasons for use of inner duct?

3. What are the reasons that fiber optic cables are not normally installed into an inner duct that has previously installed cables?

4. What is the one specific function that most hardware provides for cables?

5. True of False: all splice enclosures are essentially the same. Justify your answer.

6. Where are pre-terminated cables commonly used?

7. Why are pre-terminated cables commonly used?

PART TWO
PRINCIPLES OF INSTALLATION

PRINCIPLE

"An essential element, constituent, or quality, especially one that produces a specific effect; as, the active principle of a medicine." (Webster's New Twentieth Century Dictionary, Unabridged.) This definition applies to the information in Part 2. All the principles in this part result in the three effects, or goals, of low power loss, low installation cost and high reliability.

We have highlighted the principles, or specific effects and the methods by which the installer achieves these effects. To highlight the principle, we have used the symbol (▶) before the Principle and the symbol (▶▶) before the Method. These principles and methods support the three goals and result in the procedures in Part 4.

10 PLANNING AND MANAGEMENT ISSUES

Chapter Objective: you learn the actions to take prior to installation.

10.1 INTRODUCTION

Installers avoid many common problems through planning. These considerations include management issues for cable installation, connector installation and splicing. Such planning will include the following twelve considerations.

- Equipment and supplies required
- Equipment locations
- Product data sheets
- Potential installation techniques
- Best installation techniques
- Personnel requirements
- Personnel selection
- Testing needs
- Testing data forms
- As-built data logs
- Potential problems
- Potential safety issues

10.2 EQUIPMENT AND SUPPLIES

The installer avoids problems by reviewing the activities planned. This review will indicate the equipment and supplies needed for successful completion of these activities. These items include:

- Pulling equipment
- Pulling ropes
- Cable lubricant
- Pulleys
- Cable ties
- Epoxy curing ovens
- Epoxy
- Connector polishing films, etc.

Chapters 20-27 identify the equipment and supplies needed for the various installation activities.

10.3 EQUIPMENT LOCATIONS

After identifying the equipment required, the installer determines the equipment locations. For example, pulling equipment needs to be attached to a structure heavy enough to prevent movement (Figure 10-1). An another example, connector installation may require power for a curing oven. The installation location may not have that power.

Figure 10-1: Pulling Equipment

10.4 DATA SHEETS

The installer avoids problems with the information he obtains from data sheets for all products to be installed. For example, cable data sheets include limits on installation load, use load, and bend radii. Such information is essential to avoid damage due to violations of these limits. An another example, mechanical splice data sheets identify installation methods and cleave lengths.

10.5 INSTALLATION TECHNIQUES

The installer identifies the techniques by which the cables, connectors, and splices can be installed. Such information may be included in the data sheets. However, such information may be in application notes issued by the manufacturer. With such application notes, the installer avoids errors that damage the products. In some cases, multiple techniques are possible.

10.6 RECOMMENDED TECHNIQUES

If multiple installation techniques are possible, the installer determines the best technique. In general, the best installation technique is that indicated by the manufacturer in a data sheet or application note. However, the recommended technique may not be best suited to a specific situation. In this case, the installer determines the specific installation technique he will use.

10.7 PERSONNEL

10.7.1 REQUIREMENTS

The supervisor determines the number of installers required. For example, cable pulling into underground conduits requires a minimum of three personnel (11.3.1.6). Having one person prepare the cable ends and a second person install the connectors may reduce connector installation time.

10.7.2 SELECTION

All personnel who handle cables and connectors need some training. Even network technicians who install patch cords need to know about bend radius limitations, cleanliness of connectors, and caps on connectors and on optoelectronic ports.

The supervisor selects personnel for installation activities based on their experience. That experience should be appropriate for the activity to which installer is assigned.

Assigning inexperienced personnel to installation activities can, and often does, result in increased installation cost and reduced reliability. Here are three examples. First, a fusion splicer who installed connectors without any connector training did not know that the epoxy required a heat cure. The result was uncured epoxy, lost connectors, and wasted time.

Second, a connector installer without any experience tried, unsuccessfully, to polish connectors in a room, in which plasterers were sanding the plaster. High loss from scratches on the core resulted.

Third, a connector installer who had training on multimode connector polishing used that method for singlemode polishing. The result was scratched cores and high loss.

The safest way to ensure successful installation is to:

▶▶Train all personnel

This text provides a structure for such training.

A third method includes certification of installation personnel. This text enables trainees to pass the CFOT, CFxT, and three CFOS certification examinations offered by the Fiber Optic Association.

Another method is training and certification by an experienced fiber optic trainer. With 31 years of training experience, Pearson Technologies Inc. is such a trainer.

10.8 TESTING NEEDS

Prior to beginning an installation, the supervisor determines the testing needs. A common procedure is a check of continuity and attenuation <u>before</u> each operation. This determination results in checks:

- On cable as received or before installation
- After cable installation
- After splicing
- After all installation activities
  ▶▶Check cable before each operation

10.8.1 AS-RECEIVED TEST

The installer can test cables before installation with a bare fiber adapter (Figure 10-2) and an OTDR (15). This adapter is a connector with a clamp on the back shell. Alternative methods include using a cleave and crimp connector, a re-useable mechanical splice, and a fusion-spliced pigtail.

This check ensures that the cable has the:

- Proper length
- Acceptable attenuation rate
- Undamaged fibers

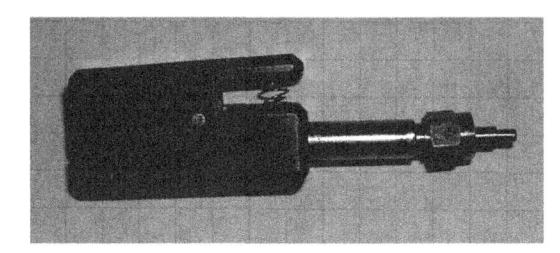

Figure 10-2: Bare Fiber Adapter

Without this test, a cable with damaged fibers discovered after installation cannot be returned to the supplier.

Alternatively, the installer can conduct a white or red light continuity test by holding a high intensity light or VFL at the end of each fiber in the cable. This test will not reveal broken fibers in a tight tube cable. Nor will this test reveal high attenuation rate.

Some installers skip this test. They rely on the consistent quality of the products they purchase. While such a skip reduces testing cost, it risks installing and pay-ing for defective cable.

10.8.2 POST CABLE INSTALLATION TEST

After installation and prior to splicing or connector installation, the installer tests the cable. It makes no sense to continue the installation on a cable with damaged fibers.

10.8.3 POST SPLICING

The installer verifies splice loss prior to link certification. The installer verifies splice loss with an OTDR (15). The installer can perform this test either after splicing or after connector installation. If connectors are not on the cable ends, the installer can perform this test as in 10.8.1.

10.8.4 POST INSTALLATION

The installer performs this testing after completion of all installation activities. Insertion loss testing is required by TIA/EIA-568-C. OTDR testing is not required but recommended by many professional installation organizations. This author recommends such testing.

10.9 CREATE DATA FORMS

The supervisor creates the data forms dictated by the testing to be done. Such forms can be used to check the status of the installation. If the testing after one step is complete, the installers can proceed to the next step.

▶▶ Create test forms

The information includes

- Test technique
- Wavelength
- Spectral width
- Equipment used
- Direction of test

In short, the data forms include all information sufficient to enable exact reproduction of the test. Without the ability to exactly reproduce the test, the installer will not know whether increases in insertion loss and OTDR test results are due to changes in the test method or to degradation of the cable, connectors, or splices.

10.10 RECORDS

The final testing results in a set of 'as-built' data records. These 'as-built' records greatly simplify troubleshooting. If future tests of the cable link reveal no changes in values from the 'as-built' values, any link problems are in the opto-electronics, not in the cable link.

Good practice for 'as-built' records is:

- Creation of three copies
- Inclusion all information on the testing methods
- Measurement of range (14.8)

The three copies are: an archive copy; an open copy that is available to any who would need it; and a project manager copy.

The test information includes all information sufficient to enable exact reproduction of the test.

The range measurement provides the installer with the maximum increase in loss that can be expected without any degradation of the link components. Without this value, interpretation of an increase in insertion loss can be difficult.

10.11 IDENTIFY POTENTIAL PROBLEMS

Before the start of the installation, the supervisor identifies potential problems. In addition, he develops plans to avoid and to solve such potential problems.

10.12 IDENTIFY SAFETY ISSUES

The supervisor reviews the entire installation process to identify potential safety issues, with the goal of avoiding and solving such issues. As optical fiber is small, and potentially dangerous, the installer uses procedures that minimize the possibility of fiber splinters.[40]

In addition, there is the possibility of eye damage from the invisible light produced by high power lasers used in testing equipment and transceivers, such as OTDRs, reflectance test sets, Gigabit Ethernet transceivers and 10 Gigabit Ethernet transceivers.

10.12.1 EYE SAFETY

10.12.1.1 ▶▶WEAR SAFETY GLASSES

The installer wears safety glasses. Should a fiber 'jump' towards his face, the glasses will prevent the fiber from landing in his eyes. Such a fiber will be very difficult to find and remove without causing eye damage.

10.12.1.2 ▶▶NO LIGHT IN FIBER

Prior to looking into a fiber with a microscopic, the installer disconnects all equipment at the opposite end. If possible, the installer 'locks out' and labels the enclosure on the opposite end, so that no one connects equipment.

Should the fiber have high power laser light while the installer is inspecting the fiber with a microscope, the laser light can burn the rods and cones of the retina. Damaged rods and cones will not regenerate. Self-imposed laser eye surgery is not part of a normal installation!

This power level can range from harmless to very harmful. While data communication LEDs have relatively low power, on the order of 0.001 mW, the VCSELs used in 1000BASE-X and 10GBASE-X can have relatively high power, on the order of 1 mW (0 dBm). In addition, the three wavelengths used in FTTH systems can have power levels from 1 mW to 100 mW (+20 dBm).

10.12.1.3 ▶▶IDENTIFY MEDICAL SERVICE LOCATION

Before beginning an installation, the installer identifies a local hospital with the capability to find and remove optical fiber. Useful knowledge is of those local hospitals that can address the problem of fiber in eyes, mouths and hands. Such knowledge will reduce the time required to correct such problems. Finally, such a time reduction can result in reduced injury caused by such problems.

10.12.2 HAND SAFETY

10.12.2.1 ▶▶BARE FIBER DISPOSAL

During fiber installation, receiving a fiber splinter is the most common danger. To avoid such danger, the installer places bare fiber in a disposable container as soon as he creates it. With this approach, the bare fibers are controlled as much as is possible and will not cause problems by becoming splinters in hands.

10.12.2.2 ▶▶USE WORK MAT

The installer uses a work mat with a dark, dull surface. The dull surface makes it easy to find fiber splinters. Use of any other surface may make the fiber difficult to see. Such fiber can be lost and find its way onto clothing or into other undesirable areas.

10.12.2.3 ▶▶WASH HANDS

The installer washes his hands before using the rest room. The reason for this principle is obvious. Just imagine the conversation at a doctor's office or hospital!

---

[40] A splinter is a fiber without the primary coating.

10.12.3 CLOTHING SAFETY

10.12.3.1 ▶▶WEAR SMOCK

The installer wears a smock that is washed separately from other clothes. Fibers that are caught on such a smock will not be washed onto other clothing. This is a common practice in cable assembly facilities.

10.12.3.2 ▶▶WEAR SMOOTH CLOTHING

When working with fiber, the installer wears smooth clothing. Smooth clothing is less likely to catch and retain fibers than rough clothing such as sweaters.

10.12.3.3 ▶▶USE LINT ROLLER

After each session in which the installer generates bare fibers, he uses a lint or pet hair roller. He cleans his arms, chest and lap to remove any bare fiber.

10.12.3.4 ▶▶ISOLATE WORK AREA

The installer isolates the work area from unauthorized or untrained personnel. While the installer may know how to avoid getting fiber splinters, others may not know either the danger of or the methods of avoiding such splinters.

10.12.4 MOUTH SAFETY

10.12.4.1 ▶▶NO FOOD IN WORK AREA

The installer does not eat, drink or smoke in any area in which he is creating bare glass fiber. All items that can find their way to the mouth should be removed from the work area. Being on a high fiber diet does not mean ingesting optical fiber!

10.12.5 CHEMICAL SAFETY

Fiber installation activities may require the use of hazardous chemicals. Safe use of such chemicals requires knowledge of their dangers and of the treatment required after exposure. The Material Safety Data Sheet (MSDS), which is available from the chemical supplier, provides this knowledge.

10.13 SUMMARY

The installer can damage fiber optic cables during their installation. This damage may be observed in the form of excess power loss. However, this damage may be hidden in the form of reduced reliability. The installer can achieve the maximum reliability possible by following installation procedures that comply with the principles in this chapter.

10.14 REVIEW QUESTIONS

1. Make a list of the items needed to ensure safety of any cable and connector installation.

2. Is the power in an FTTH fiber more or less dangerous than that in a data communication network?

3. Is the power in a DWDM fiber higher or lower than that in a data network?

4. Prior to inspecting a connector with a microscope, what should the installer do?

5. What is the first action that an installer should take after creating a short length of bare fiber?

6. How can an installer implement eye safety?

7. What three steps can an installer implement for hand safety?

11 CABLE INSTALLATION PRINCIPLES

Chapter Objectives: you learn the principles for installation and preparation of the ends of fiber optic cables.

11.1 INTRODUCTION

There are four goals of cable installation:

- Avoid breakage
- Avoid reduced power at the receiver
- Avoid reduced reliability
- Proceed in a safe manner (10.12).

To meet these four goals, the installer needs to recognize and respect certain principles. We organize these principles into four groups:

- Environmental limits
- Installation limits
- NEC compliance
- End preparation

11.2 ENVIRONMENTAL LIMITS

The network designer is responsible for specification of a cable proper for the intended environment. With this definition of responsibility, the installer might not need to be concerned with environment limitations. However, network designers may not recognize all relevant environmental conditions. In order that the installer recognize and avoid common field problems, we provide this first principle.

▶Respect environmental limits

The installer installs the cable in environmental conditions that are within its limits. Should the cable be exposed the conditions in excess of its limits, the cable can fail to protect the fibers. Such failure can lead to fiber breakage and increased attenuation rate.

While not exhaustive, five environmental conditions represent most of the conditions[41] that have resulted in installation problems:

- Moisture
- Operating temperature range
- Bend radii
- Crush load
- Use load and vertical rise distance

11.2.1.1 MOISTURE

An installer can install a cable that is not moisture resistant in an environment that contains moisture. Such a cable can:

- Channel moisture into electronics
- Develop increased attenuation rate and fiber breakage due to frozen water
- Experience reduced fiber strength and breakage due to attack from the chemicals in ground water

11.2.1.2 TEMPERATURE OPERATING RANGE

An installer can install a cable in an environment with a temperature outside its operating range. Such a cable can exhibit:

- Increased attenuation rate
- Degradation of cable materials

In the former case, there may be insufficient power at the receiver for proper link operation. In the latter case, the fibers may break.

11.2.1.3 BEND RADII

During installation, the installer can bend a cable to less than either of the two bend radii (4.6.2). In this case, the cable can develop:

- -Increased attenuation rate
- Fiber breakage
- Reduced fiber strength

Such a situation can occur in an underground conduit path, in which the conduit sweep, or elbow, has a radius less than either bend radius of the cable. In this situation, the cable path (its' environment) is forcing a violation of the bend radius.

---

[41] This is a statement of the author's experience, not a statement with statistical support.

11.2.1.4 CRUSH LOADS

If installer installs a cable in an environment that imposes either a long or short-term crush load in excess of the limit of the cable, the cable can experience increased attenuation rate and fiber breakage. Such a situation can occur when an indoor, tight tube cable is directly buried.

11.2.1.5 USE LOAD

An installer can install a cable with a long term, or use, load on the cable in excess its rating. Such a condition can occur when the cable is installed between widely spaced buildings, widely spaced telephone poles, widely spaced power transmission towers, and up a long vertical rise. Two problems can occur: increased attenuation rate and fiber breakage.

▶▶Limit use load

The installer limits the long-term load to a value less than or equal to the rating of the cable. To do so, he obtains the use load from on a data sheet, a web page, or the manufacturer.

▶▶Limit vertical rise distance

The use load and the vertical rise distance are different statements of the same characteristic. The vertical rise distance is the distance to which the cable can be installed vertically without support. The installer limits the cable to a vertical distance less than the cable's rating. To do so, the installer obtains the vertical rise limit from the data sheet. The installer can achieve a total vertical rise distance in excess of this limit by supporting the cable at a separation no greater than this limit.

Indoor loose tube cables can allow the fibers to slide out of the cable. Thus, the installer

▶▶ Installs service loops in loose tube cables installed vertically

11.3 INSTALLATION LIMITS

There are two types of cable installation:

➢ Pulling the cable into its path

➢ Placing the cable in its location

The installer installs the cable with conditions that are within the limits of the cable. The principle is:

▶Respect installation limits

11.3.1 PULLING CABLE

During pulling of the cable, the installer respects five installation limits:

➢ No twisting
➢ Installation load
➢ Installation bend radius
➢ Installation temperature range
➢ Storage temperature range

11.3.1.1 ▶NO TWISTING

Avoidance of fiber breakage requires that fiber cables be installed without twisting.

▶▶Use swivel

In order to avoid twisting, the installer uses a pulling swivel between the pull rope and the cable (

Figure 11-1). The installer attaches the pull rope and the cable to the pulling swivel.

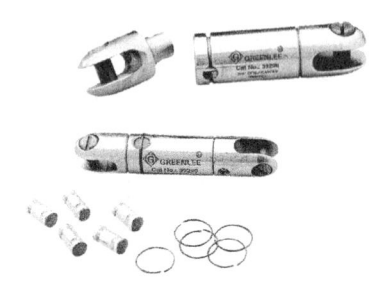

Figure 11-1: Swivel With Shear Pins (Greenlee Textron)

11.3.1.2 INSTALLATION LOAD

In order to comply with the installation load limit, the installer must know the load limit and have a method for limiting the load applied to the cable. The installer will learn the limit from the cable data sheet, creating two principles.

▶Know installation load limit
▶Limit installation load

There are three methods by which an installer limits the short-term, or applied installation. In addition, there are two

Professional Fiber Optic Installation 11-3

methods for reducing the applied installation load.

The three methods have the advantage of providing concrete evidence that the installation load has not been exceeded. These methods are use of a:

▶▶ Pulling eye with a swivel with a shear pin (

Figure 11-1)

▶▶ Pulling device with a slip clutch (Figure 11-2)

▶▶ Pulling device with a load gage (Figure 11-4)

Figure 11-2: Puller With Slip Clutch (Greenlee Textron)

The eye has a shear pin, which is rated at a load less than the cable maximum installation load. For example, a cable with a rating of 600 pounds-force requires a shear pin rated at 500-550 pounds-force. If the installer exceeds the rating, the shear pin, not the cable, breaks.

The installer can use a pulling eye with a swivel without a shear pin if he has some other method of limiting the load applied to the cable. Pullers have two methods to provide such limitation. These methods are a slip clutch and a load gage.

An installer can set the puller slip clutch (Figure 11-2) to a level less than the installation load of the cable. Should the applied load exceed the level set, the clutch slips, preventing fiber damage.

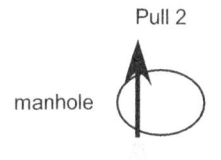

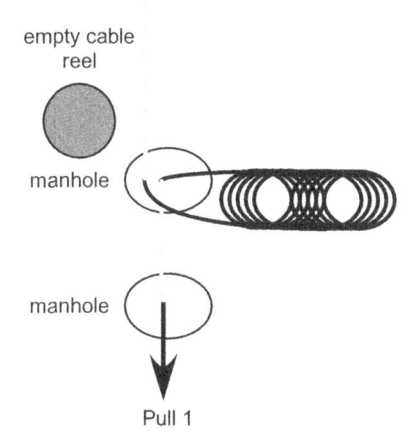

Figure 11-3: A Mid Pull

Figure 11-4: Puller With Load Gage[42]

An installer can set the puller load gage (Figure 11-4) to a level less than the cable installation load. Should the applied load exceed the level set, the load gage stops

---
[42] Courtesy of Condux International.

the pulling motor without fiber breakage or damage. Some load gages have an additional advantage: they allow attachment of a chart recorder, which provides proof that the installer did not load the cable in excess of it's rating.

These three methods limit the load but do nothing to reduce the load. There are two methods for reducing the load applied to the cable. The first method is use of a fiber optic cable lubricant.

▶▶Use fiber optic cable lubricant

The installer uses a cable lubricant to reduce friction and load on the cable. While copper cable lubricants exist, the installer uses a fiber optic cable lubricant, as such a lubricant is matched to the jacket of the cable.

The second method for reducing the load is pulling the cable by hand in multiple, reduced length pulls. Breaking a long pull into multiple pulls of reduced length reduces the load on the cable. If the installer installs the cable in multiple pulls, he stores the cable at intermediate locations in a 'Figure 8' pattern. The advantage of this method is reduced load. The disadvantage is increased labor cost.

▶▶Use 'figure 8' installation

During a 'Figure 8' pull, the installer pulls the cable into the first manhole and out of a subsequent manhole (Figure 11-5). The subsequent manhole may be the next manhole along the cable path or the nth manhole along the path. The installer determines this manhole by the load he is able and willing to impose on the cable.

As the installer pulls the cable from the manhole, he places the cable on the ground in a 'Figure 8' pattern (Figure 11-6). This pattern can be 12 feet high and 4-6 feet wide. For practical reasons, the pattern is rarely higher than 24 inches.

When the installer has pulled all the cable out of the manhole, he and several helpers pick up the 'Figure 8' and flip it over so that the cable end is on top.

The installer pulls the cable back into the same manhole and out another manhole along the cable path (Figure 11-7). At this manhole, the installer repeats the 'Figure 8' pattern on the ground. The installer can repeat this process as many times as desired, until the cable is installed along the entire path. This author is aware of installation by this method to a distance of 40,000 feet.

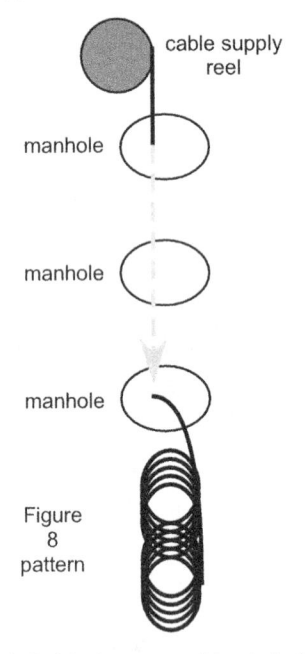

Figure 11-5: Underground Install, Step 1

The installer can use the Figure 8 method for both unidirectional pulls (Figure 11-5 and Figure 11-7), and mid pulls, (Figure 11-3). In a bi-directional pull, the installer pulls the cable in one direction, ('pull 1', Figure 11-3). He places the remaining cable in a figure 8 pattern on the ground. Finally, he pulls the cable from the figure 8 pattern in the opposite direction ('pull 2', Figure 11-3).

Some professional installers use a fourth method for limiting the load they apply during installation. This method requires pulling a 600 pound-force cable by hand. The assumption is that it is essentially impossible to create a 600-pound load by hand pulling in a horizontal axis. This method has the disadvantage of providing no evidence that the installers have not exceeded the installation load.

PROFESSIONAL FIBER OPTIC INSTALLATION 11-5

Figure 11-6: Cable Storage In A Figure 8 Pattern

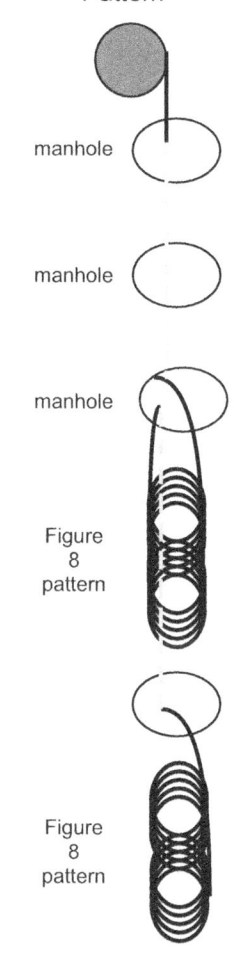

Figure 11-7: Underground Install, Step 2

All four methods require attachment of a pull rope to the cable. The installer attaches a pull rope to a cable so that the cable strength members support the load and that no load is imposed on the fibers.

▶ Do not load fibers

The installer has at least five attachment possibilities:

- ➢ Attachment of the pull rope around the outside of the cable jacket
- ➢ Attachment of the pull rope to a Kellems grip (Figure 11-8) that grips the cable through the jacket
- ➢ Attachment of the pull rope to central strength members
- ➢ Attachment of the pull rope to strength members outside the loose buffer tubes
- ➢ Attachment of the pull rope to strength members between multiple jackets

With all these possibilities, the installer needs another principle. The best method of attachment is the method recommended by the cable manufacturer.

▶▶ Use recommended attachment method

Figure 11-8: Kellems Grip

The manufacturer has designed the cable so that it can be installed without damage. He has done so by assuming that the installer will attach the pull rope to specific strength members. Attachment of a pull rope to any cable structural element other than these strength members may result in a load imposed on the fibers. Such a load can result in fiber breakage.

There is one final method that some installers follow. The installer uses a loose tube cable design for installations in which the installation load will be high. The excess fiber (4.2.1.1) in this type of cable allows the fiber to move to reduce the stress on the fiber. This movement can be for hundreds of feet from the high stress area.

An example of a high load installation: a professional installer applied a 1200 pound-force load to a loose tube cable rated at 600 pounds-force without experiencing fiber breakage or residual increase in attenuation rate.

© PEARSON TECHNOLOGIES INC.

▶▶Use loose tube cable for high load installations

11.3.1.3 BEND RADIUS

In order to avoid breakage, the installer limits the bend radius of the cable to above the minimum value, such as those values in Equations 11-1 and 11-2. This limitation means that each deviation from a straight path requires some form of control. A pulley or sheave provides such control. To ensure compliance with these limits, the installer obtains the bend radii from the cable data sheet.

Short term bend radius = 20 x cable diameter

Equation 11-1

Long term bend radius = 10 x cable diameter

Equation 11-2

▶Limit bend radii

A cable that complies with TIA/EIA-568-C meets these requirements. When so indicated on a cable data sheet, a cable can withstand a bend radius smaller than that indicated by Equation 11-1.

11.3.1.4 SUPPLY REEL

The installer monitors the supply reel to avoid bend radius violations due to:

- Improper winding of cable on reel
- Loosening of cable
- Cable wrapping around the shaft supporting the reel
- Back wrapping when the pull stops

If the cable is improperly wound on the reel (Figure 11-9), the cable will attempt to pull a lower layer from under an upper layer. This situation results in the cable making a right angle as it leaves the reel- an obvious bend radius violation!

Figure 11-9: Cable Wound Incorrectly

▶▶Monitor supply reel

If the cable loosens during the pulling process, it can ride over the top of the flange of the reel and wrap around the shaft that supports the reel, another example of bend radius violation.

If the installer does not stop the reel from rotating at the end of the pull, the cable can wrap backwards around the reel, a third example of bend radius violation.

11.3.1.5 COMMUNICATION DURING PULLING

The installer communicates and coordinates the pulling actions. In advance of a stop, the installer in charge of the pulling equipment informs the installer at the supply reel of his intent. With this advance notice, the installer at the supply reel can don the heavy work gloves needed to avoid splinters as he grabs the rapidly spinning supply reel flange in order to stop the reel.

▶▶Communicate and coordinate pulling

11.3.1.6 PEOPLE PER PULL

The installer in charge of the pulling equipment can forget to alert the installer at the cable supply reel. Because of this possibility, each installation requires at least three installers. The third installer coordinates the activity of the other two.

▶▶At least 3 people per pull

11.3.1.7 PULLEY MONITORING

An installer monitors each pulley location (11.3.1.3). This monitor ensures that the cable does not cable jump from the pulley. Should the cable jump, the installer can have two problems: fiber damage and strength member damage.

▶▶Monitor all pulleys

Should the cable jump from a pulley, the bend radius is no longer under control. In addition, the cable may scrape against a sharp edge at the entrance of the conduit. Should this sharp edge cut through the jacket, it may damage strength members under the jacket. Such damage will reduce the installation load capability of the cable, increasing the likelihood of fiber damage.

11.3.1.8 PULL CABLE

During installation, the installer pulls the cable. Pushing can cause a violation of the bend radius.

▶▶Always pull cable

11.3.1.9 INSTALLATION TEMPERATURE RANGE

The installer installs the cable at a temperature within the range specified on the data sheet. At excessively low temperatures, the cable materials may be brittle enough to crack. At excessively high temperatures, the cable materials may stretch excessively, resulting in many problems.

▶▶Comply with installation temperature range

11.3.1.10 STORAGE TEMPERATURE RANGE

The installer stores the cable at a temperature within the storage temperature range specified on the data sheet. Storage of a cable at a temperature outside of its storage range will cause the material problems (11.3.1.9).

▶▶Comply with storage temperature range

11.3.2 CABLE PLACEMENT

In this section, we address the issue of cable placement. By placement, we mean installation without pulling. Such placement occurs in cable trays, troughs and raceways.

11.3.2.1 LONG TERM BEND RADIUS

After the installer completes the pulling, he limits the long-term bend radius to at or above the minimum value (Equation 11-2).

▶▶Limit long term bend radius

11.3.2.2 ▶▶BUNDLE CABLES

The installer bundles cables running along same path. Bundling reduces the risk of bend radius violation. In addition, bundling simplifies circuit tracing.

11.3.2.3 ▶▶HAND TIGHTEN TIES

If he uses cable ties, the installer tightens these by hand, not with a cable tie tool. Excessively tight cable ties can deform the cable jacket, resulting in a localized bend radius violation and excess power loss.

11.3.2.4 ▶▶USE VELCRO™ BANDS

The installer uses Velcro™ bands to bundle cables. Velcro bands minimize the risk of bend radius violation.

11.3.2.5 ▶▶SEGREGATE FIBER CABLES

This is the author's opinion. Whenever copper and fiber cables reside in the same tray, the installer places the fiber cables in inner duct. By segregating the two types of cables, the installer reduces the risk of bend radius and crush load violations of the fiber cables.

Such segregation avoids a bend radius violation during installation of additional cables in the tray. If an installer is not aware of bend radius concerns, he may use cable ties to attach new cables to those in place. In this situation, a violation of bend radius may result.

Segregation avoids a crush load violation that can occur when heavy copper cables are installed on top of fiber cables in ladder trays. In this situation, the fiber cables can experience localized bend radius violations.

11.3.2.6 ▶▶LEAVE SERVICE LOOPS

The installer leaves service loops throughout the link. Service loops are lengths of excess cable that can be pulled into problem areas. Service loops are inexpensive insurance: they are less expensive than replacing the entire segment!

For indoor links, a common practice is a 10-12' service loop near both ends of each link. This service loop coil may in the back of an enclosure, on a cable tray, or above a suspended ceiling.

For outdoor links, a common service loop practice has three parts:

- A 100' service loop for each 1000' of cable
- A 100' service loop for each street crossing
- No more than 200' feet of service loop per 1000' of cable

Finally, there will be a service loop of at least 10-12' at each cable end. Often this service loop is 50'. This service loop is in addition to the length of cable that will be stripped for termination. For aerial cable systems, service loop holders (Figure 9-9) hold the service loop.

11.3.2.7 ▶▶MARK CABLE

In any location in which the cable can be accessed, the installer marks it as "Fiber Optic Cable". There are two reasons for such marking. First, when the cables are so marked, electricians will not be tempted to cut, reroute and splice them with black tape! Second, installers will not be tempted to used such marked cables in a manhole as a ladder step.

11.3.2.8 ▶▶LEAVE SAG

Installers install outdoor cables with sag to allow for thermal expansion and contraction. A 2.5-foot sag for 150-foot span is a common practice.

11.4 NEC COMPLIANCE

Indoor cables must comply with the National Electric Code (NEC) and local electrical codes (4.4). Horizontal cable runs require OFN or OFC-rated cables. Cable runs between floors require OFNR, OFNP, OFCR, or OFCP-rated cables. Cable runs in air handling plenums require OFNP or OFCP-rated cables. In addition, firewall penetrations must meet the requirements of the applicable fire code.

To determine compliance, the installer reads the printed information on the cable. The NEC requires that all cables be printed with their rating and test number.

11.5 END PREPARATION

After pulling and placing the cable, the installer prepares the cable ends for splicing or connector installation (21.5).

11.5.1 PRINCIPLE

The installer protects buffer tubes and fibers. This principle means that the jacket ends inside the enclosure, not outside the enclosure. The buffer tubes and fibers are not designed to withstand exposure to the working environment.

▶ Protect buffer tubes and fibers
▶▶ Seal cable ends with fiber optic cable sealant

The installer seals the ends of grease filled, gel blocked cables with a fiber optic cable sealant. Water blocking compounds can, and will, flow from even a small vertical drop. Such flow will cause maintenance problems. An improper sealant can attack primary coating of the fiber.

It is this author's understanding that some silicone sealants cure to produce acetic acid while others cure to produce water. Acetic acid can attack the primary coating of the fiber. Thus, the installer should use a fiber optic cable sealant.

11.5.2 COSMETICS

Some installers put heat shrink tubing over the end of the jacket. This conceals any uneven jacket preparation, sealant, and cable materials not trimmed flush with the jacket.

11.6 REVIEW QUESTIONS

1. An installation team is planning to pull a cable through an underground conduit system. The system consists of conduits between manholes. There are no sweeps at locations at which the

cable path changes direction or at which the conduit changes elevation. What must the installer do at these changes of direction?

2. Regarding Question 1: the manholes in this system have a cover diameter of 36". With the assumption that the installers have single piece pulleys only, what is the maximum cable diameter that the installers can install without violation of the short term bend radius?

3. The installer is planning to place a service loop of cable inside the back of an enclosure mounted on a 19" rack. The loop will be mounted vertically in the enclosure. The enclosure fills the full width of the rack. What is the maximum cable diameter that the installer can place in this enclosure?

4. Regarding Question 3: How high must the inside height of this enclosure be?

5. Installers are installing the cable system shown in Figure 11-10. The cable changes direction by exiting one conduit and entering another in each of the manholes indicated. The supervisor believes that the cable can be installed in a single pull. A swivel-pulling eye with a shear pin will control the load. The swivel is between the cable and the pull rope. A truck will be used to pull in the cable. The supervisor needs to determine the minimum number of installers he needs to send. Including the truck driver, what is this number?

6. You observe your installer tying a pull rope to the strength members of a fiber cable. Do you see any potential problems with this activity?

7. Prior to preparing the end of a filled and blocked cable, an installer stops at Home Depot for silicone sealant. What should you say to him?

8. What is the installation rule for cable bend radius?

9. What is the bend radius rule for storage of a coil of cable?

10. To what part of cable is a pull rope with swivel attached?

11. How is the cable attached to the swivel?

12. What three steps can an installer use to minimize pulling load on a cable during installation?

13. How tight should cable ties be attached the a cable?

14. What method of attachment is preferred to cable ties?

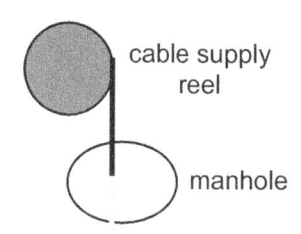

cable supply reel

manhole

800 feet

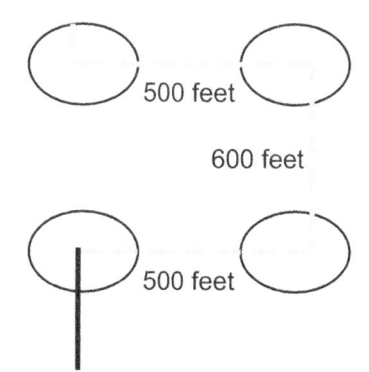

500 feet

600 feet

500 feet

Figure 11-10: Map For Question 5

12 CONNECTOR INSTALLATION PRINCIPLES

Chapter Objectives: you learn the principles for installation of fiber optic connectors. These principles result in the procedures of Part 4.

12.1 INTRODUCTION

In this chapter, we present the principles that result in the procedures for connector installation by four methods:

- Epoxy and polish
- Quick cure adhesive
- Hot Melt™ adhesive
- Cleave and crimp

These methods are commonly used and represent a majority of connectors installed in North America.

Taken as a group, these methods include combinations of the following six steps:

- End preparation of the cable
- Injection of adhesive
- Insertion of the fiber into the connector
- Crimping
- Curing of adhesive
- End finishing

12.2 CABLE END PREPARATION

Cable end preparation involves:

- Removal of the jacket(s)
- Removal of water blocking compounds
- Trimming of strength members
- Removal of buffer tube and primary coating (Figure 12-1, Figure 12-2, and Figure 12-3)

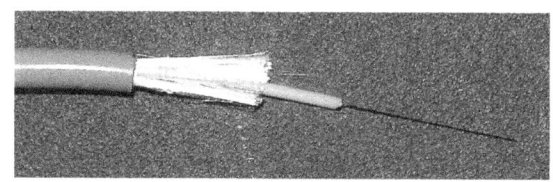

Figure 12-1: Single Fiber Cable End Prepared For Connector Installation

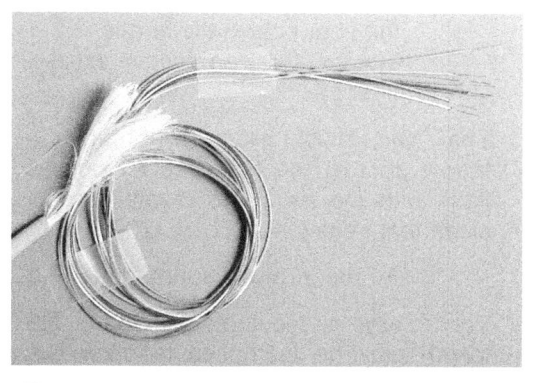

Figure 12-2: Eight Fiber Premises Cable End Prepared For Connector Installation

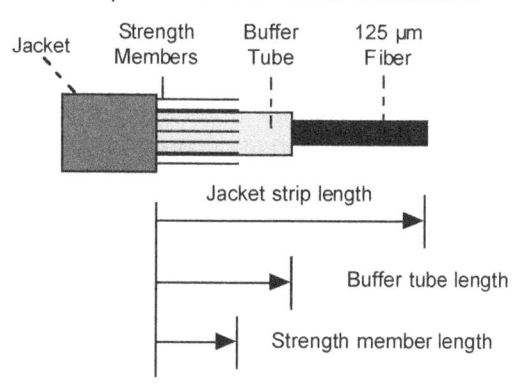

Figure 12-3: End Preparation Dimensions For Fibers With Individual Jackets

12.2.1 DIMENSIONS

The cable end preparation dimensions depend on three factors:

- The nature of the installation
- The specific connector
- The specific enclosure into which the cable is to be installed

The installation can be either patch cord assembly or enclosure installation.

12.2.1.1 PATCH CORD ASSEMBLY

For patch cord assembly, the connector to be installed determines the end preparation dimensions. Six principles determine these dimensions (Figure 12-3).

12.2.1.1.1 FIBER AND FERRULE

For connectors requiring polishing, two lengths, the jacket removal length and the bare fiber length, are long enough that:

- The bare fiber protrudes through the fiber hole in the ferrule
- The installer can remove excess fiber easily

The fiber must protrude through the ferrule so that the fiber can be polished flush with the ferrule to create a round, clear, featureless, flush core (20.4.1).

▶ Fiber protrudes through ferrule

For epoxy connectors, the bare fiber length may be increased to allow easy removal of the excess fiber. However, if the bare fiber length is increased excessively, the installer may break the fiber hen installing the connector into a curing oven.

For some connector styles, such as the SC and the LC, the jacket removal length is increased. Such an increase enables a slight amount of flexing of the fiber under the boot. This flexing occurs when the connector is inserted into a patch panel. This flexing enables pull proof and wiggle proof behavior (5.3.4.2.1 and 5.3.4.2.2).

Excessive jacket removal length can result in two problems:

- long bare fiber
- short jacket.

Excessive bare fiber length can result in breakage during insertion through the ferrule. In some cases, this broken fiber can become jammed in the fiber hole, resulting in loss of the con-nector.

When the connector has quick cure adhesive, excessive fiber length can result in premature adhesive curing and bare fiber inside the connector. Bare fiber inside a connector results in reduced reliability.

The SC connector installed with epoxy or quick cure adhesive provides an example of such reduced reliability: if there is bare fiber inside the connector or if the buffer tube is short, so that the buffer tube is not immersed in the epoxy or adhesive, bare fiber remains inside the connector. The SC connector requires that the fiber in the back shell flex slightly whenever the connector is inserted into a patch panel. Repeated flexing of bare fiber results in breakage. During training programs, we have observed such failure after 5-10 insertions.

The boot will not support a short jacket. A cable that is not covered by the boot is a condition of reduced reliability (Figure 12-4 and Figure 12-5).

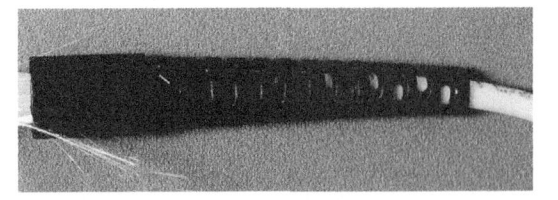

Figure 12-4: Proper Jacket Removal Length

Figure 12-5: Excessive Jacket Removal Length

12.2.1.1.2 STRENGTH MEMBER

The strength member must extend beyond the end of the jacket to be gripped by the crimp sleeve.

Some connectors have no crimp sleeve. Instead the strength members fold back over the jacket, which fits into the back shell of the connector. The back shell of the connector is crimped to both the strength member material and the jacket.

Excessive strength member length can interfere with proper operation of the connector. For example, excessive strength member length will prevent sliding of the outer housing of an SC connector over the inner housing (Figure 5-3). Such sliding is necessary for both insertion and removal of the connector.

▶ Strength member length must be proper
▶ A common strength member length is 3/16" - 5/16"

12.2.1.1.3 BARE FIBER

The buffer tube length must be long enough to butt against the inside of the ferrule, or to be completely immersed in an adhesive that hardens so the adhesive prevents motion of bare fiber (Figure 12-6). Bare fiber in a connector is a condition that can result in reduced reliability.

▶ No bare fiber in connector

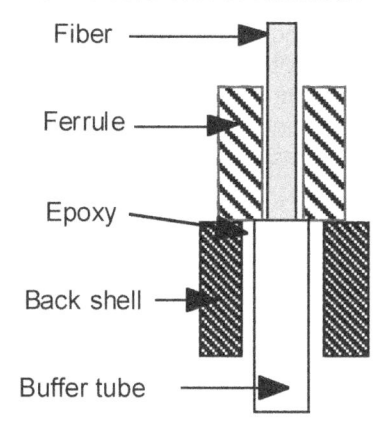

Figure 12-6: Buffer Tube Butts Against Ferrule

In order to comply with the principles in 12.2.1.1.1-12.2.1.1.3, the installer takes two steps:

▶▶ Determines dimensions
▶▶ Uses template

He uses a template for each connector type to control the stripping dimensions. This template has the form of Figure 12-3.

12.2.1.1.4 CRIMPING

This section applies to connectors that require adhesive. Connectors that require no adhesive and fiber cleaving require a crimper unique to the connector. Crimping of such a connector grips fiber and buffer tube.

The installer performs crimping to:

> Attach the cable strength members to the connector back shell
> Grip the jacket

12.2.1.1.5 CRIMPER NEST

An oversized crimp nest will not grip the strength members, allowing the connector to be pulled from the cable easily. An undersized crimp nest tends to destroy the crimp sleeve. In addition, the tolerance of the crimp sleeve is relatively tight. For example, a 0.141" crimp nest may result in a loose crimp sleeve, while a 0.137" nest results in a tight sleeve. Close is not good enough!

To avoid such problems, the installer uses the exact crimp nest size(s) recommended by the connector manufacturer (Figure 12-7).

▶▶ Use proper nest size

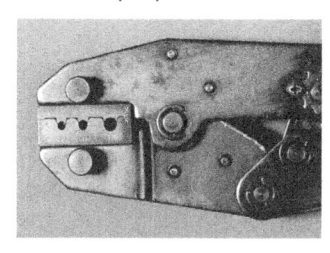

Figure 12-7: Crimp Nest In A Crimper

The large diameter of the crimp sleeve attaches the strength members to the connector. Once the crimp sleeve is attached to the connector, the connector cannot fall off the cable. The small nest attaches the jacket to the connector. This attachment prevents the jacket from shrinking back along the cable, resulting in the reduced reliability condition shown in Figure 12-5.

▶▶ Large crimp first

12.2.1.2 ENCLOSURE INSTALLATION

For installation of connectors onto a cable that is installed in an enclosure, both the enclosure and the connector determine the end preparation dimensions. These dimensions are determined by four principles:

▶ Sufficient buffer tube length
▶ Strength member length proper
▶ No bare fiber outside of connector
▶ Fiber protrudes through ferrule

For all connectors, the jacket removal length is long enough to enable the cable to be attached to the enclosure at a specific location (Figure 12-8 and Figure 12-9), to allow the buffer tube to extend outside the front of an enclosure for connector installation, and to allow the end of the buffer tube to reach the patch panel that is integral to the enclosure. Excess buffer tube length will be coiled inside of the enclosure (Figure 12-10).

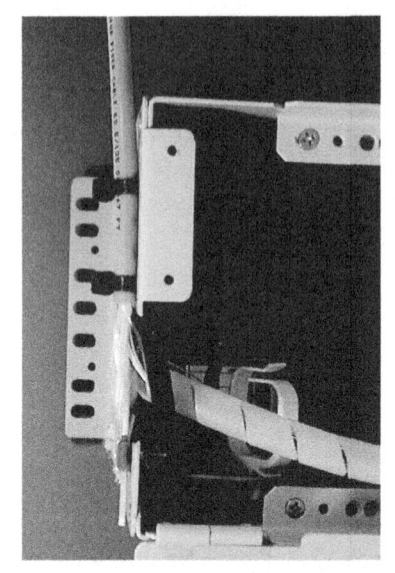

Figure 12-8: Indoor Cable Attachment Location

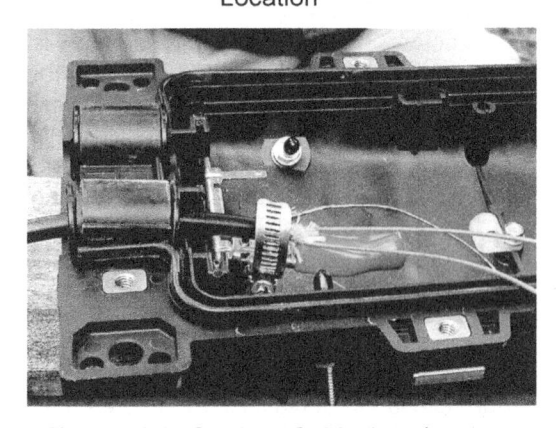

Figure 12-9: Outdoor Cable Attachment Location

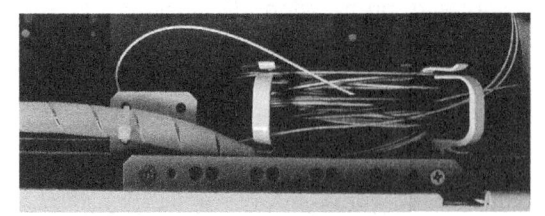

Figure 12-10: Buffer Tube Coiled In Indoor Enclosure

This length must not be excessive, as excessive length can result in damage to the buffer tube when the volume of buffer tubes exceeds the capacity of the enclosure. As a practical matter, a practical tolerance on buffer tube length is ±1-2".

The enclosure determines the strength member length. The enclosure is designed so that the strength members can be attached at a specific location (Figure 12-8 and Figure 12-9). In general, excessive strength member length is a cosmetic or workmanship concern, but not a performance or reliability concern.

▶Strength member length must be sufficient

The third principle, no bare fiber, is almost automatic. Without a jacket, there is nothing to prevent the fiber from entering the connector until the buffer tube butts against the inside end of the ferrule.

The fourth principle, fiber protrudes through ferrule, is controlled by the bare fiber length. The fiber will protrude beyond the tip of the ferrule, as long as the installer

▶▶Determines dimensions

▶▶Uses a template

12.2.2 FIBER PREPARATION

In order to install connectors, the installer avoids damaging and contaminating the cladding.

12.2.2.1 CLADDING DAMAGE

Because the installer removes the primary coating, the cladding can become scratched or otherwise damaged. Such damage can result in fiber breakage. The installer

▶Avoids damage

Avoiding damage requires cleaning with of lens grade tissues, which are not abrasive.

▶▶Use lens grade tissues
▶▶Strip, clean and insert fiber without delay

If the cladding is exposed to air for a significant length of time, moisture in the air can reduce the fiber strength. By itself, water will not reduce fiber strength. Moisture in the air is not pure water. As such, it can attack the fiber.

12.2.2.2 CONTAMINATION

Low power loss requires a small clearance between the cladding and the inside of the fiber hole. This small clearance requires prevention of contamination of the cladding surface.

▶Avoid contamination

The installer cleans the fiber with lens grade, lint free tissues to remove contamination. After cleaning the fiber, the installer does not place it down or against any surface. To avoid contamination, the installer

▶▶Inserts the fiber in the connector immediately after cleaning
▶▶Uses 98 % isopropyl alcohol

70 % isopropyl alcohol contains water and oil. Both will contaminate the cladding. Oil will prevent the adhesion of an epoxy or adhesive.

12.3 ADHESIVES

We use the term 'adhesive' to mean any material that is used to glue, or fix in place, the fiber inside the connector. Fiber connectors use three adhesive types:

- Epoxy
- Quick cure adhesive
- Hot melt adhesive

12.3.1 EPOXY

We define the term 'epoxy' to mean a one or two part chemical system that requires a minimum cure time of two minutes with heat or a minimum cure time of 6 hours without heat.[43]

12.3.1.1 THREE FUNCTIONS

Epoxy provides three functions:

- A strong bond between the fiber and the ferrule
- A strong bond between the buffer tube, or primary coating, and the connector back shell
- Support of the fiber protruding beyond the end of the ferrule during polishing

---
[43] This definition is arbitrary, but fits the experience of the author.

12.3.1.2 TYPES

We organize fiber optic epoxies into two main groups, each of which has three subgroups. The two main groups are:

- Heat cured epoxies
- Room temperature cured epoxies

The three subgroups are based on curing time:

- Fast cure (1-5 minutes)
- Medium cure (5-60 minutes)
- Slow cure (> 60 minutes)

Heat reduces cure time, but increases the possibility of thermal cracking of the fiber. As cure temperature increases, the likelihood of cracking increases. Cracking can result when a heat cured connector cools to room temperature. The different thermal expansion rates of fiber, epoxy and ferrule can result in excessive compression of the fiber. This compression causes cracking. Cracks can divert light from its proper path, causing increased connector loss. At the time of this writing, thermal cracking is uncommon.

Heat curing can be used, as long as the combination of temperature, fiber core diameter, ferrule material and epoxy does not result in cracking. The installer avoids cracking by:

▶▶Use proper time and temperature combination
▶▶Use no heat

12.3.1.3 EPOXY STRENGTH

The curing time and temperature must be sufficient to achieve adequate strength between the fiber and the ferrule. With insufficient strength, tension on the fiber can result in the fiber withdrawing into the ferrule. Pressure on the fiber can result in the fiber protruding from the ferrule. Both types of motion, called '**pistoning**', are undesirable.

▶Sufficient epoxy strength required

The combination of curing temperature, fiber core diameter, ferrule material and epoxy results in sufficient strength.

Whenever practical, the installer uses an epoxy that cures at room temperature. Such an epoxy will not exhibit thermal cracking. In addition, there will be no thermostat to malfunction. Such a malfunction can result in cracking.

Of course, there are situations in which use of heat is desirable. Small field installations and cable assembly operations use heat to reduce cure time.

12.3.1.4 INJECTION

The installer injects epoxy into the back shell of the connector with a syringe. The amount he injects is critical to proper operation of the connector.

12.3.1.4.1 EPOXY INTERFERENCE

The epoxy is injected so that the epoxy does not interfere with proper operation of the connector.

▶No Interference

Excessive epoxy can create connector malfunction due to expansion and displacement. When heated, excessive epoxy in the back shell expands. Such expansion can cause the epoxy to flow out of the back shell onto other areas of the connector. Such epoxy can cause the connector to fail to function.

For example, the SC and LC connectors contain a tube inside the back shell (Figure 12-11 and

Figure 12-12). This tube is mechanically independent of the back shell. Should epoxy flow between the tube and the back shell, the connector will lose its pull proof and wiggle proof performance (5.3.4.2.1 and 5.3.4.2.2). In addition, such excessive, expanding epoxy can wick under a jacket via the strength members, making the cable rigid and sensitive to handling.

When the fiber and buffer tube are inserted into the back shell, excess epoxy will be displaced from the back shell. In the case of the SC and LC connectors, such displaced epoxy can flow between the central tube and the back shell. As stated earlier, the connector will lose its pull proof and wiggle proof performance.

Figure 12-11: SC Inner Tube

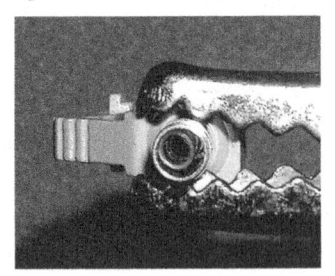

Figure 12-12: LC Inner Tube

To avoid excessive epoxy, the installer injects enough epoxy to fill the fiber hole in the ferrule and one additional drop in the bottom of the back shell (Figure 12-13). The additional drop bonds the buffer tube to the connector. To avoid interference, the installer

▶▶Minimizes epoxy

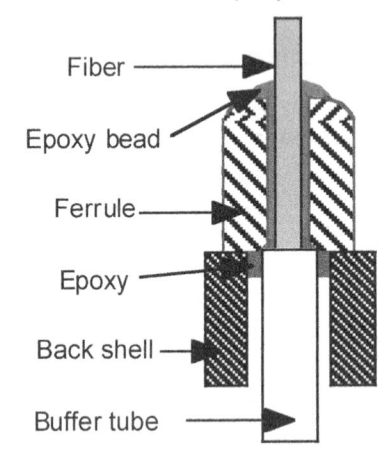

Figure 12-13: Sufficient Epoxy In Back Shell

12.3.1.4.2 BEAD SIZE

Epoxy connectors have a bead on the ferrule tip (Figure 12-13). During polishing, this bead supports the fiber. With this support, fiber shattering and mechanical cracks can be avoided.

▶Control bead size

There are three methods of controlling bead size. The first is to inject epoxy until it appears through the fiber hole in the ferrule. Immediately, the installer removes the syringe from the connector. When the installer inserts the fiber through the ferrule, the fiber will force additional epoxy onto the ferrule tip. This method results in the a bead size.

The second method requires the installer to wipe the epoxy from the ferrule tip before inserting the fiber. The size of the bead will be controlled by the epoxy forced through the fiber hole by the fiber. As long as the bare fiber length is consistent, there will be a consistent bead size.

The third method requires insertion and retraction of the fiber until it is flush with the ferrule. The installer wipes all epoxy from the tip of the ferrule and reinserts the fiber. This method results in an small bead, which is appropriate for hand polishing by experienced installers only.

With a toothpick, inexperienced installers can add epoxy to the tip of the ferrule. This additional epoxy enables polishing without damaging the fiber.

The bead size determines polishing time and process yield through avoidance of shattering. As this size increases, polishing time and cost increase and the incidence of shattering decreases. Obviously, there is a trade off.

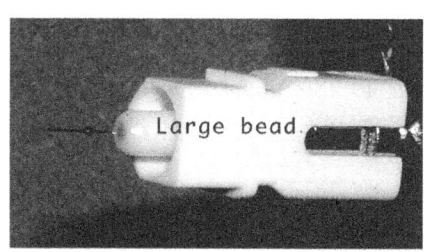

Figure 12-14: Large Bead For Inexperienced Installers

This trade off depends on the installer experience: the more experience the installer has, the smaller the bead can be. A practical strategy is a bead of approximately 0.020-0.030" high (Figure 12-14) for novice installers and a 0.015" bead (Figure 12-15) for experienced installers. As a visual reference, installers can use a paper clip. Paper clips have a diameter of approximately 0.030".

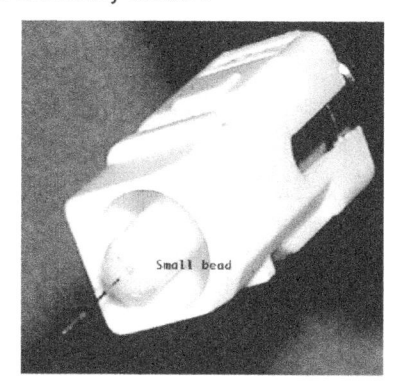

Figure 12-15: Small Bead For Moderately Experienced Installers

12.3.1.5 CURING

Epoxy curing has three variables: expiration date, time, and temperature. Epoxies may not cure after their expiration date. In addition, some epoxies will not cure if they have been exposed to freezing temperatures.

Epoxies will cure properly as long as the curing time exceeds a minimum time. Additional time will not cause degradation of the bond.

Temperature is a different matter. Low temperature does not result in full bond strength between the fiber and the ferrule. Without such a bond, the fiber can move into and out of the ferrule ('pistoning'). Such movement can result in link failure due to insufficient power at the receiver.

Excessive curing temperature can result in degradation of the bond strength and in fiber cracks.

The installer controls the temperature at that recommended by the epoxy manufacturer. To ensure such control, he monitors the oven temperature with an accurate thermometer. To ensure proper curing, the installer:

▶▶Controls oven temperature
▶▶Monitors oven temperature
▶▶Cures for a minimum time

To ensure the best possible reliability, the installer chooses an oven with a thermostat that fails in the 'full off' mode. An oven without such a thermostat overheats, resulting in cracked fibers.

12.3.2 QUICK CURE ADHESIVES

We define 'quick cure adhesive' as a one or two part chemical system that cures without heat in a relatively short time, typically less than two minutes.

These adhesives have expiration dates. After the expiration date, the adhesive cures slowly. Eventually, the adhesive fails to cure. Therefore, prior to use, the installer:

▶▶ Checks the expiration date.

12.3.2.1 TYPES

Quick cure adhesives can be a two part liquid product or a one-part gel product. In general, both are edge-filling adhesives. Edge-filling adhesives harden when the adhesive fills a narrow space, such as that between the outside of the fiber the inside of the fiber hole in the ferrule.

12.3.2.2 INJECTION

Excess quick cure adhesive inside the connector creates the same problems as does excess epoxy. In addition, two part, quick cure adhesives do not cure inside the back shell. Thus, excess adhesive inside the connector has no benefit.

By itself, two part, quick cure adhesive can cure slowly. As a result of this characteristic, the installer need not rush the insertion of the fiber after adhesive injection out of concern for premature curing of the adhesive. After being injected, the adhesive may not cure for five minutes. Therefore, the installer

▶▶ Need not rush

12.3.2.3 PRIMER APPLICATION

The second part of a two-part, quick cure adhesive systems, is called a primer, accelerator, or hardener. This part is applied to the fiber prior to its insertion into the connector. This application can be by spraying, dipping or brushing. The best method of application is

▶▶ Brushing of primer

We recommend against spraying, as spraying can create an acetone mist, a carcinogen.

Dipping would seem to provide the most consistent and complete coverage of the fiber. However, dipping is not convenient when the primer bottle is partially empty. Brushing has proven convenient and without any problems.

12.3.2.4 FIBER INSERTION

After brushing the primer onto the fiber, this author has observed a problem with insertion of the fiber. If the time between application of the primer and insertion of the fiber is excessive, the fiber may not fit into the fiber hole. After brushing the primer onto the fiber, the installer:

▶▶ Inserts the fiber without delay

The adhesive starts curing as soon as the fiber is inserted. Quick-cure adhesives cure in less than two minutes, and, occasionally, in as little as 30 seconds. Because of this short cure time, the installer inserts the fiber into the connector quickly. To do otherwise would allow the adhesive to cure prior to full insertion of the fiber. Premature curing results in bare fiber inside the connector, a condition of low reliability. To avoid premature curing of the adhesive, the installer

▶▶ Inserts the fiber quickly

12.3.2.5 TIP BEAD

Insertion of the fiber forces the quick cure adhesive from the fiber hole onto the ferrule tip. As the adhesive cures only when in thin sections, application of the primer to the fiber or to the tip of the ferrule results in a very small bead (Figure 12-16).

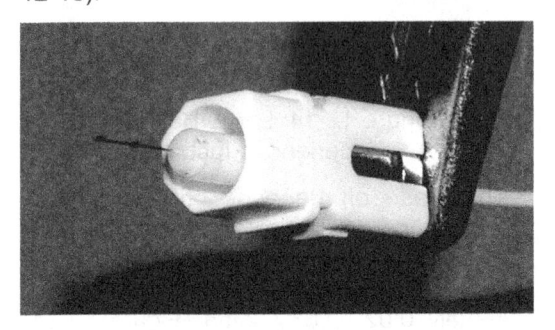

Figure 12-16: Bead Without Application Of Additional Adhesive

Repeated application of adhesive and primer to the tip of the ferrule does increase the bead size. Increased bead size increases the resistance to fiber shattering during polishing (Figure 12-17). However, the increase in bead size is not significant. In addition, application of additional adhesive to the ferrule tip increases connector installation time, thus reducing the advantage of this method.

Figure 12-17: Bead Size With Three Adhesive Applications

Curing begins as soon as the fiber is inserted into the connector. The older the adhesive is, the longer the cure time will be. Since 'old' adhesive cures slowly and provides increased time for fiber insertion, the installer in training can

▶▶Use old adhesive

This author's experience is that quick cure adhesives can be used to several years after the expiration date. Eventually, such adhesive will fail to cure.

To verify full curing, the installer pulls on the cable. If the fiber protruding beyond the ferrule tip does not move into the ferrule, the adhesive has cured fully.

12.3.3 HOT MELT ADHESIVE

We define the term 'hot melt adhesive' to mean a one part adhesive with two characteristics: it is preloaded in the connector and requires connector pre-heating prior to insertion of the fiber. When the connector cools, the adhesive grips the fiber. 3M pioneered and is the sole source of this system.

12.3.3.1 ABILITY TO REHEAT

Because the adhesive can be reheated, it is possible to salvage and repair damaged connectors. This characteristic results in reduced installation and maintenance costs. Such repair is possible under two conditions:

➢ Reheating must be done prior to crimping of the crimp sleeve
➢ There is extra fiber inside the connector

To provide extra fiber inside the connector, the installer inserts the fiber fully into the connector. He withdraws the fiber by approximately 1/16". After the installer re-heats the connector, he can push this extra fiber through the ferrule for a second polishing to repair a damaged core.

▶▶Leave extra fiber inside connector

If a fiber is so badly shattered that it cannot be re-polished, installer reheats the connector, removes the fiber, prepares a new cable end and reinserts the new end into the connector. This method can result in 100% yield and reduced maintenance cost.

This method works well with connectors installed on premises cable. This method can work once or twice with jacketed cable and the ST-™ compatible connector. This method will not work when installed on a jacketed cable with the SC connector.

12.3.3.2 POLISH TIME

The viscosity of the hot melt adhesive is so high that it tends to stick to the fiber as the installer inserts the fiber. The experienced installer can reduce polishing time by increasing the bare fiber length by 1/2". By increasing the fiber length, the installer reduces the bead size and polishing time to 20-30 seconds for multimode connectors. However, if the fiber length is excessive, there will be so little adhesive on the tip of the ferrule that any breakage of fiber during polishing requires connector rework or replacement. To reduce polishing time, the installer

▶▶Increases bare fiber length

12.3.4 FIBER INSERTION

Regardless of the method of installation, the installer inserts a fiber into a connector. Low loss requires a small clearance between the cladding and the fiber hole in the ferrule. In addition, clearance and tolerances of singlemode connectors are smaller than those in multimode connectors. The installer exercises more care inserting fibers into singlemode connectors than into multimode connectors.

12.3.4.1 FIBER BREAKAGE

The installer's objective is insertion of the fiber without breakage. If the fiber breaks, the installer may loose the connector. If the fiber does not bend, it cannot break. Therefore, during fiber installation, the installer

▶Does not bend the fiber

Rotation of the connector during fiber installation allows the fiber to slip past lips or steps inside the connector. To avoid such bending, the installer

▶▶Rotates the connector

Difficulty during fiber insertion may be due to dirt in the fiber hole. To clean dirt from a fiber hole, the installer can

▶▶Flush the connector with a syringe filled with isopropyl alcohol[44]

12.4 END FINISHING

End finishing is the process of removing excess fiber from the end of the ferrule and creating an optical grade surface on the fiber end. When properly performed, this process results in a core that is round, clear, featureless and flush with the surface of the ferrule (20.4). This process includes three steps:

> Excess fiber removal
> Air polishing
> Pad polishing

12.4.1 FIBER REMOVAL

The objective of fiber removal is removal of excess fiber without causing it to break below the surface of the ferrule (Figure 12-18 and Figure 12-19). If the fiber breaks below this surface, the installer may not be able to polish the fiber so that the core is defect free and low loss. The installer removes the fiber by scribing it.

---

[44] This method of cleaning does not work on Hot Melt connectors.

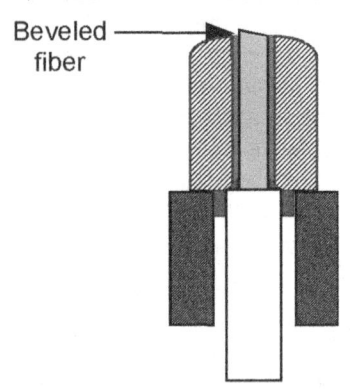

Figure 12-18: Fiber Broken Below Ferrule

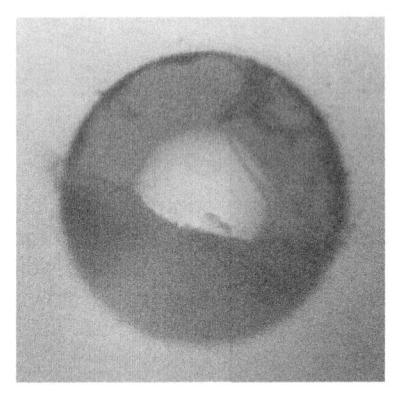

Figure 12-19: Broken Fiber Appearance

Scribing is the placement of a single, small scratch on the fiber at the tip of the bead of epoxy or adhesive (Figure 12-20). Scribing is not sawing the fiber or breaking the fiber.

▶Scribing is scratching, not sawing

The installer uses a scriber with a wedge shape (Figure 12-21), not one with a point. The installer aligns the edge of a wedge scriber to the fiber more easily than the point of a 'pencil' type scriber. After scribing, the installer removes the fiber by pulling the fiber away from the tip of the ferrule.

▶▶Use wedge scriber

During scribing, the installer rests his hands together. By resting his hands together, the installer will not accidentally hit and break the fiber with the scriber. If he hits the fiber, he can break the fiber below the surface of the ferrule (Figure 12-18).

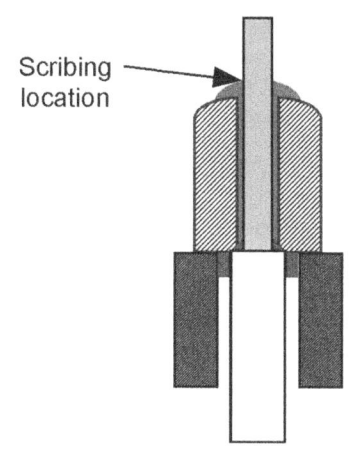

Figure 12-20: Location Of Scribe

Figure 12-21: Wedge Scribers

▶▶Scribe once

The installer scratches the fiber once, and only once, with a light pressure. With a single scratch, the fiber breaks on a single plane. If the fiber breaks on multiple planes, the fiber may break below the surface of the ferrule.

▶Pull fiber

After scribing the fiber, the installer pulls the fiber away from the tip of the ferrule. To do so without breaking the fiber below the ferrule, the installer avoids bending bend the fiber to the side. To avoid bending, the installer slides his fingers up the ferrule, onto the fiber. He pulls the fiber along its axis.

12.4.2 AIR POLISHING

The goal of air polishing is creation of a fiber end that is flush with the bead of epoxy or adhesive (Figure 12-22) and has no sharp edges (Figure 12-23). If the installer achieves both these characteristics, he will not snag the fiber on the polishing film. If he does not snag the fiber, he will not break it below the end of the ferrule. Air polishing requires rubbing of a coarse polishing film against the fiber. The film is 12μ or 15μ. No pad is used.

▶Flush fiber, dull edge

▶▶Always air polish before pad polishing

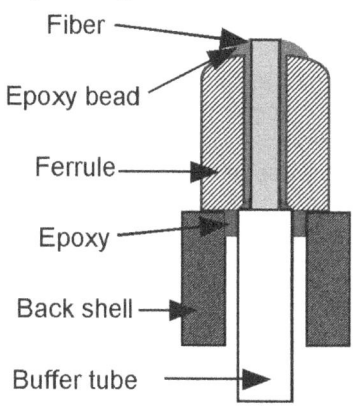

Figure 12-22: Desired Air Polish Condition

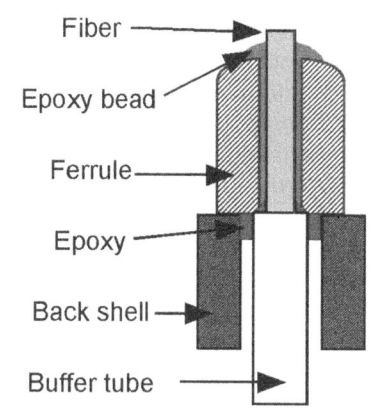

Figure 12-23: Undesired Condition

12.4.3 PAD POLISHING

12.4.3.1 OBJECTIVE

Pad polishing has two objectives:

- ➢ The creation of a lens grade, or optical grade, surface on the fiber end
- ➢ The end is flush with the ferrule

Usually, achieving these objectives requires two or three polishing steps from coarse to fine films.

We polish on rubber pads to allow the fiber to conform to the radius of curvature of the ferrule, resulting in low loss and low reflectance. Glass and hard plastic polishing plates produce a flat fiber end, high loss and high reflectance. Such plates are used only on legacy non-contact connectors, such as Biconic and SMA connectors.

12.4.3.2 TYPES

There are two types of pad polishing: hand polishing and machine polishing. Hand polishing results in the desired low loss. Cable assembly facilities use machine polishing in place of hand polishing, as it results in reduced cost. Machine polishing of singlemode results in consistent low loss and low reflectance.

Machine polishing of 24 connectors to low loss and low reflectance can take as little as three minutes. In comparison, field polishing of a single connector can take thirty seconds (multimode) to three minutes (singlemode).

Machine polishing has one aspect that is different from that of hand polishing. All machine polishing requires grinding through the ferrule to produce a round cladding. Hand polishing may not produce a round cladding. In order to remove ceramic ferrule material, all machine polishing films, except the final film, are diamond films. Diamond films are more expensive than multimode polishing films.

12.4.3.3 FILMS

Both hand and machine polishing are performed with relatively fine abrasive film(s) placed on a resilient rubber pads.

Fine films have grit sizes from 5μ to 0.05μ.

▶▶To achieve low loss and good appearance on multimode connectors, the final film is at least as fine as 1μ.
▶▶To achieve low loss and low reflectance on singlemode connectors, the final film is at least as fine as 0.5μ.

Some singlemode polishing procedures require a polish solution, which contains suspended fine abrasives.

All polishing films are harder than the fiber. Some films are harder than some ferrules. Diamond films are harder than the hardest ferrule material, which is ceramic. Alumina (aluminum oxide) films are harder than LCP, or composite, and stainless steel ferrules but softer than the ceramic ferrules.

Diamond film polishing removes fiber and ferrule. The fiber stays nearly flush with the ferrule (Figure 12-24). Polishing ceramic ferrules with alumina films removes fiber but not ferrule. Excessive polishing with alumina film results in undercutting, which creates an air gap, high loss and high reflectance (Figure 12-25 and Figure 5-13).

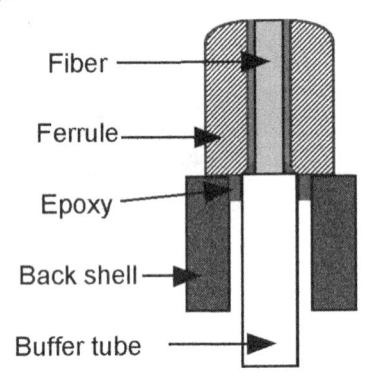

Figure 12-24: Fiber Flush With Ferrule

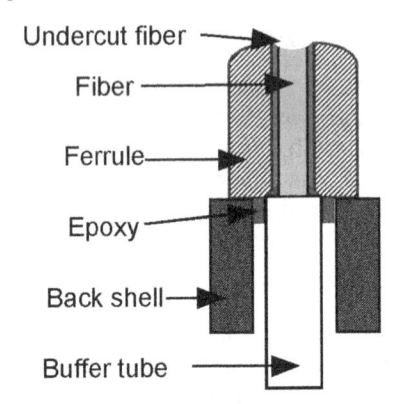

Figure 12-25: Undercut Fiber

12.4.3.4 POLISHING TOOL

A polishing tool, also known as a puck and fixture aligns the connector to the pad (Figure 12-26).

Figure 12-26: Polishing Fixtures

12.4.3.5 POLISHING LIQUID

The installer can perform wet or dry polishing. Cable assembly polishing is wet. For wet polishing, the installer uses distilled or reverse osmosis (RO) water. These types of water are free from particles that can cause scratches on the core. Such scratches divert the light from its proper path and can result in increased loss and reflectance. In addition, wet polishing increases film life, as the liquid flushes polishing materials from the films.

▶▶Use distilled or RO water for wet polishing

In environments with dirty air, the installer uses dry polishing. Many field environments have significant amounts of particles in the air. In such environments, wet polishing is not the best choice, as the liquid can attract particles from the air.

▶▶ Use no liquid for field polishing

12.4.3.6 EQUIPMENT CLEANING

One objective of polishing is the removal of scratches. Scratches can be from contamination or from previous polishing steps. Scratches on the core can divert and block light from its normal path, resulting in increased connector loss.

▶Avoid and remove scratches on core

Prior to use, the installer cleans the connector, the polishing pad, the polishing films, and tool.

▶▶Prior to use, clean equipment

In addition, the installer cleans the connector and tool before he changes to a film with grit finer than that of the previous film.

▶▶Clean connector and tool when changing to finer film

The installer can clean with lens grade tissues and one of the following: lens grade gas, distilled water or 98% isopropyl alcohol. Gas is convenient, but relatively expensive. Water and isopropyl alcohol are less convenient and relatively inexpensive.

12.4.3.7 POLISHING MOTION

The installer polishes the connector with a 'Figure 8' motion. This motion avoids creation of a bevel on the end of the fiber (Figure 12-27). Such a bevel creates an air gap between mated connectors. An air gap results in high loss and high reflectance.

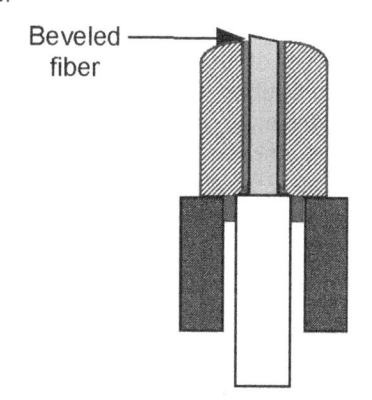

Figure 12-27: Bevel On End Of Fiber

▶▶Use 'Figure 8' Polish

12.4.3.8 POLISHING PRESSURE

The installer performs polishing without breaking the fiber below the surface of the ferrule. Should the bead of epoxy or adhesive be sheared from the tip, the fiber will not conform to the radius.

▶▶Use light pressure on bead

The installer uses a light polishing pressure until the fiber is flush with the ferrule. Once the fiber is flush, the installer cannot shatter the fiber below the ferrule. The fiber is flush when the bead is gone.

Hot melt connectors require less pressure than do epoxy connectors. Hot melt adhesive is gummy or rubbery. If the installer uses excessive pressure during polishing, the adhesive compresses but the fiber does not. In this case, the fiber can protrude beyond the bead of adhesive (Figure 12-23), snag on the polishing film, and break below the surface of the ferrule (Figure 12-18 and Figure 12-19).

Hot melt and quick cure adhesives can create a very small bead on the tip of the ferrule. This bead provides less support of the fiber during polishing than does a bead of epoxy. Because of this difference in support, the installer performs the air polish and the first pad polishing with less pressure than he would use on an epoxy connector. This reduced pressure is required until the fiber is flush with the ferrule.

▶▶Use reduced pressure on hot melt and quick cure connectors

Connectors with the 1.25 mm ferrule (LC, LX.5 and MU) have a cross section area that is 25% of that of 2.5 mm ferrules (ST™-compatible, SC, FC). In order to avoid excessive pressure and broken fibers, the installer reduces his polishing pressure to 25% the pressure he would use when polishing the 2.5 mm ferrules.

▶▶Use reduced pressure on 1.25 mm ferrules

During polishing, the installer may detect 'scratchiness'. Scratchiness is caused by glass fiber with a sharp edge protruding beyond a bead on the tip of the ferrule (Figure 12-23). This fiber can snag on the polishing film, breaking below the tip of the ferrule (Figure 12-18 and Figure 12-19).

The installer starts polishing all connectors slowly, with very light pressure and with a 1/2" high 'Figure-8' pattern. He maintains light pressure and a slow polish until the scratchiness stops.

▶▶Scratchiness requires slow, light pressure polishing

12.4.3.8.1 POLISHING TIME

When polishing ceramic ferrules with films other than diamond, the installer polishes for a minimum time. Over polishing with alumina films can result in undercutting, high loss and high reflectance (Figure 12-25).

▶▶Minimize polish time

12.4.3.9 REPOLISHING

To remove minor damage from the core of a fiber in a ceramic ferrule, the installer uses diamond polishing films. These films remove damage by removing both fiber and ferrule.

▶▶Salvage ceramic ferrules with diamond films

To remove minor damage from the core of a fiber in a LCP or stainless steel ferrule, the installer can use either diamond or alumina polishing film. Because alumina films are much less expensive than diamond films, they are preferred for salvaging connectors with these two ferrule materials.

▶▶Salvage other ferrules with alumina films

Note that it is possible to salvage ceramic ferrules with alumina films. However, the number of polishing strokes is limited by avoidance of undercutting. In other words, polishing with alumina may work. If the number of strokes is excessive, high loss and reflectance will result.

12.5 CLEAVE AND CRIMP INSTALLATION

The 'cleave and crimp' connector has a preinstalled fiber stub that the manufacturer has pre-polished. This stub has index matching gel on the inside end of the fiber (Figure 12-28). Because of this structure, the connector is actually a mechanical splice in a connector.

Installation by this method requires five steps:

- ➢ Cable end preparation
- ➢ Fiber end preparation
- ➢ Cleaving
- ➢ Insertion
- ➢ Crimping

The first two steps are the same as for all installation methods. The dimensions are different from those of other methods.

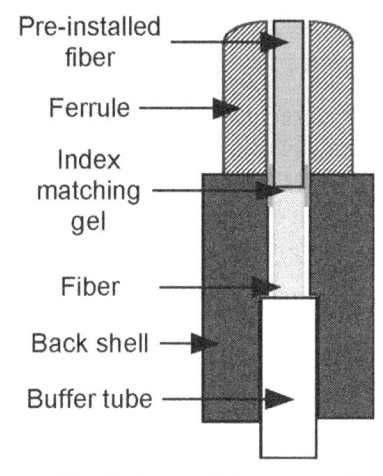

Figure 12-28: Internal Structure Of The Cleave And Crimp Connector

12.5.1 CLEAVING

The performance of this connector is strongly determined by the quality of the cleave on the fiber. Obtaining low and consistent cleave angles requires a high quality cleaver. Such a cleaver can cost $1200-$1600. Justification of this expense is easy: the installer saves more than the additional cost of this cleaver from improved yield and reduced labor cost. To achieve the best results, the installer will:

▶▶Use a high quality cleaver

Cleave and crimp connectors grip the fiber in two places: the cladding and the primary coating or buffer tube. A long cleave length may prevent the crimp from creating a proper grip through the primary coating and/or buffer tube. Without a proper grip, the connector can exhibit high loss or slippage of the fiber from the connector. A short cleave length may prevent the cleaved end from contacting the pre-installed fiber.

▶▶Control cleave length tightly

The installer cleaves the fiber to the cleave length and tolerance indicated in the connector instructions. A common cleave length tolerance is ± 0.5 mm.

12.5.2 INSERTION

Contamination of the fiber end results in increased loss. Such contamination can block light from its proper path or create a gap between the fiber ends. There are two methods for avoiding such contamination.

▶▶Minimize delay

After cleaving the fiber, the installer inserts the fiber into the connector without delay. Delay can allow dirt and dust to collect on the end of the fiber, resulting in high loss.

During insertion of the fiber, the installer does not touch the cleaved end against anything. Doing so may contaminate the fiber end, resulting in high power loss. In addition, the fiber should not be placed on any surface. Such action will result in contamination.

A common misconception is that the fiber should be cleaned after cleaving. Such cleaning is more likely to contaminate than to clean the cleaved end.

▶▶Avoid contacting cleaved end with anything

After inserting the fiber, the installer verifies that the fiber has not slipped from its fully inserted position. Such slippage creates a gap between the fiber and the preinstalled fiber. This gap results in high loss (Figure 5-13).

▶▶Verify full contact of fiber with internal fiber

12.5.3 CRIMPING

Crimping of a cleave and crimp connector is similar to that of any other connector. However, the crimper is unique to the product. Finally, there are two crimps: one to grip the cladding and the second to grip the fiber through the primary coating or buffer tube.

12.6 SUMMARY

The information, principles and methods presented in this chapter indicate the need for attention to detail. Such attention is necessary to address the subtleties of achieving low loss, low reflectance, high reliability, and low installation cost.

12.7 REVIEW QUESTIONS

1. Organize the following polishing films in the proper sequence of use: 3 µ, 12 µ, 0.5 µ, and 1 µ.

2. Organize the following polishing films in the proper sequence of use: alumina, diamond.

3. Is tap water the best source of liquid for wet polishing? Justify your answer.

4. Is cracking common in connectors using quick cure adhesive? Justify your answer.

5. Is wet polishing preferred for all polishing? Justify your answer.

6. Does oven curing of quick cure adhesive takes less than 2 minutes? Justify your answer.

7. Should the strength members be cut longer than necessary if a connector requires that the strength members be crimped to the back shell? Justify your answer.

8. Is undercutting is desirable? Justify your answer.

9. Should installation of a fiber into all connectors be done as slowly as possible to avoid breakage and other problems? Justify your answer.

10. What single action almost completely eliminates fiber breakage during insertion?

11. Your technician states that the cause of cracking in his room temperature cured connectors was the epoxy. Do you believe him? Justify your answer.

12. Your technician is terminating a 12 fiber premises cable with cleave and leave connectors. He has just finished cleaving all 12 fibers, which he has placed on a work mat. He is about to insert the first fiber into a connector. Is there anything wrong with this procedure? If so, what?

13. Your technician is polishing both multimode and singlemode connectors. To simplify his process, he plans to use the same polishing films. He plans to finish the polishing with a 1 µ film. Is there anything wrong with this procedure? If so, what?

14. Your technician states that the cause of high loss in his hot melt adhesive connectors was pistoning. Do you believe him? Justify your answer.

15. You are to make a decision on which cleaver to purchase for your cleave and crimp connectors. You plan to install 1200 connectors. The inexpensive cleaver costs $300. The expensive cleaver costs $1200. The connectors cost $15. You have been led to believe that you will lose 5 % of the connectors with the expensive cleaver and 13 % with the inexpensive cleaver. Which cleaver should you purchase?

13 SPLICING PRINCIPLES

Chapter Objectives: you learn the principles of both fusion and mechanical splicing. These principles apply to both indoor and outdoor cables and to both mid-span and pigtail splicing. These principles result in the procedures of Chapters 25.

13.1 INTRODUCTION

The principles in this chapter are relatively independent of the enclosure, the mechanical splice used, and the specific fusion splicer used. As such they are general principles that apply to most products.

The process of splicing requires eleven steps:

- Ensure compatibility of the cable with the enclosure
- End preparation of the cable
- Fiber grouping
- Attachment of the cable to the enclosure
- Tray preparation
- Fiber preparation
- Making of the splice
- Placement of the splice and fiber into a tray
- Placement of the tray into an enclosure
- Testing
- Closing and sealing the enclosure

13.2 CABLE-ENCLOSURE COMPATIBILITY

While uncommon, some cables are incompatible with some enclosures. This incompatibility results from two limitations: buffer tube bend radius and fiber bend radius.

The buffer tube incompatibility is a result of the enclosure forcing the buffer tube into a bend radius that is too small for the buffer tube. At a small radius, the buffer tube may kink (Figure 13-1) and break the fibers.

This kinking can result from incompatibility between the enclosure and cable or from improper buffer tube dimensions, improper buffer tube processing, or deformation of the buffer tube. Such kinking can occur many months after the installation of the splices.

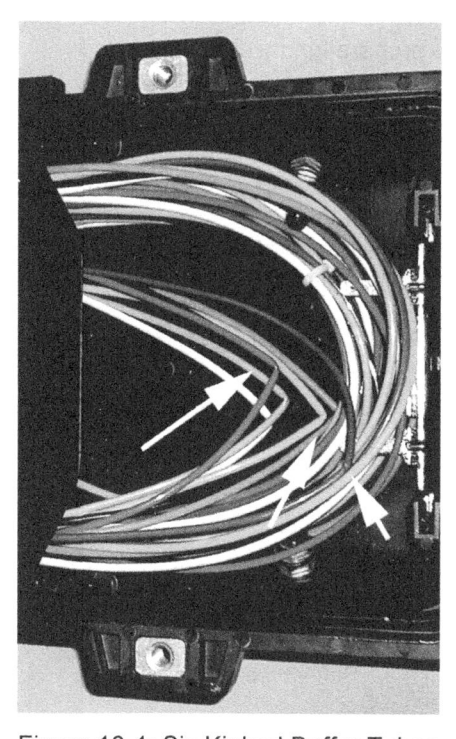

Figure 13-1: Six Kinked Buffer Tubes

While unlikely, fiber incompatibility can occur. This incompatibility is a result of the dimensions of the enclosure and/or the splice tray. These dimensions force the fiber into a bend radius that results in increased fiber attenuation. Such increased attenuation will be evident from splice loss at a long wavelength that is higher than that at a short wavelength. Examples of this incompatibility are:

- A multimode splice with a loss higher at 1300 nm than at 850 nm
- A singlemode splice with a loss higher at 1550 nm than at 1310 nm
- ▶Ensure Compatibility

To avoid kinking, fiber breakage, and increased splice loss, the installer ensures that the cable and enclosure are compatible.

13.3 END PREPARATION

The installer prepares the cable end to dimensions (Figure 13-2) that enable attachment of all parts of the splice system at the proper locations (Figure 13-3). The enclosure and the splice trays determine these dimensions.

This preparation involves:

- Removal of cable jacket(s), armor, and moisture blocking greases
- Trimming of strength members
- Removal of some of the buffer tube length
- Removal of water blocking gels from fibers

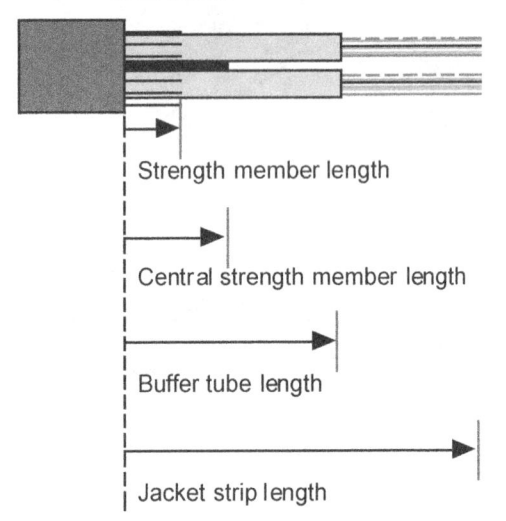

Figure 13-2: Cable End Preparation Lengths

The installer prepares the cable end to dimensions appropriate for the enclosure and trays. The lengths of the strength members beyond the end of the cable jacket allow the installer to attach these members to the enclosure so that the cable cannot be pulled out (Figure 13-4 and Figure 13-7).

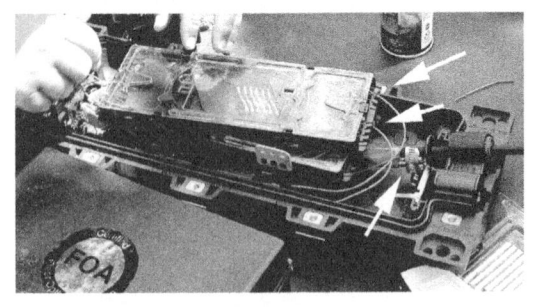

Figure 13-3: Enclosure Attachment Locations

▶ Match dimensions to enclosure

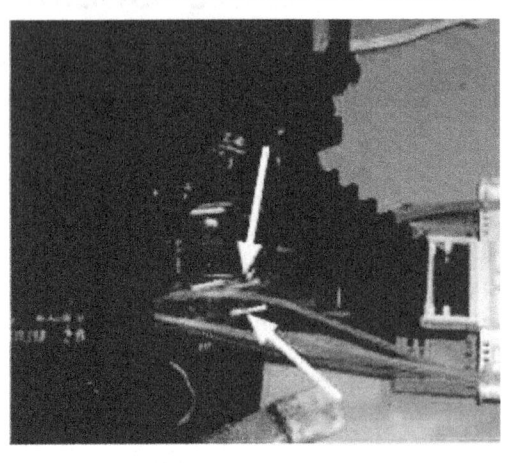

Figure 13-4: Strength Member Attachment

The lengths of the buffer tubes beyond the end of the cable jacket allow them to enter the splice tray when the tray is properly located in the enclosure. This buffer tube length is defined by the enclosure in that the buffer tubes must make a specific number of loops in the enclosure (Figure 13-3). In addition, this buffer tube length allows the installer to place the trays outside the enclosure during splicing (Figure 13-5).

The length of the fiber beyond the end of the buffer tube allows the splice to rest in the splice holder in the center of the splice tray. As with the buffer tube length, the fiber must make a specific number of loops in the splice tray. In addition this fiber length allows the splicing technician to route the fibers outside the tray and into the splicer (Figure 13-5) or into the mechanical splicing tool (Figure 13-6).

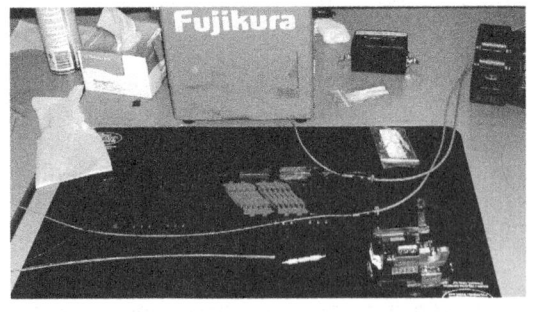

Figure 13-5: Buffer Tubes Out Of Enclosure

Figure 13-6: Mechanical Splice Tool

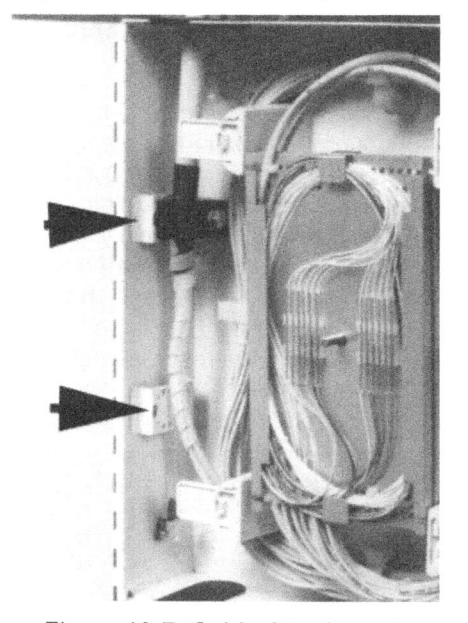

Figure 13-7: Cable Attachment

These lengths depend on the enclosure. For example, the jacket removal length can be 96". The buffer tube length can be 48". The central strength member and strength member lengths can be 2-4". In general, required buffer tube and fiber lengths for indoor enclosures are shorter than those for outdoor enclosures.

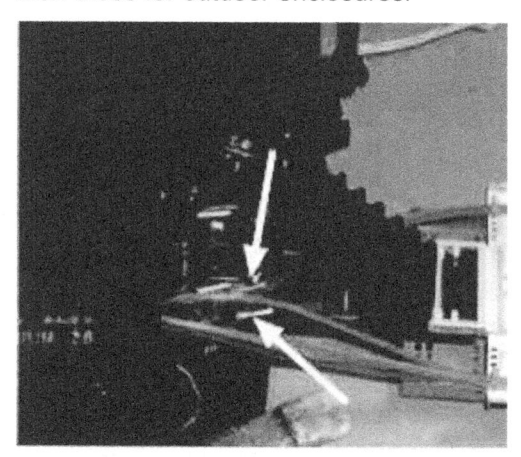

Figure 13-8: Strength Member Attachment

13.4 FIBER GROUPING

After the installer prepares the cable end, he may need to group fibers for routing to a specific location in the tray. Some designs, such as the central buffer tube, ribbon, and premises do not provide such grouping. For these designs, the installer creates groups of fibers or buffer tubes with spiral wrap or tubing, so that each group can be separately routed to its tray (Figure 13-9).

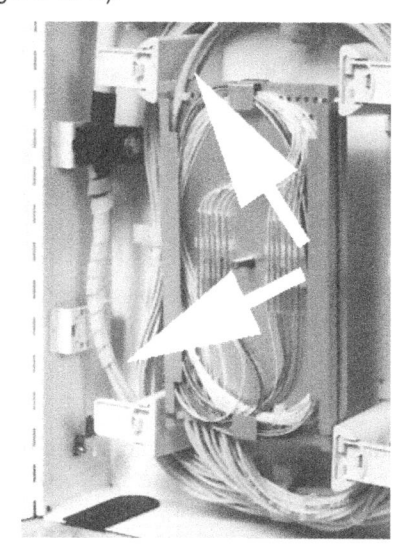

Figure 13-9: Fibers Grouped And Routed With Tubing

For example, the central buffer tube cable design may have 216 fibers in 18 groups of 12. With such a cable, grouping with tubing or spiral wrap will be essential for a reliable enclosure. If necessary, the installer installs fibers into tubing or spiral wrap in order to group them. Some cable designs, such as the multiple fiber per tube design (4.3.1), provide this grouping.

13.5 ATTACHMENT

The technician attaches the cable to the enclosure to prevent the cable being pulled from the enclosure. The method of attachment is specific to the enclosure. For example, the cable in Figure 13-4 is attached with a metal clamp, while the cables in Figure 13-7 and Figure 13-10 are attached with cable ties (top arrow) and a metal clamp (bottom arrow).

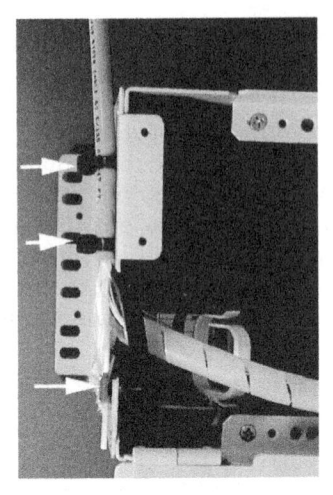

Figure 13-10: Indoor Enclosure Cable Attachment

13.5.1 CONFIGURATIONS

There are two cable-splicing configurations possible:

- Butt
- In-line

In a butt configuration, all cables enter the enclosure from the same end of the enclosure (Figure 13-11). This configuration is most common. In an in-line configuration, the cables enter the enclosure from the opposite ends (Figure 13-12). In-line fiber optic splicing is not recommended.

The installer uses the butt configuration to attach the cables to the enclosure in accordance with manufacturer's instructions. The butt configuration is preferred as its use simplifies installation of trays into the enclosure. Finally, the installer uses the butt configuration to attach the buffer tubes to the splice tray (Figure 13-13).

▶▶Use butt configuration

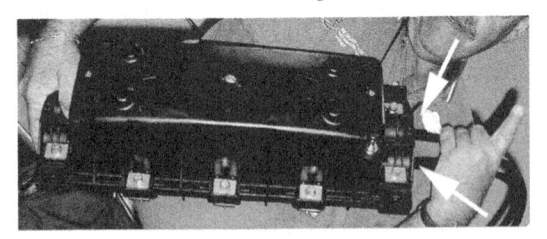

Figure 13-11: Butt Configuration

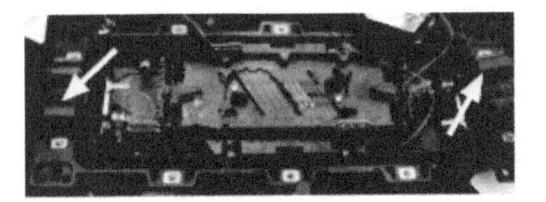

Figure 13-12: In-Line Configuration

13.5.2 GROUNDING AND BONDING

If an outdoor cable has conductive components, such as stainless steel armor, steel strength members, or conductors, the installer bonds the cable. The installer bonds the cable by connecting together the conductive components of the two cables. Bonding requires hardware appropriate for the enclosure (Figure 13-14). Bonding requires removal of the insulating materials from the conductive elements in the structure of the cables.

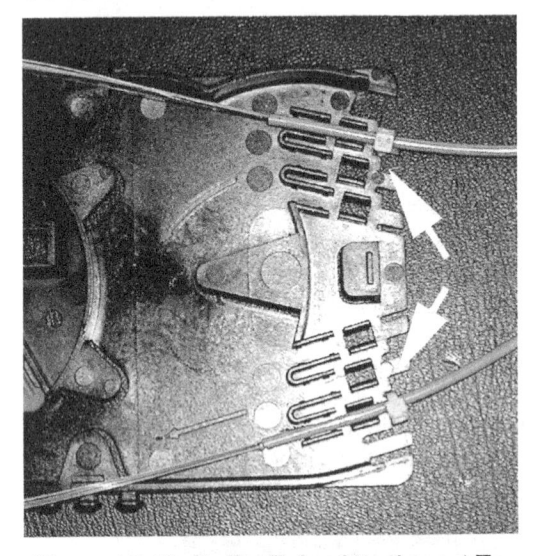

Figure 13-13: Buffer Tube Attachment To Splice Tray

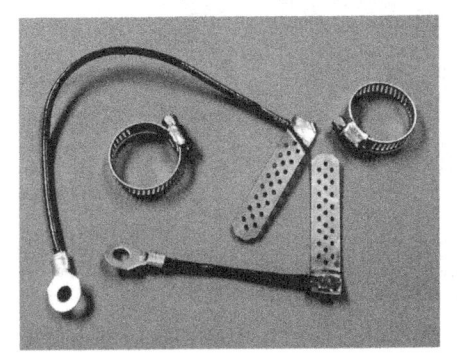

Figure 13-14: Grounding and Bonding Hardware

▶Ground conductive cables

If an outdoor cable terminates indoors, the installer grounds the conductive elements in accordance with the requirements of the relevant electrical code (Figure 13-15).

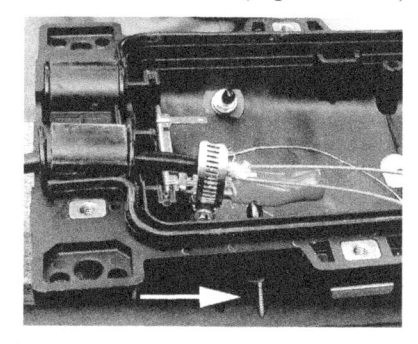

Figure 13-15: Enclosure Grounding Location

13.6 TRAY PREPARATION

Some splice trays require insertion of splice holders (Figure 13-16). Some trays do not require such installation, as they have permanently-installed splice holders.

Figure 13-16: Splice Holder Inserts

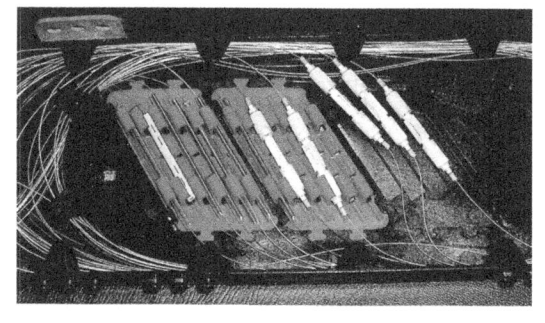

Figure 13-17: Incorrect Splice Holders

Some splice holders accept heat shrink and adhesive splice covers but not mechanical splices (Figure 13-17). The installer inserts the holders appropriate to the splice type and splice cover type. Use of the incorrect insert can result in splices loose in the tray (Figure 13-17), a condition of reduced reliability. The installer:

▶▶Uses proper inserts

13.6.1 BUFFER TUBE ATTACHMENT

The technician attaches the buffer tubes to a tray. The method of attachment can be cable tie (Figure 13-13), a press fit into a slot, or crimping lip of the tray onto the tube.

If done improperly, such attachment could result in increased optical power loss due to bend radius violation. Such violation is less likely on loose tubes than on tight tubes, as the rigidity of the loose buffer tubes resists transfer of force to the fibers.

At the location of the attachment, the buffer tube enters the tray for a short distance, approximately one inch. If the buffer tube enters the tray by an excessive distance, the fibers may experience a bend radius violation as they exit the tube. In addition, the excessive tube length may make placing fibers in the tray difficult.

▶Avoid increased loss conditions

The installer avoids any conditions that can increase loss. Such conditions include twisting and violations of bend radius and crush load.

▶Avoid violating tube bend radius

For mid-span splicing, the technician attaches one buffer tube from each cable to the same end of the tray (Figure 13-13). The installer uses this butt configuration, as it simplifies coiling of the buffer tube in the enclosure. With the butt configuration, attachment of more than one buffer tube per cable per tray results in ease of placement of the tray in the enclosure.

The technician routes the buffer tubes from the enclosure to the splice tray without twisting or crossing them. Twisting and crossing will complicate the placement of a completed tray into the enclosure.

▶Avoid twisting and crossing tubes

13.7 FIBER PREPARATION

The technician takes the following five steps:

- Routes these tubes out of the tray (Figure 13-13)
- Separates one fiber from the others all the way back to the buffer tube
- Places the heat shrink cover on one fiber
- Strips the primary coating, or tight tubes and primary coatings, to expose the fiber for cleaving

13.7.1 SPLICE COVER

Before fusion splicing, the installer slides a heat shrinkable splice cover (Figure 13-18) on one fiber. Alternatively, after splicing, he installs an adhesive cover (Figure 13-19).

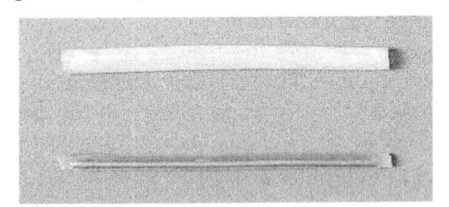

Figure 13-18: Heat Shrink Splice Cover[45]

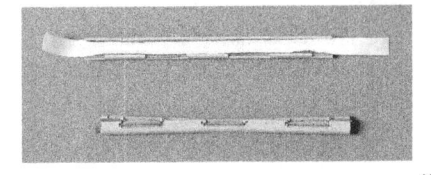

Figure 13-19: Adhesive Splice Cover[46]

13.7.2 STRIP LENGTH

The cleaver determines the length of bare fiber. This length will be sufficient for the cleaver to produce a flat and perpendicular end. Such an end is essential for low loss.

The installer strips the primary coating or the buffer tube and the primary coating to the length required by the cleaver. For the Fujikura CT04 and CT07 cleavers, this strip length is 1.5-1.75". Other cleavers can require strip lengths up to 4".

---
[45] Preshrink, Top; Post shrink, Bottom
[46] Open, Top; Closed, Bottom

▶▶ Strip fiber to proper fiber length

If the strip length is long, the fiber may bend in the cleaver. If the strip length is short, the cleaver cannot grip the fiber on both sides of the scribe. Both situations result in high angle cleaves and high loss splices.

After stripping the fiber, the technician cleans the fiber. He and inspects for debris. If necessary, he cleans the fiber repeatedly until it is clean.

13.7.3 CLEAVER FUNCTION

All steps with the fiber and cleaver have two goals:

- A flat, perpendicular fiber end
- An end without contamination

The cleaver produces a low loss, flat and perpendicular end. It does so by holding the fiber perpendicular to and at a tightly controlled distance from a scribing blade. The cleaver holds the fiber on both sides of the scribing blade, perpendicular to the path of the scribing blade. The cleaver places a single, narrow scratch of controlled depth. The breaking arm places a controlled bend on the fiber so that the fiber breaks at the scratch. The principle of cleaving is creation of a

▶ Low angle, uncontaminated end

In order that the cleaver holds the fiber properly, the splicing technician cleans each fiber immediately before cleaving. As dirt or dust on the fiber can change the angle at which the fiber rests in the cleaver, the installer cleans each fiber with isopropyl alcohol and lens grade tissues..

▶ After cleaving, the splicing technician does not clean the fiber, since such cleaning can contaminate the end face of the fiber.

However, cleaning of the fiber after cleaving is not necessarily fatal for a fusion splice. Prior to splicing, fusion splicers emit a pre-fuse arc across both fibers. This arc creates a electrostatic charge on both the fiber and any dirt on the fiber. As like-charged materials repel each other, this charge results in removal of dust and dirt from the fiber. Achievement of a low angle cleave requires that the technician

> ▶▶Before the first use and periodically there-after, he cleans the cleaver

This author recommends that the technician clean the cleaver with lens grade gas. He can use isopropyl alcohol and lens grade tissues. The environment in which the technician performs the splicing determines the frequency for cleaning the cleaver. As the dust level in the air increases, the frequency of cleaning increases.

Immediately after cleaving, the technician places the fiber into a splicer or into a mechanical splice. He does not place the fiber on any surface. Doing so can contaminate the cleave. Avoiding contamination requires the technician to

> ▶▶Move the fiber immediately to the splicer or the mechanical splice

13.7.4 CLEAVE LENGTH

Two factors determine the cleave length:

- The fusion splicer or the mechanical splice
- The coated diameter of the fiber

The cleave length must be long enough that the bare fiber fits into v-groove holders that position and grip of the fiber (Figure 13-20). If the cleave length is short, this final positioning will be on the primary coating or tight tube. With such positioning, the splicer may not have be able to align the fibers.

Figure 13-20: Precision V-Grooves

For a 250 µ fiber, a common cleave length is 12 mm for a fusion splicing. For the same fiber, a common cleave length is 7-12 mm for a mechanical splice.

For a fiber in a 900 µ buffer tube, a common cleave length is 15-20 mm for both fusion and mechanical splices. To avoid problems in splicing or with the splice cover, the technician

> ▶▶Uses recommended cleave length

The technician determines the cleave length from the splicer or splice instructions.

13.7.5 CLEAVE QUALITY

For both fusion and mechanical splices, the cleaved fiber end must be smooth and perpendicular (Figure 13-21). If the fiber end has missing glass (Figure 13-22) or extra glass (Figure 13-23), or a large angle on the end (Figure 13-24), the splice will be high loss due to an air gap or a reduced core diameter in the splice. Prior to fusion splicing, the installer

> ▶▶Evaluates cleave quality

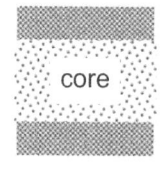

Figure 13-21: Acceptable Cleave

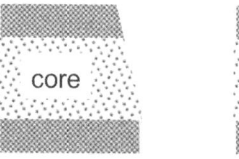

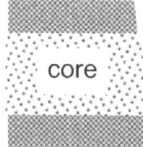

Figure 13-22: Cleaves With Missing Glass

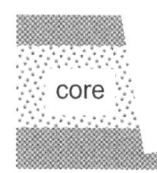

Figure 13-23: Unacceptable Cleave Due to Extra Glass

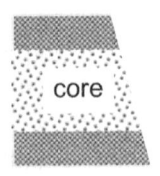

Figure 13-24: Unacceptable High Angle Cleave

Most splicers measure and display the cleave angle. For such splicers, the technician need only ensure that the maximum angle setting is appropriate. This author's experience is that cleave angles less than 1.5° result in low loss. Cleave angles greater than 2°-3° can result in high loss. High quality cleavers produce consistent cleave angles below 1°.

When making a mechanical splice, the technician does not inspect the cleave. Doing so can contaminate the fiber end.

The cleaver scribing blade does become dull with use. When such dullness occurs, many cleavers have a scribing blade that can be rotated to a new position. While not always true, many scribing blades are rated at 1000 cleavers (500 splices) per position. This author's experience is that scribing blades can work well to 2000-2500 cleaves.

13.8 FUSION SPLICING

13.8.1 FIBER PLACEMENT

The technician places the fibers into the splicer and crudely aligns them. The precision of the initial position of the is not critical, as most splicers examine this placement to ensure sufficient range of motion for both fibers. However, there are two principles of placement.

▶ Place each fiber on its side of electrodes
▶ Position fibers near electrodes

Each fiber is placed into the splicer on its side of the electrodes (Figure 13-25 and Figure 13-26).

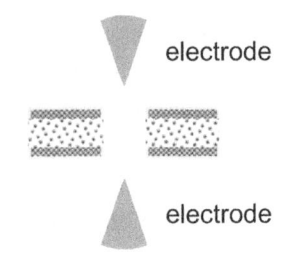

Figure 13-25: Fibers Correctly Positioned

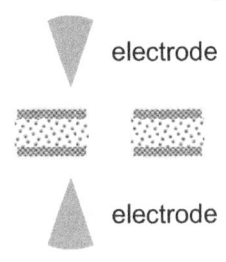

Figure 13-26: Fibers Incorrectly Positioned

For splicers with active alignment, the technician need not align the fibers precisely. Current generation splicers have active alignment. With such active alignment, the splicer will compensate for some misalignment due to dirt on the cladding or in the grooves of the splicer.

However, this compensation is limited. If excessive dirt is on the fiber or in the splicer grooves, the splicer will be unable to achieve the precise alignment required for low loss. In summary, fiber cleanliness is required to enable fiber alignment by the splicer.

Some splicers have fiber holders that are integrated with the cleaver and splicer. With such integration, the holder controls the initial fiber placement. The technician has minimal concern for placement.

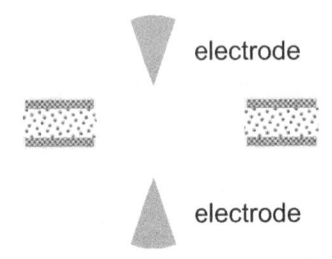

Figure 13-27: Excessive Fiber Separation

The installer positions both fibers close to the electrodes (Figure 13-25). Such positioning places the fibers within the range of motion of the splicer.

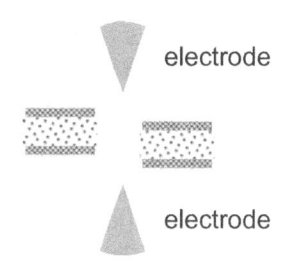

Figure 13-28: Fibers Misaligned In Splicer

If misalignment is evident (Figure 13-28), the technician must clean the fibers and

▶▶Clean grooves

If required, the technician cleans the fibers and the splicer grooves repeatedly so that the fibers align (Figure 13-25).

13.8.2 SPLICER OPERATION

The splicer fuses together the fibers with a program determined by the type of fiber being spliced. This program is defined by a 'recipe' or 'menu', which is built into all current-generation splicers. Prior to splicing, the technician chooses the appropriate menu.

Most automatic splicers, such as PAL splicers manufactured by Alcoa-Fujikura, Corning Cable Systems, Ericcson and Sumitomo, perform eight operations:

- Check of z-axis motion
- Cleaning of fibers
- Measurement of the cleave angles of both fibers in two axes
- Alignment of fibers
- Control of arc current, arc time, and overrun (6.3.3.1)
- Estimation of splice loss
- Splice strength check
- Shrinking of splice cover

13.8.2.1 Z AXIS CHECK

The z-axis is along the fiber length. The z-axis check ensures that the fibers can be moved sufficiently to provide adequate overrun (6.3.3.1).

13.8.2.2 OVERRUN SETTING

Overrun is the amount by which the ends of the two fibers move past one another (Figure 13-29). The technician sets a proper overrun with his choice of fiber in the settings menu. If the z-axis movement is insufficient, the splicer stops the process and alerts the technician. A correctly set overrun results in

▶Uniform diameter and low loss

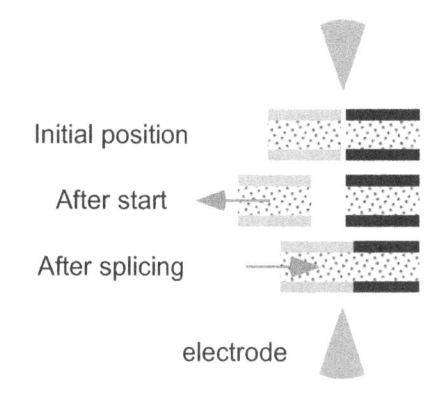

Figure 13-29: Overrun

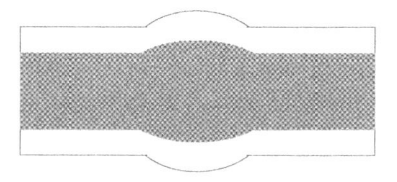

Figure 13-30: Bulged Splice Due Excessive Overrun

A bulge, results from excessive overrun (Figure 13-30), An hourglass shape results from insufficient overrun (Figure 13-31). Both conditions result in high loss.

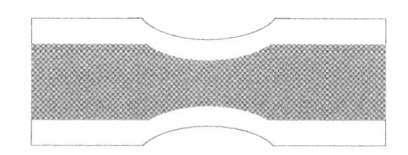

Figure 13-31: Hourglass Splice Due Insufficient Overrun

13.8.2.3 FIBER CLEANING

Prior to inspection of the cleave angles, the splicer passes a low level electrostatic arc across the fibers. This pre-fuse arc cleans the fibers of dust and dirt.

13.8.2.4 ANGLE MEASUREMENT

After this cleaning action, the video microscope in the splicer creates two digitized images of the ends of the fibers. The images are at 90° to each other. The images are analyzed for cleave angles. The splicer measures and compares the cleave angles to a maximum acceptance angle. If either cleave angle is greater than the acceptance angle, the splicer stops the process and alerts the technician. If the cleaver is in good condition, a common maximum cleave angle is less than 1.5°.

13.8.2.5 ACTIVE FIBER ALIGNMENT

For singlemode fibers in an active alignment splicer, the splicer analyzes the core boundaries of both fibers, which are visible due to the collimated light that the splicer produces. For multimode fibers, the splicer analyzes and aligns the cladding of both fibers. The splicer adjusts the fiber positions so that the fibers are optimally aligned. With this alignment method, the splicer can achieve minimum splice loss even if the fibers that have different core or cladding diameters. This splicer performs this alignment in two planes at 90° to each other. For passive alignment splicers, the V-grooves provide alignment in the X- and Y-axes.

13.8.2.6 FUSING

The splicer positions both fibers in their initial locations. The splicer moves one or both of the fibers out of and slowly into the arc. During this movement, the splicer controls the arc current (Figure 13-32), arc time, and overrun. These three values depend upon the fiber being spliced and the splicer. The arc current may or may not be ramped from zero to a maximum value (Figure 13-32, left).

The splicer controls the arc current and time as long as the electrodes are in good condition. These electrodes wear and need periodic replacement. The splicer instruction manual provides the replacement interval. Some splicers record the number of arcs and inform the technician that electrodes need replacement.

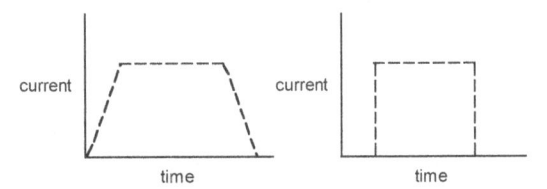

Figure 13-32: Ramped (Left) And Non-Ramped (Right) Splicing Currents

However, this author's experience indicates that the electrodes last longer than indicated by the replacement interval recommended by splicer manufacturers. One of our fusion splicers had a recommended electrode replacement frequency of 1000 arcs. We have experienced consistent, 0.05 dB singlemode splices up to 2500 arcs!

Arcing produces a low level, consistent sound. When the electrodes become oxidized and in need of replacement, the sound level increases and is not constant; i.e., the sound has the character of uneven hissing.

In addition, oxidized electrodes have a dull, white color. New electrodes have a shiny metal appearance.

If the splicer controls the arc current and time properly, the location of the singlemode splice is not visible after splicing. Multimode splices may have a faint indication of location. During splicing, should the glass temperature be low, the joint will be visible after splicing.

Occasionally, the splicing and end face conditions combine to create a gas bubble in the fiber. Such a bubble is unacceptable. It can cause high loss, high reflectance and low splice strength.

13.8.2.7 LOSS ESTIMATION

After completing the fusing process, most current-generation splicers make an estimation of splice loss. The splicer makes this estimation by analyzing the microscopic image of the core alignments in two axes at 90° to one another. This splice loss is a calculated and estimated value. It is not the true value of the splice. The true value of the splice loss is the average the OTDR splice loss measurements in both directions (15.7.4.3).

This estimation allows the technician to place the fiber into the tray with confidence that the splice has acceptable loss. Without this estimation, the splicer must wait for a splice loss measurement from an OTDR technician, who has been sitting idly, waiting for the splicing technician to make the splice. With an acceptable estimated value, the technician can dirt from the fiber. Achievement of a low angle cleave requires that the technician

▶▶ Before the first use and periodically there-after, he cleans the cleaver

This author recommends that the technician clean the cleaver with lens grade gas. He can use isopropyl alcohol and lens grade tissues. The environment in which the technician performs the splicing determines the frequency for cleaning the cleaver. As the dust level in the air increases, the frequency of cleaning increases.

Immediately after cleaving, the technician places the fiber into a splicer or into a mechanical splice. He does not place the fiber on any surface. Doing so can contaminate the cleave. Avoiding contamination requires the technician to

▶▶ Move the fiber immediately to the splicer or the mechanical splice

13.7.4 CLEAVE LENGTH

Two factors determine the cleave length:

- The fusion splicer or the mechanical splice
- The coated diameter of the fiber

The cleave length must be long enough that the bare fiber fits into v-groove holders that position and grip of the fiber (Figure 13-20). If the cleave length is short, this final positioning will be on the primary coating or tight tube. With such positioning, the splicer may not have be able to align the fibers.

Figure 13-20: Precision V-Grooves

For a 250 µ fiber, a common cleave length is 12 mm for a fusion splicing. For the same fiber, a common cleave length is 7-12 mm for a mechanical splice.

For a fiber in a 900 µ buffer tube, a common cleave length is 15-20 mm for both fusion and mechanical splices. To avoid problems in splicing or with the splice cover, the technician

▶▶ Uses recommended cleave length

The technician determines the cleave length from the splicer or splice instructions.

13.7.5 CLEAVE QUALITY

For both fusion and mechanical splices, the cleaved fiber end must be smooth and perpendicular (Figure 13-21). If the fiber end has missing glass (Figure 13-22) or extra glass (Figure 13-23), or a large angle on the end (Figure 13-24), the splice will be high loss due to an air gap or a reduced core diameter in the splice. Prior to fusion splicing, the installer

▶▶ Evaluates cleave quality

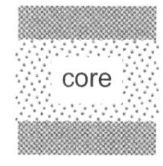

Figure 13-21: Acceptable Cleave

Figure 13-22: Cleaves With Missing Glass

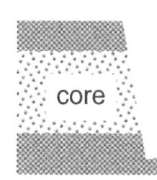

Figure 13-23: Unacceptable Cleave Due to Extra Glass

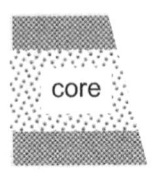

Figure 13-24: Unacceptable High Angle Cleave

Most splicers measure and display the cleave angle. For such splicers, the technician need only ensure that the maximum angle setting is appropriate. This author's experience is that cleave angles less than 1.5° result in low loss. Cleave angles greater than 2°-3° can result in high loss. High quality cleavers produce consistent cleave angles below 1°.

When making a mechanical splice, the technician does not inspect the cleave. Doing so can contaminate the fiber end.

The cleaver scribing blade does become dull with use. When such dullness occurs, many cleavers have a scribing blade that can be rotated to a new position. While not always true, many scribing blades are rated at 1000 cleavers (500 splices) per position. This author's experience is that scribing blades can work well to 2000-2500 cleaves.

13.8 FUSION SPLICING

13.8.1 FIBER PLACEMENT

The technician places the fibers into the splicer and crudely aligns them. The precision of the initial position of the is not critical, as most splicers examine this placement to ensure sufficient range of motion for both fibers. However, there are two principles of placement.

> ▶ Place each fiber on its side of electrodes
> ▶ Position fibers near electrodes

Each fiber is placed into the splicer on its side of the electrodes (Figure 13-25 and Figure 13-26).

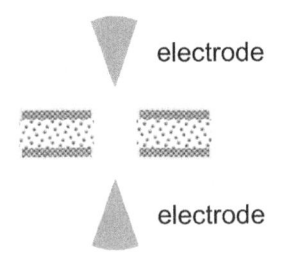

Figure 13-25: Fibers Correctly Positioned

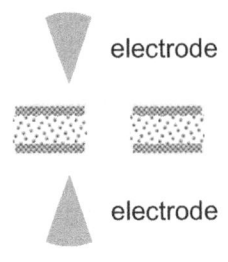

Figure 13-26: Fibers Incorrectly Positioned

For splicers with active alignment, the technician need not align the fibers precisely. Current generation splicers have active alignment. With such active alignment, the splicer will compensate for some misalignment due to dirt on the cladding or in the grooves of the splicer.

However, this compensation is limited. If excessive dirt is on the fiber or in the splicer grooves, the splicer will be unable to achieve the precise alignment required for low loss. In summary, fiber cleanliness is required to enable fiber alignment by the splicer.

Some splicers have fiber holders that are integrated with the cleaver and splicer. With such integration, the holder controls the initial fiber placement. The technician has minimal concern for placement.

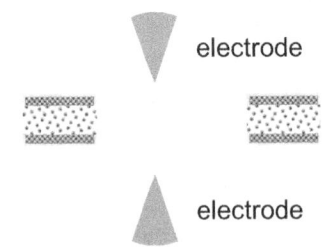

Figure 13-27: Excessive Fiber Separation

The installer positions both fibers close to the electrodes (Figure 13-25). Such positioning places the fibers within the range of motion of the splicer.

▶▶Place splice in tray based on estimated loss

The installer uses the estimated loss as an indicator of acceptance. The installer will use the OTDR splice loss for final acceptance.

13.8.2.8 STRENGTH TEST

One principle of splicing is:

▶The splice is stronger than the fiber

Sophisticated splicers pull on the two fibers to check the strength of the splice. If the splice is sufficiently strong, it will not break.

13.8.2.9 SPLICE COVER APPLICATION

A principle of splicing is

▶Leave no bare fiber

When exposed to air, bare fiber can experience a significant reduction in strength. With this reduction, bare fiber reduces network reliability. To avoid this reduction, the technician

▶▶Places fiber in splice cover

After fusing, the splicing technician removes the splice from the splicer, and takes one of two steps. He centers a heat shrinkable splice cover over the center of the bare glass (Figure 6-8). He places the splice cover in a heating oven, which shrinks the cover. This shrinkage starts from the center of the splice and moves towards the ends. After shrinking the splice cover, the splicing technician places the fiber in the tray. Alternatively, the technician will place an adhesive splice cover on the splice (Figure 6-9).

13.9 MECHANICAL SPLICING

To install a mechanical splice, the technician inserts cleaved fibers into the splice, one at a time. The mechanical splice provides both functions of alignment and of a cover.

The installer inserts the first fiber to a specific location, which depends upon the splice. For example, the installer installs the first fiber into a 3M Fibrlok™ splice until it stops. As another example, the installer installs the first fiber into a Siemon Company ULTRAsplice™ splice until the fiber is in the center of the splice. The center of the ULTRAsplice™ splice has a glass capillary tube, which allows the technician to see the position of the first fiber.

Figure 13-33: Equal Bow Results In Low Loss 3M Fibrlok™ Splice

The installer installs the second fiber in a manner that depends upon the splice. For example, the installer installs the second fiber into a 3M Fibrlok™ splice until the amount of bow of both fibers is the same (Figure 13-33).[47] As another example, the installer installs the second fiber into a Siemon Company ULTRAsplice™ splice until it is in contact with the first fiber.

The technician 'closes' the Fibrlok™ splice so that the splice grips the fibers properly. Closing may be by crimping, as in the 3M Fibrlok™ splice, or by sliding or rotating a cover or collar, as in the ULTRAsplice™. Most mechanical splices grip the fiber by compression.

Some mechanical splices may require tuning. Tuning is the rotation of one or both fibers in order to achieve either an acceptably low loss or the lowest possible loss. The need for tuning is uncommon.

Tuning can be done three ways, with:

➢ An OTDR
➢ An insertion loss test set
➢ A feature finder or fault finder (18.3.2)

When the technician uses an OTDR or insertion loss test set, he tunes the splice to the lowest loss. When the technician uses a feature finder or faultfinder, he tunes the splice until the visible light from the feature finder disappears.

---

[47] This is not the complete procedure.

13.10 FIBER PLACEMENT

The technician places the fiber in the splice tray by coiling the fiber. He lays the fiber along the long sides of the tray from one end to the other. He crosses, or twists, the fibers at the end opposite the buffer tubes. He lays the fiber along the long sides of the tray back towards the end with the buffer tubes. At this end, he again twists the fiber, but in the direction opposite to the direction of the first twist (i.e., a 'reverse twist'). The technician repeats this process until all the fiber is in the tray. To avoid problems, the technician uses the method of a

▶▶Reverse twist

The reverse twist prevents an accumulation of the twisting, which causes the fiber to pop out of the tray. In addition, the technician avoids repeated twisting in the same direction, which can result in fiber breakage. If the technician has prepared the fiber lengths properly, the splice cover, or mechanical splice, will be in the proper position for placement in the splice holder (Figure 13-34).

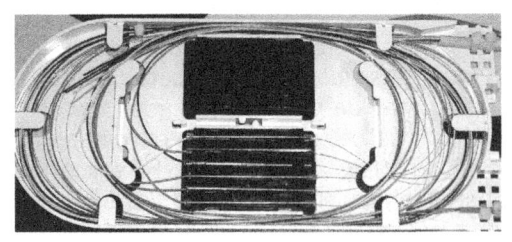

Figure 13-34: Coiled Fiber

After placing all the splices in the splice holder, the technician places a cover on the tray. If the splicing is pigtail splicing, the installer attaches the pigtails to the end of the tray (Figure 13-35).

13.11 TRAY PLACEMENT

The installer attaches the trays to the inside of the enclosure. This attachment can be by bolts, by Velcro straps (Figure 13-36) or by other mechanisms.

The installer attaches trays to prevent damage due to tray movement.

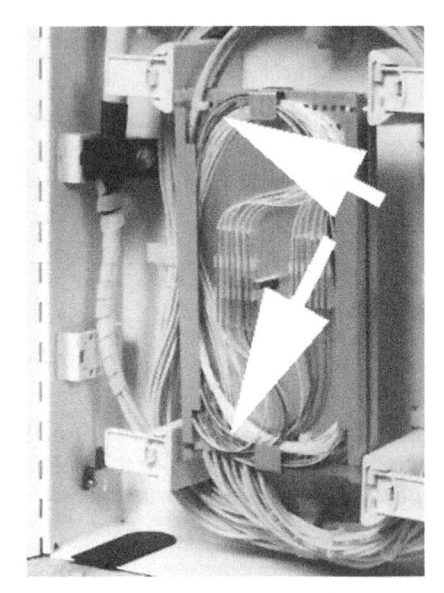

Figure 13-35: Pigtails and Tubing Attachment Locations

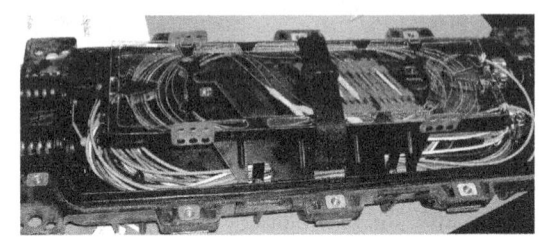

Figure 13-36: Velcro Strap Attachment

13.12 TESTING

There is no single rule for the sequence of splice testing. However, there are two common methods.

> Test all splices at end
> Test each splice after it's made

The most meaningful time for testing is after the enclosure is closed and placed. At this time, there are no additional actions that could result in increased loss. Finally, testing at this time reduces testing cost.

Some organizations test all splices after the enclosure is closed but before the enclosure is placed in its final location. With this testing time, there is small chance that a cable placement error will result in increased loss that will not be detected.

Testing prior to closing of the enclosure risks bend radius violation of the buffer tubes. Such violation can result in increased loss that will not be detected.

If the splicer provides automatic splice loss estimation, the OTDR technician can test all splices after all trays have been placed in the enclosure but before the enclosure is closed. Should a high loss splice be found before the installer closes the enclosure, the cost of repair is low.

While uncommon, a splicer may not make a loss estimation. In this case, an OTDR technician tests each splice after it is placed in the tray. Should a high loss splice be found, the cost of repair is low. The disadvantage of this method is increased labor cost, as the OTDR technician must make a final test after the enclosure is closed.

13.13 ENCLOSURE CLOSURE

Closing the enclosure requires cleaning of all gasket grooves, placing gaskets into the grooves, and installing moisture seals (Figure 13-37). Moisture seals are required on outdoor enclosures but not on indoor enclosures. This closing process may include checking the seals by pressurizing the enclosure (Figure 13-38).

Some outdoor enclosures do not require seals, but have weep holes so that condensed moisture can drip from the enclosure. Such enclosures are used in environments that do not experience freezing. Such enclosures are placed with the weep holes on the bottom.

When the technician completes all steps, he places the enclosure in its final location (Figure 13-39).

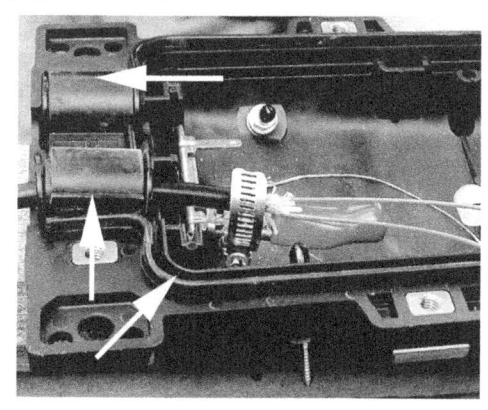

Figure 13-37: Moisture Seals And Gasket Grooves

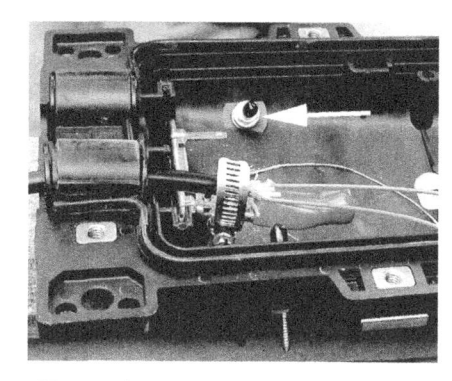

Figure 13-38: Pressurization Valve

Figure 13-39: Enclosure Placement

13.14 SUMMARY

The process of splicing requires following a large number of principles and methods. Each principle, by itself, is relatively simple. However, the number is large. When properly completed, the splice enclosure (Figure 13-39 and Figure 13-9) will be reliable and require no maintenance.

13.15 REVIEW QUESTIONS

1. A splicing supervisor must make a decision: rent a splicer without the loss estimation feature at a reduced cost or rent a splicer with this feature at an increased cost. The difference between the two rental fees is $500. He owns an OTDR so there is no additional cost. He expects the splicing to take 3 man-days at eight hours per day. He expects that an OTDR technician could test all the fibers in no more than one day. The total loaded labor rate of the OTDR technician is $50/hour. Which piece of equipment should he rent?

2. Put the following steps in sequence:

A. Place tray in enclosure
B. Pressurize enclosure
C. Clean grooves for gaskets
D. Place splice in splice holder
E. Attach buffer tube to tray
F. Test splice
G. Splicing
H. Cleaving
I. Coil fiber in tray
J. Place splice holder in tray
K. Attach cable to enclosure

3. Are strength members cut flush with the jacket of the cable? Justify your answer.

4. What is/are the purpose(s) of having several feet of fiber exposed beyond the end of the buffer tube?

5. What is/are the purpose(s) of having several feet of buffer tube exposed beyond the end of the jacket?

6. What determines the length of bare fiber?

7. What determines the cleave length?

8. Choose the most accurate statement. Justify your choice.

A. Cleaning of the fiber is more important to fusion splicing than to mechanical splicing.

B. Cleaning of the fiber is more important to mechanical splicing than for fusion splicing.

C. Cleaning of the fiber is equally important to fusion splicing and to mechanical splicing.

PART THREE
PRINCIPLES AND METHODS OF TESTING AND INSPECTION

14 INSERTION LOSS PRINCIPLES AND METHODS

Chapter Objectives: you learn the principles and methods of insertion loss testing of links in LANs, telephone networks and CATV networks.

14.1 INTRODUCTION

14.1.1 CHARACTERISTICS TO BE TESTED

As you learned from Chapter 3, light experiences two changes as it moves through a fiber:

- Pulses spread or disperse
- The intensity is reduced

Thus, we might expect to need to perform testing of both dispersion and power loss. Sophisticated fiber systems, such as DWDM networks and networks transmitting above 2.5 Gbps, do require both types of tests.

However, data networks do not require dispersion tests because of the manner in which data standards are developed. Fiber optic LANs are designed and implemented within limits on transmission distance based on the performance properties of the link components. In setting these limits, the standard committees have determined the maximum dispersion that can occur under worst-case distances and properties. As long as the network is implemented within these limits, dispersion will not be excessive and signal accuracy will be acceptable. In addition, installation errors cannot result in increased dispersion, as all factors that determine dispersion cannot be degraded by the installer.

However, installation errors can, and often do, result in increased power loss. Thus, power loss tests of installed links are required to verify that there is power at the receiver sufficient to enable the it to accurately convert the optical signal to an electrical signal.

14.1.2 POWER LOSS TESTS

There are two types of power loss tests: insertion loss and OTDR. The insertion loss test is required by the data standards and included in TIA/EIA-568-C.

The ideal goal of any test procedure is simulation of the manner in which optical power is lost between a transmitter and a receiver. With simulation, the loss measured will be close to the loss experienced between a transmitter and a receiver.

With one modification to the launch conditions, the singlemode insertion loss test comes close to simulation. However, as the reader will see, simulation of multimode conditions requires multiple launch conditions, because multimode transmitters produce significantly different conditions.

14.2 TEST PRINCIPLE

The basic principle of insertion loss testing is:

- Measurement of the difference between power input at the transmitter (Figure 14-1) and power delivered to the receiver (Figure 14-2).
- Thus, insertion loss testing requires two measurements.

Two measurements closely simulate the loss of a link (Figure 14-3). The test leads (Figure 14-2) replace and simulate the patch cords in the link (Figure 14-3).

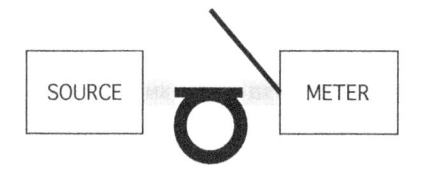

Figure 14-1: Input Power Measurement

14.3 SINGLEMODE TESTING

While simple in concept, singlemode testing is slightly more complicated than the basic principle (14.2) in two aspects. The first aspect is the loop on the test lead connected to the source (Figure 14-1). Singlemode laser diodes launch power into the cladding (Figure 8-1). Such power must be removed from the test conditions, as this power is not included in the OPBA of the transceivers (8.4.1). In addition, this cladding power will be lost either in the first connector pair after the transmitter or by attenuation in the fiber near the transmitter. The 1.2" diameter loop in Figure 14-1 removes this cladding power. Testing without this loop results in increased loss measurements.

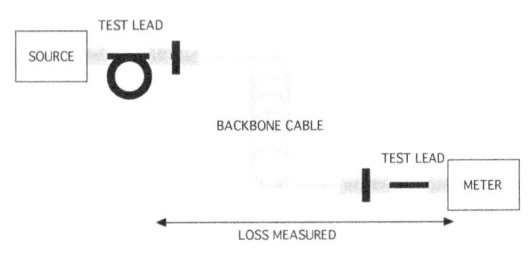

Figure 14-2: Output/Receiver Power Measurement

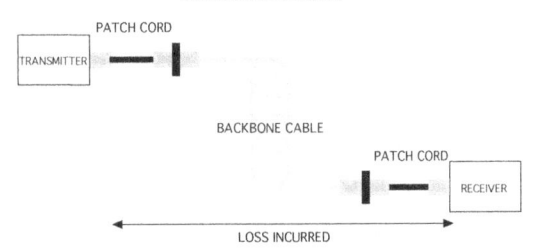

Figure 14-3: Loss In Link

The second aspect is the use of a single test lead in the measurement of the input power (Figure 14-1). This test method is called Method A.1 in the singlemode insertion loss test standard, EIA/TIA-526-7A.

The use of a single lead results in a loss measurement that includes the loss of two connector pairs, as occurs in a typical link (Figure 14-3). Were the input power measurement made with two test leads (Figure 14-4), the loss measurement (Figure 14-2) includes the measurement of only one additional connector pair, which is the difference between the number of pairs shown in Figure 14-4 and that in Figure 14-2.

A multimode insertion loss measurement made with a two lead reference is known as 'Method A'. Thus, the use of a single test lead provides a loss measurement that is close to the loss that a transmitter-receiver pair experiences. A simple analysis shows that the loss measured with a two lead reference (Figure 14-4) will be one connector pair lower than that made with a one lead reference.

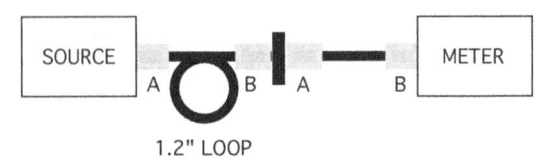

Figure 14-4: Input Power Measurement With Two Leads

14.4 MULTIMODE TESTING

Like singlemode insertion loss testing, measurement of input power level of multimode testing is performed with a single reference lead. This method is known as Method B, or the one lead reference method in EIA/TIA-526-14B, the multimode insertion loss test standard. However, insertion loss testing of the multimode link is more complicated than that of the singlemode link. The reason for this complication is the fact that loss in multimode fibers depends on the manner in which the transmitter launches light into the core. Different multimode transmitters launch light into the core with different power distribution. In contrast, all singlemode transmitters launch light with the same power distribution.

14.4.1 CAUSE OF DIFFERENCES IN MULTIMODE LOSS MEASUREMENTS

If the characteristics of the multimode testing light source are the same as those in the transmitter, the testing will simulate network operation. However, multimode transmitters have two different light sources: LEDs and VCSELs.[48] These two sources launch power into the core in significantly different power distributions. These differences create differences in

> Attenuation in the fiber
> Power loss at connectors

These differences in transmitter characteristics are:

---
[48] Although the data standards allow the use of singlemode transmitters on multimode fibers, we ignore such use, as such use is rare and more expensive than other options.

- LEDs overfill the core diameter and the NA.
- VCSELs under fill both

Thus, testing with one type of source will not accurately indicate the loss that occurs with the other type.

14.4.1.1 ATTENUATION RATE VS. DIVERGENCE ANGLE

The attenuation of light depends upon the travel path of the rays of light. The longer the travel path is, the higher the attenuation will be. Thus, rays of light traveling at a large angle to the axis travel paths that are longer than those that travel parallel to the axis or at low angles to the axis. The higher the percentage of rays traveling at high angles to the axis, the longer will be the travel paths and the higher will be the attenuation rate. The technical term that indicates this behavior is 'differential modal attenuation' (DMA).

With this understanding, we can compare attenuation rates of the same multimode fiber with LEDs and VCSELs. LEDs have a large portion of the optical power at high angles to the fiber axis (Figure 14-5). VCSELs have a small portion of the optical power at high angles of the fiber axis (Figure 14-6).

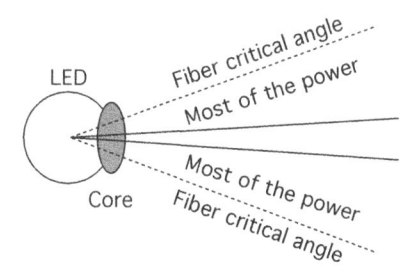

Figure 14-5: LED Power Distribution

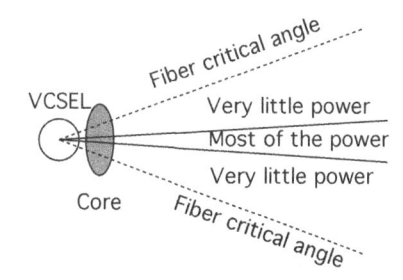

Figure 14-6: VCSEL Launch Power Distribution

- From this comparison, we would expect attenuation rates experienced by a fiber with an LED transmitter to be higher than those measured with a VCSEL transmitter. Such is the case.

14.4.1.2 LOSS AT CORE BOUNDARY

Power loss at connections is strongly influenced by the power at the core boundary. The power at the core boundary is influenced by the angle of divergence, as indicated above, and by the spot size of the transmitter.

Spot sizes can be larger or smaller than the core diameter. If the transmitter spot size is larger, that transmitter is said to 'overfill' the core. An overfilled core has a significant fraction of the optical power at the core boundary.

- Since the single largest cause of connection power loss is lateral core offset, a large amount of optical power at the core boundary results in a relatively large power loss (Figure 14-7).

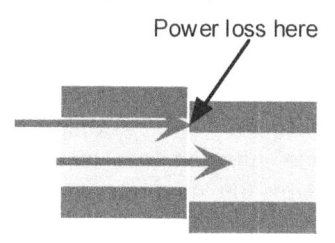

Figure 14-7: Increased Power Loss At Connection With Overfilled Core

850 nm LEDs have spot sizes larger than the core and a significant amount of power at the core boundary. VCSELs have spot sizes smaller than the core, 30μ. With such a small spot size, there is relatively little optical power at the core boundary (Figure 14-8).

- From this comparison we would expect connection losses with an LED source to be higher than those measured with a VCSEL source. Such is the case.

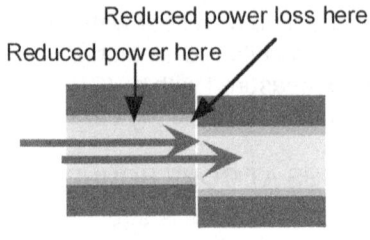

Figure 14-8: Reduced Power Loss At Connection With Under filled Core

Simulation of operating conditions in the insertion loss test procedure is desirable. However, multimode simulation has two disadvantages. Firstly, simulation requires multiple tests with different light sources. Multiple tests make simulation testing difficult, expensive, and unpopular. Secondly, multiple tests create a problem of interpretation if one test result is acceptable and the other is not. Finally, simulation testing has not been required by TIA/EIA-568-C or its predecessor, TIA/EIA-568-B.

14.4.2 MULTIMODE TEST PROCEDURES

In this section, we examine the four multimode insertion loss test procedures required by the next version of EIA/TIA-526-14B.[49] These procedures are:

- Test with an EF compliant source
- Test with an EF compliant mode conditioner
- Test with a mandrel
- Test without a mandrel

These procedures are those that are expected to be approved in the near future (14.4.2.1) and those required by existing standard (14.4.2.2).

14.4.2.1 FOUR CURRENT PROCEDURES

The current multimode insertion loss draft standard, ANSI-TIA-526-C d3, identifies four measurement situations (Table 14-1).

The draft states that encircled flux (EF) launch conditions be used for situation 1 and *allows* use of High Order Mode Loss

---

[49] This standard is expected to be approved in late 2014 or early 2015.

(HOML) launch conditions for the remaining three conditions, unless the 'channel loss' budget of the optoelectronics is less than 3 dB. EF compliant launch conditions *may* be used for all four situations.

Situation	Core Diameter	Wavelength
1	50μ	850 nm
2	50μ	1300 nm
3	62.5μ	850 nm
4	62.5μ	1300 nm

Table 14-1: Multimode Insertion Loss Situations

High order modes are those at high angles to the axis (Figure 14-5). These modes travel near the core boundary. HOML conditions result in a restriction of most of the power to the center of the core.

EF Launch Conditions

EF launch conditions produce test results that are close to those experienced by typical launch conditions of an 850 nm, VCSEL. These conditions must comply with the template in Figure 14-9. Note that all the power is within a 30μ diameter, with very little in the central 9μ of the core.

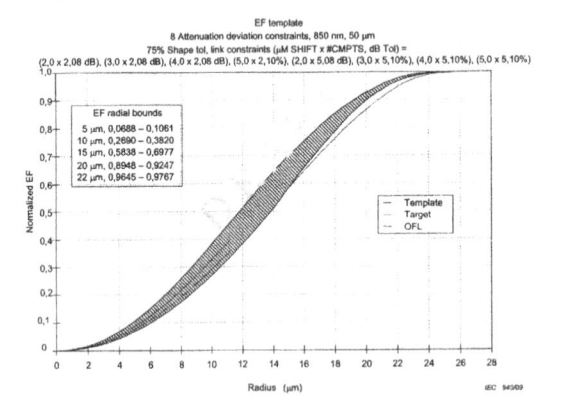

Figure 14-9: Encircled Flux Template For Core Power Distribution

There are two methods for achieving EF launch conditions: use of an EF compliant light source (Figure 14-10 and Figure 14-11); or use of a test lead, aka 'mode controller', that provides such compliance (Figure 14-12 to Figure 14-14).

At the time of this writing, Kingfisher, AFL, and EXFO are manufacturers of EF compliant 850 nm test sources. Arden

Photonics appears to be the only manufacturer of mode controllers.

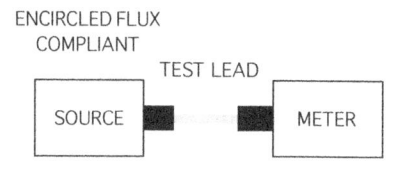

Figure 14-10: Measurement Of Input Power With EF Compliant Source

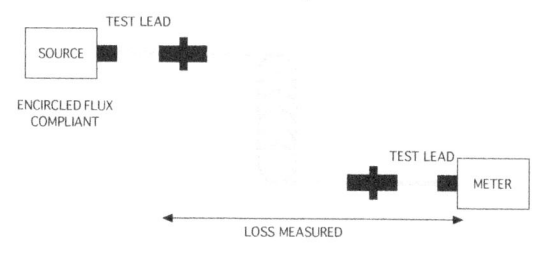

Figure 14-11: Loss Measurement With EF Compliant Source

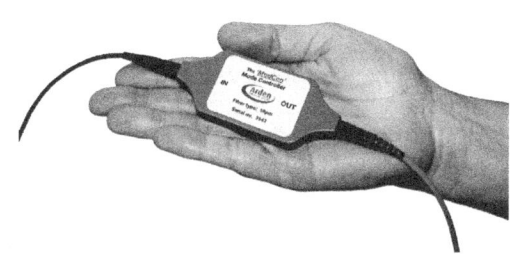

Figure 14-12: EF Mode controller (Courtesy Arden Photonics)

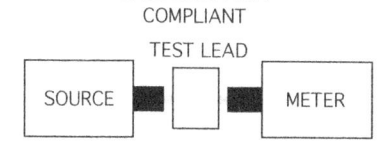

Figure 14-13: Measurement Of Input Power With EF Compliant Test Lead

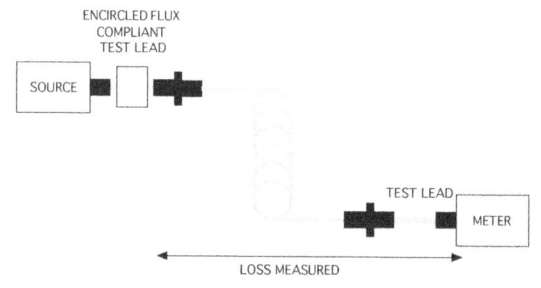

Figure 14-14: Loss Measurement With EF Compliant Test Lead

HOML Launch Conditions

According to The current multimode insertion loss draft standard, ANSI-TIA-526-C d3, situations 2-4 require meeting HOML launch conditions. These conditions require qualifying the source and test cable. The source and meter are connected with a qualified test lead (Figure 14-15). The power level is measured. The 3mm test lead is wrapped around a 0.87" mandrel (Figure 14-16). The power level is measured. Other mandrel diameters apply for other core diameter-jacket diameter combinations (Table 14-2).

If the loss with the mandrel is 0.1-0.6 dB, the installer removes the mandrel and tests as shown in Figure 14-17 and Figure 14-18.

If the loss with the mandrel is >0.6 dB, the installer leaves the mandrel in place and tests as shown in Figure 14-19 and Figure 14-20. Note that the mandrel required for multimode testing in TIA/EIA-568-B has a diameter of 0.9", almost the same as that required in Figure 14-19.

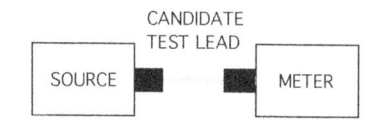

Figure 14-15: Measurement Of Input Power For HOML Qualification

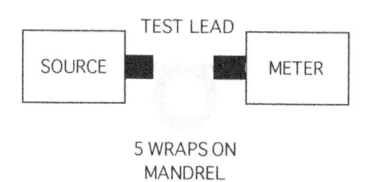

Figure 14-16: Measurement Of Output Power For HOML Qualification

Core μ	900 μ	1.6 mm	2.0 mm	2.4 mm	3.0 mm
50	25 mm	24 mm	23 mm	23 mm	22 mm
50	0.98"	0.94"	0.91"	0.91"	0.87"
62.5	20 mm	19 mm	18 mm	18 mm	17 mm
62.5	0.79"	0.75"	0.71"	0.71"	0.67"

Table 14-2: HOML Mandrel Diameters

Figure 14-17: Measurement Of Input Power, HOML 0.1-0.6 dB

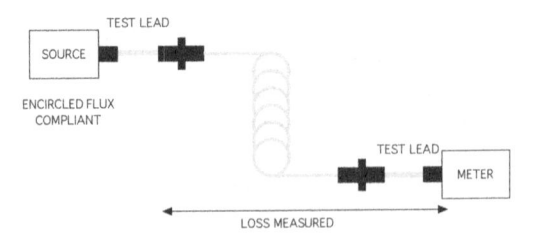

Figure 14-18: Measurement Of Insertion Loss

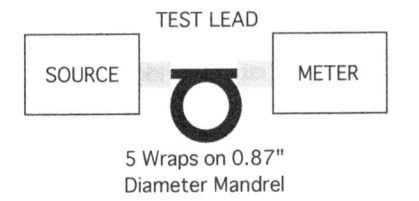

Figure 14-19: Measurement Of Input Power, HOML >0.6 dB

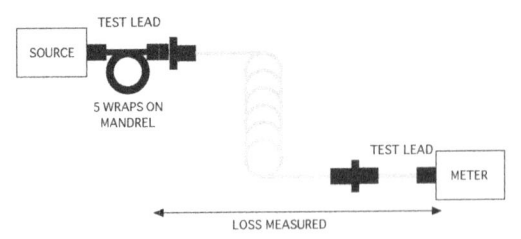

Figure 14-20: Measurement Of Insertion Loss

If the loss with the mandrel is <0.1 dB, the installer repeats the qualifying shown in Figure 14-15 and Figure 14-16 with a different source and/or test lead until he qualifies a new combination.

14.4.2.2 PREVIOUS PROCEDURE

The previous multimode insertion loss measurement procedure for all four measurement situations (Table 14-1) is the EF procedure shown in Figure 14-10 and Figure 14-11 or that shown in Figure 14-13 and Figure 14-14.

14.5 MULTIPLE WAVELENGTH TESTS

To this point, we have implied that testing is to be performed at the wavelength of the transmitter. With one exception, TIA/EIA-568-C requires that all links be tested at both operating wavelengths.

The exception, which is stated in Section 5.3.3.3 of TIA/EIA-568-C, is for short horizontal links, known as 'Level III cabling link segments'. These segments may be tested at one wavelength. Such segments are deemed short enough that insertion loss difference due to a wavelength difference are insignificant.

Even if testing at two wavelengths was not required, it should be performed. The reason for testing at both wavelengths is stress sensitivity. Stress can result from bending, crushing, excessive temperature, or any violation of a cable performance parameter. The key technical fact that leads to testing at both wavelengths is:

> As the wavelength increases, the fiber becomes increasingly sensitive to stress.

This sensitivity results in increase in loss. In other words, loss due to improper installation may not be apparent if the test is performed only at a short wavelength. However, such increased loss can be detected at a long wavelength.

This author has observed this relationship repeatedly In field-testing of both multimode and singlemode cables. In some situations, fibers that functioned properly at a short wavelength would not function properly at the long wavelength.

14.6 BI-DIRECTIONAL TESTING

TIA/EIA-568-C allows testing in one direction. While the insertion loss will be different in opposite directions, the difference is not significant.

These directional differences have six potential causes:

> Core diameter differences
> NA differences
> Differential modal attenuation
> Core offset
> Cladding non-circularity
> Fiber offset in connectors

Core diameter differences can create directional effects. When light travels from a small core to a large core there will be little power loss. However, when light travels in the opposite direction, there can be increased power loss.

NA differences can create directional effects. When light travels from a small NA fiber to a large NA fiber, there will be little power loss. However, when light travels in the opposite direction, there will be an increased loss.

Differential modal attenuation (DMA) occurs in multimode fibers. DMA is the mechanism by which the ratio of the power in the center of the core to that at the core boundary increases as the light travels further along a fiber. In other words, the optical power becomes increasingly focused at the center of a multimode fiber as the light travels further along a fiber.

This mechanism is directly related to the travel path of the rays of light. The longer the travel path, the more attenuation the rays experience. As defined by the NA, the critical angle rays travel the longest path and experience the highest attenuation in the core. The critical angle rays will spend more 'time' near the core boundary than in the center of the fiber. The axial rays travel the shortest path and experience the lowest attenuation. The axial rays will comprise much of the power in the center of the core.

As light travels along a fiber, the fraction of the total power in the central region of the core increases while the fraction of the total power at the core boundary drops. Connector loss is related to the amount of power at the core boundary. As the distance between the source and a connector pair increases, the power level at the core boundary becomes less and the loss of the connectors becomes less. In other words, multimode connectors exhibit reduced loss as the distance from the source increases.

One consequence of DMA is the directional effect of insertion loss measurements. To see this mechanism, consider a two-segment link with mid span connectors that are not equidistant from the ends (Figure 14-21). When the testing light source is on the end of the link that is close to the mid span connector pair, there will be some DMA which reduces the power at the core boundary of the connector pair. When the testing light source is on the end of the link that is far away from the mid span connector pair, there will be increased DMA which reduces the power at the core boundary to a level below the level with the source at the opposite end. Thus, the measured insertion loss will be higher in the direction with connectors closer to the source than in the opposite direction.

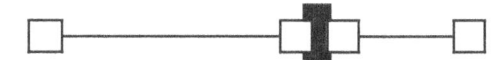

Figure 14-21: Connectors At Unequal Distances From Link Ends

These directional differences are usually small. For data links within buildings, these directional differences are small, on the order of several tenths of a dB. For links between campus buildings, directional differences are small. The differences that cause directional effects tend to cancel out with increases in length and number of connections.

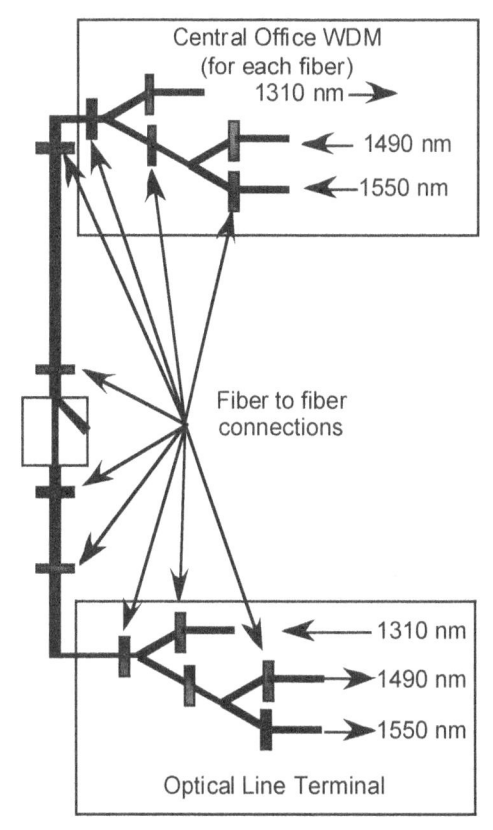

Figure 14-22: Minimum Configuration Of FTTH/PON Network

For an FTTH/PON network, these directional differences can be significant. This significance results from the large number of optical components in the link. A minimum configuration (Figure 14-22) includes at least five passive devices and at least ten fiber-to-fiber connections. If each passive device and each connection exhibited a small directional effect of 0.1 dB, the worst-case, combined directional difference could be 1 dB.

14.7 BIMM TESTING

Bend insensitive, multimode (BIMM) fibers have a core profile that is different from that of non-BI multimode fibers. As a result, the insertion loss testing must be performed with test leads that are not bend insensitive.

14.8 RANGE TESTS

Regardless of the test method, this author recommends that installers make multiple tests on multiple fibers to determine a 'range' value for the network.

> - The range is the maximum increase in loss that could occur between successive measurements of the same link, when there is no degradation of any link components.

The range value enables proper interpretation of an increase in loss that might occur between multiple tests of the same link. Multiple tests could be performed in order to troubleshoot link problems or as part of a preventive maintenance program.

The range value is used as follows:

> - If the increase in loss between measurements made at different times is less than the range, the link exhibits no evidence of degradation.
> - If this increase exceeds the range, degradation of link components has occurred. Additional troubleshooting is required.

Determination of range is simple process. The installer makes 4-6 tests on one link. From these tests, he calculates the difference between the maximum and minimum loss measurements. This is the link range.

The installer repeats this process on 4-6 fibers, resulting in 4-6 link range values. If all range tests have been made properly, the values will be close to one another, within 0.2 dB. The largest of the link range values is the range value for the network.

Occasionally, one of the range values is significantly higher than all other values. The author's experience is that this situation results from dirt on the connectors. Either dirt was introduced onto or dirt fell off the connectors. In such a case, the installer cleans the connectors and the test leads and repeats the range test.

If the range is not known, the installer can estimate the range from the connector repeatability. Connector manufacturers specify the repeatability of their products. A typical repeatability for keyed and contact connectors is 0.2 dB. With this value, the installer might expect that the range, which involves disconnecting and reconnecting two ends, would be twice the repeatability, or 0.4 dB.

However, this author finds measured ranges to be smaller than double the repeatability. Range testing for multimode, keyed, and contact connectors indicates a typical range of less than 0.2 dB. Finally, such testing indicates repeatability closer to 0.10 dB than to the 0.20 dB value found in connector data sheets.

14.9 SUMMARY OF TEST STANDARDS

This chapter is based on the data and test standards existing at the time of this writing. Unless qualified with the word 'opinion', all information is consistent with these standards.

14.9.1 TIA/EIA-14-B

This standard, also known as OFSTP-14, defines multimode insertion loss testing. Past standards required use of Method B, a one lead test. The current version of this standard, TIA/EIA-526-14 B, allows use of the one lead method (Method B), the two lead method (Method A) and three lead test method (Method C). IEC 61280-4-1, the international standard that is the equivalent of TIA/EIA-526-14 B, has replaced this standard.

14.9.2 TIA/EIA-7-A

This standard defines singlemode insertion loss testing. TIA/EIA-568-C requires testing to Method A.1, a one lead reference method.

14.9.3 IEC 61280-4-1

This international, multimode test standard has been adopted as a replacement for TIA/EIA-526-14-A. This test standard includes three reference methods with the input power level set with 1, 2, or 3 reference leads. The one lead method is equivalent to Method B of TIA/EIA-526-14-A for multimode and Method A.1 of TIA/EIA-526-7-A for singlemode. The two lead method is equivalent to Method A of TIA/EIA-526-14-A. The three lead method is used to test multimode link loss by excluding connector loss.

This standard allows OTDR testing instead of insertion loss testing as long as there is a 'launch' cable on both ends of the link under test. This author does not recommend such tests as OTDR tests do not produce the same loss values as do insertion loss tests.[50]

14.10 EQUIPMENT REQUIREMENTS

Insertion loss testing requires the following five types of equipment:

- Stabilized light source
- Calibrated power meter
- Qualified test leads
- Low loss barrels
- Multimode HOML mandrel or mode conditioner

A source and meter create an 'optical loss test set' (OLTS). Leads, barrels, mandrels, and an OLTS form a loss test kit.

14.10.1 SOURCE

The testing light source must be stabilized. A stabilized source provides constant optical power. With constant power, the installer can perform a large number of insertion loss tests with the same the reference, or input, power.

The light source must have a wavelength close to that of the network transmitter. Of course, if the network will be used at more than one wavelength, as is the case in most data networks and in FTTH/PONs, the installer will test links at all wavelengths. Many sources have multiple wavelengths in a single unit (Figure 14-23 and Figure 14-24).

According to 61280-4-1, the multimode source must meet the EF distribution of Figure 14-9. Alternatively, the input reference lead can incorporate a device that results in this distribution. This device is a mode controller (Figure 14-25).

Testing done prior to the adoption of 61280-4-1 required a Category 1 source with a mandrel. This source overfills the core and the NA. The mandrel removes light from the core to create a consistent power distribution, regardless of the source used.

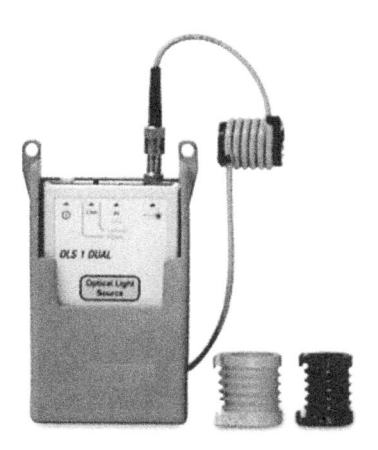

Figure 14-23: Category 1 Multimode Source With Mandrel[51]

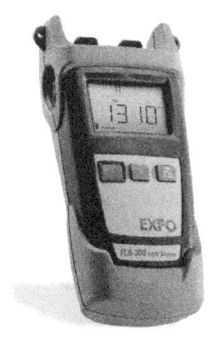

Figure 14-24: Three Wavelength Singlemode Source[52]

---

[50] http://thefoa.org/tech/ref/testing/5ways/fiveways.html

[51] Courtesy Noyes/AFL Limited
[52] Courtesy EXFO

Figure 14-25: Mode Conditioner For Multimode Testing[53]

14.10.2 METER

Three features are common to most power meters. The first feature, a requirement, is calibration traceable to the National Institute of Standards and Technology (NIST). The calibration must be at each wavelength in the unit. With multiple calibrations, the installer can use a single unit to test multiple wavelengths.

The second feature is an 'offset' or 'zeroing' function. This function allows the installer to set the input power level, or reference power level, to 0 dB. With a 0 dB input power level, the meter reading is the link loss.

A third feature is changeable adapters (Figure 14-26). Changeable adapters allow the meter to be used to make measurements by the one lead method. Measurements made by this method reduce measurement uncertainty.

Figure 14-26: Meter With Adapters

A common misunderstanding is the location of power measurement. All power measurements are at the detector (Figure 14-27).

---
[53] Courtesy Arden Photonics

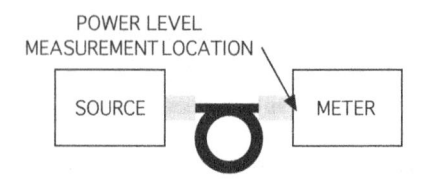

Figure 14-27: Location of Power Measurement

14.10.3 REFERENCE LEADS

14.10.3.1 GENERAL REQUIREMENTS

Test leads are variously called 'reference leads', 'source' or 'input leads', 'meter' or 'output leads'. We shall use the term 'test' leads. Where necessary, we will use the terms 'input' for the lead at the source and 'output' for the lead at the power meter. When testing, the installer uses at least two test leads. The method used determines the number of leads.

According to 61280-4-1, these leads must meet six requirements:

- ➢ Be qualified
- ➢ Be 2-5 m long
- ➢ Have the same core diameter
- ➢ Have the same NA
- ➢ Multimode leads not be bend insensitive (BIMM)
- ➢ Have the same connector type as that on the cables to be tested

In addition, the output end of the multimode input test lead shall have the power distribution of the EF template (Figure 14-9).

"Qualified" means that all connectors of the reference leads are low loss connectors. Qualified means that the connectors that mate to the connectors under test 'should' be 'reference grade'. Note that use of such reference grade connectors is not a requirement, but a recommendation. According to 61280-4-1, 'reference grade' is defined as having loss ≤ 0.1 dB when mated to another reference grade lead.

Unqualified reference leads may have high loss connectors. Such connectors will produce high losses when used to test low loss links. If the installer uses such high loss reference connectors, he risks rejection of low loss, properly installed connectors and cables.

This author's experience is that if the connectors on the test leads are not reference grade, they will have a loss ≤0.5 dB. Most values have been 0.2-0.4 dB. This latter value is consistent with the requirement of TIA/EIA-568-B, the predecessor standard. The process of qualifying reference leads is defined in 61280-4-1.

The process of qualification creates a matched pair of reference leads. After qualification, this pair can be used for testing, but only as a pair. The prudent installer will create a second, or back up, pair for use when the primary pair becomes damaged.

The qualification is a single end test. That is, the measurement is of the loss of the connector on one end of the test lead.

14.10.3.2 SINGLEMODE QUALIFICATION

The installer starts with two singlemode candidate leads. The singlemode qualification process is defined in TIA/EIA-568-C. The input test lead requires a 1.2" loop. This loop removes power in the cladding. Such power can increase loss measurements. The installer measures the input power as shown in Figure 14-28.

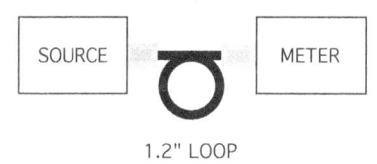

Figure 14-28: Singlemode Input Power Measurement

The installer connects the second candidate test lead, lead B, to the first, as shown in Figure 14-29.

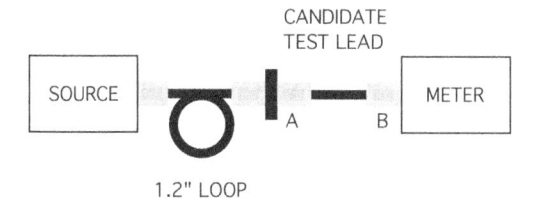

Figure 14-29: First Reference Lead Test

If the connectors on the test leads are 'reference quality', the power meter should read a power level above -0.1 dB If the connectors on the test leads are not 'reference quality', the power meter should read a power level above -0.5 dB.

The installer reverses the ends of candidate B, as shown in Figure 14-30.

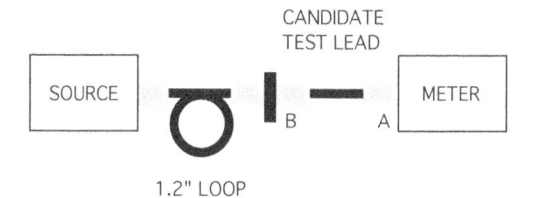

Figure 14-30: Second Reference Lead Test

To be acceptable according to TIA/EIA-568-C, the power level in Figure 14-30 must be above -0.75 dB. However, this author recommends a value above -0.5 dB. This reduced value complies with the measurement principle that reference leads have loss lower than that of the connectors under test.

If the connector interface on the source is the same as the connector type under test, the installer can perform a complete qualification. To complete the qualification, the installer repeats the tests in Figure 14-28 to Figure 14-30 with the ends of lead attached to the source reversed. Then he may repeat these same tests with the second candidate lead at the source. Finally, he repeats these same tests with the ends of lead B reversed.

To become a qualified test pair, all eight 'single end' loss measurements must meet the same requirement, as stated above.

14.10.3.3 MULTIMODE QUALIFICATION

The process of qualification creates a matched pair of reference leads. This process requires two tests as described in this section. This qualification test is a single end test, in that the loss is that of the connector on one end. The insertion loss tests in 14.3 and 14.4 are double end tests, in that the loss includes that of the connectors on both ends.

As indicated in Table 14-1, there will be four multimode test situations. Once the installer identifies the situation, he chooses an input power measurement indicated by one of the following figures:

- Figure 14-10
- Figure 14-13
- Figure 14-17
- Figure 14-19

The installer measures the input power level, as indicated by one of the figures listed above. He attaches the second candidate test lead to the first and measures the loss. He reverses the ends of the second candidate lead, and measures the loss, as shown in Figure 14-31, Figure 14-32, Figure 14-33, or Figure 14-34. If the loss of the second candidate lead is less than the maximum allowed, the two candidates become a qualified test lead pair.

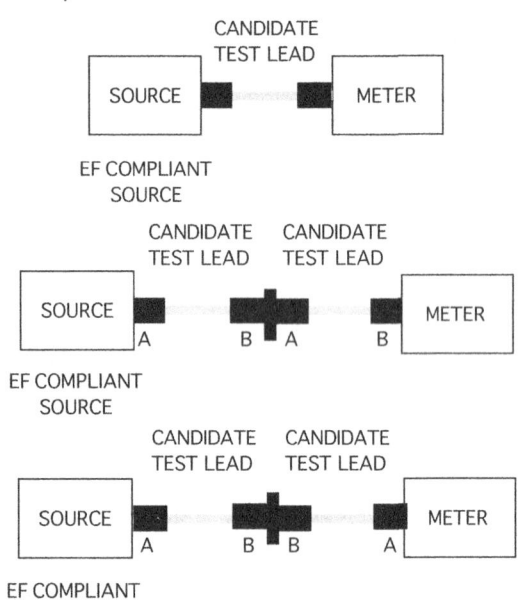

Figure 14-31: Lead Qualification With EF Compliant Source

The maximum allowed loss can vary. If the leads are 'reference quality' leads, the maximum is 0.1 dB. TIA/EIA-568-C.0 allows the test leads to be as high as 0.75 dB.

However, test leads should have loss lower than the connectors being tested. This author's opinion is that the maximum loss value for test leads be 0.5 dB. This author's experience is that ceramic ferrule multimode connectors have loss less than this value. Additionally, most of the hundreds of test leads qualified for use in training have exhibited single end losses of between 0.1 and 0.4 dB.

Figure 14-32: Lead Qualification With EF Compliant Test Lead

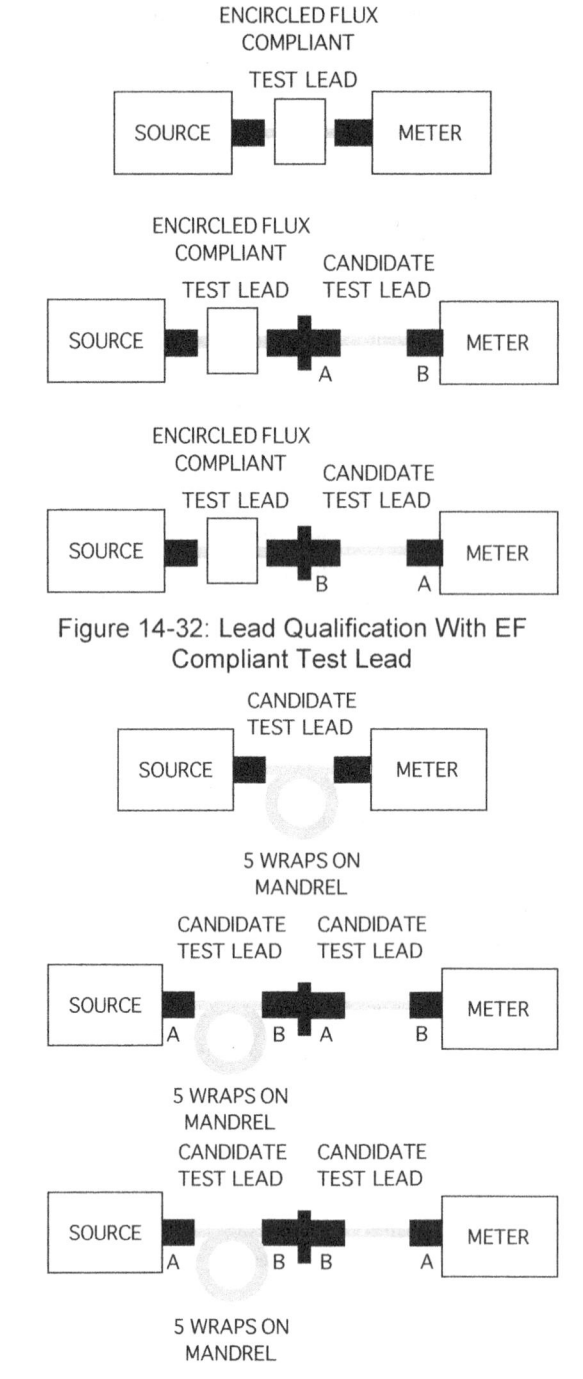

Figure 14-33: Lead Qualification With HOML Between 0.1 dB and 0.5 dB

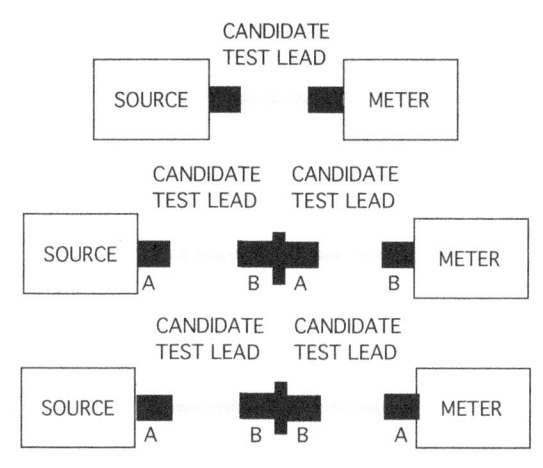

Figure 14-34: Lead Qualification With HOML>0.5 dB

14.10.3.4 LEAD MAINTENANCE

Prior to testing, the installer cleans and inspects all reference leads with a microscope. Cleaning can restore reference leads to low loss and low reflectance.

During testing, the installer cleans the connectors on reference leads frequently to remove small particles of dirt and dust, which can cause high loss measurements.

14.10.3.5 INFLUENCE OF REFERENCE LEADS ON LOSS

To some extent, loss measurements are arbitrary, as they are relative to reference leads with unknown, but low, loss. This arbitrary loss is deliberate, if only because absolute, known measurements would be inconvenient and expensive.

To obtain reference leads with known loss, the installer would need reference leads, in which the core diameter, NA, cladding diameter, core offset, cladding non-circularity, fiber hole diameter, and fiber hole offset were all specified to be within tight tolerances. Such leads would be expensive and would have a limited lifetime. In addition, such leads would not improve the primary functions of the test: to indicate proper operation of the opto-electronics and proper link installation.

In spite of this characteristic, most reference leads with low loss connectors will produce loss measurements within ±0.2 dB of one another. Measurements with this low variability will fulfill their primary function.

14.10.4 BARRELS

The insertion loss test requires a pair of low loss barrels (Figure 14-35). The barrels must be precise, not dirty or worn. Worn barrels result in insertion loss measurements that are higher than reality. Worn or dirty barrels create the risk of rejection of properly installed links. If the installer uses the same barrels in qualification of the reference leads as he uses in testing, the lead qualification process results in barrel qualification.

All barrels have precision alignment sleeves. These sleeves create the alignment required for low loss. Alignment sleeves are of three materials:

> Plastic or polymer
> Copper beryllium
> Ceramic

Figure 14-35: Barrels

This author does not recommend use of plastic alignment sleeves, as they can wear after a few insertions. Copper beryllium sleeves are acceptable for multimode patch panels, but not for testing. Such sleeves wear after 500-1000 insertions. This wear can result in loss increases of 0.5 dB. In addition, wear produces particles that result in increased loss and increased frequency of cleaning of reference leads.

Ceramic sleeves do not appear to wear or create particles that cause increased loss until over 2000 insertions. Ceramic sleeves are required for singlemode testing.

14.11 ADVANTAGES

The three advantages of the insertion loss test are:

> Low-cost equipment
> Test procedure simplicity
> Simplicity of interpretation of the test results

14.12 DISADVANTAGE

The disadvantage of the insertion loss test is its blindness to conditions of reduced reliability. Conditions of reduced reliability can result in acceptable insertion loss values.

Examples of such conditions are:

- Low loss connectors with high attenuation rate cable
- Low attenuation rate cable with high loss connectors
- Low loss connectors, low attenuation rate cable and a violation of the cable performance parameter; i,e., violation of bend radius, crush load, temperature range, use load, or a break in tight tube cable

These conditions may not be revealed by an insertion loss test. Thus, we need another test to verify proper and reliable installation. We examine that test, the OTDR test, in the next chapter.

14.13 REVIEW QUESTIONS

1. True or false: all networks are tested for both power loss and dispersion.
2. True or false: all data networks are tested for both power loss and dispersion.
3. Is there a test that is required in an FTTH/PON network but not in a data communication network? If so, what is the test?
4. List six testing wavelengths in order of increasing sensitivity to micro bending.
5. What are the five types of equipment required for an insertion loss test?
6. What are the two requirements a light source must meet?
7. What is the function of these two requirements?
8. What are the two basic requirements for an optical power meter?
9. What are the five requirements that reference leads must meet?
10. An alternative name for the Method B insertion loss test is the _____ reference lead test.
11. True or false: insertion loss measurements are always exactly the same in both directions. Explain your answer.
12. Advanced Question: explain the two reasons that singlemode fibers have power in the cladding. Hint: the answers are in Chapters 3 and 8.
13. Advanced Question: true or false: the insertion loss test tells you that all components in a link have been properly installed.
14. What does a range test enable?
15. What does EF mean?
16. What is the advantage of EF testing?
17. Which source would result in a lower link loss, an LED test source or a VCSEL test source,?
18. What is the wavelength of an EF compliant source?
19. What does HOML mean?
20. HOML testing is done in what two situations?
21. Why is the loop on the source test lead in the singlemode insertion loss test required?
22. Would a singlemode insertion loss with a loop be higher or lower than one without a loop?
23. Using the figures like those in the text, draw a multimode insertion loss measurement of input power level by Method B. Label all equipment: light source, launch test lead, receive test lead, cable under test, barrel(s), power meter.

24. Using the figures like those in the text, draw a multimode insertion loss measurement of output power level by Method B. Label all equipment: light source, launch test lead, receive test lead, cable under test, barrel(s), power meter.

25. What is the minimum number of connector pairs in a Method B test of an installed link?

26. What three characteristics of a transmission system can be determined with a power meter?

27. After making an insertion loss test, comparing that value to the OPBA and calculating the maximum and typical link losses, what three independent statements can the installer?

15 OTDR PRINCIPLES AND METHODS

Chapter Objectives: you learn how to make and interpret OTDR traces. In addition, you learn how to make measurements from these traces. Finally, you learn how to differentiate between defective product and defective installation.

15.1 INTRODUCTION

The disadvantage of the insertion loss test is its blindness to the location and distribution of loss. For example, an acceptably low insertion loss can result from at least four situations:

- Nominal loss connectors and nominal attenuation rate cable
- Low loss connectors and high attenuation rate cable
- High loss connectors and low attenuation rate cable
- Low loss connectors, low attenuation rate cable, and a bend radius violation, or some other violation of the cable performance parameters

Since the insertion loss test is blind to the location and distribution of power loss, it cannot be used to indicate proper installation of all components of the link. A 'component' is a cable segment, splice, or connector pair.

At best, the insertion loss test provides a crude inference of proper installation. In a strictly technical and legal sense, a low insertion loss measurement does not prove low loss or proper installation of all components in a link. To provide indication of proper installation, another test is needed. That test is optical time domain reflectometer test (OTDR).

- Optical time domain reflectometry is a test procedure that allows testing of **almost** every component in a fiber optic link.

With such testing, the installer can verify proper installation of each component. If each component in a link is properly installed, the link will have the maximum reliability possible.

- This, **maximum reliability,** is the key advantage of, and reason for, OTDR testing.

15.2 TYPES

There are three types of OTDRs:

- Mainframe OTDRs (Figure 15-1)
- Mini- OTDRs (Figure 15-2 and Figure 15-3)
- OTDR modules (Figure 15-4)

The AC-powered mainframe OTDRs, with their high-speed processors, are used in fiber and cable manufacturing facilities. The high speed OTDRs justifies their high cost.

A mini-OTDR meets the needs of most field installation work. This OTDR has a reduced-speed processor for extended battery life. In addition, it has small size and light weight, for convenience during field testing.

OTDR modules are cost effective when the purchaser already has a computer or notebook computer than can be used as an OTDR. Modules are controlled and powered by an external computer, such as a notebook or netbook computer.

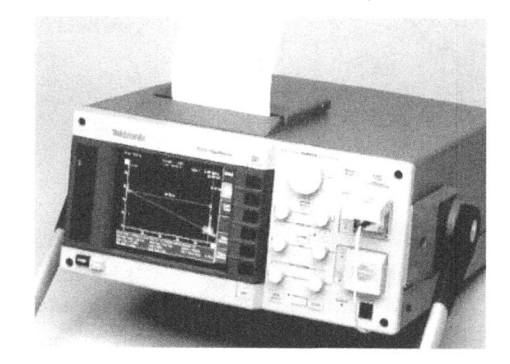

Figure 15-1: Mainframe OTDR (Tektronix)

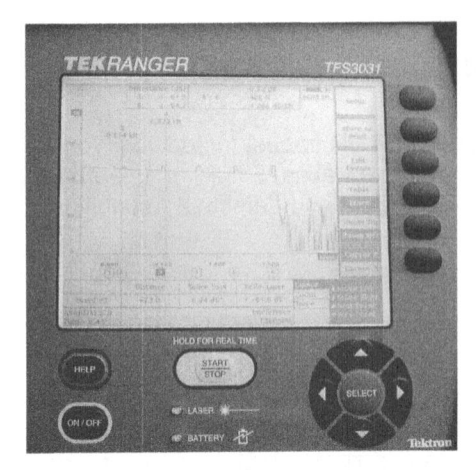

Figure 15-2: Tektronix Mini-OTDR

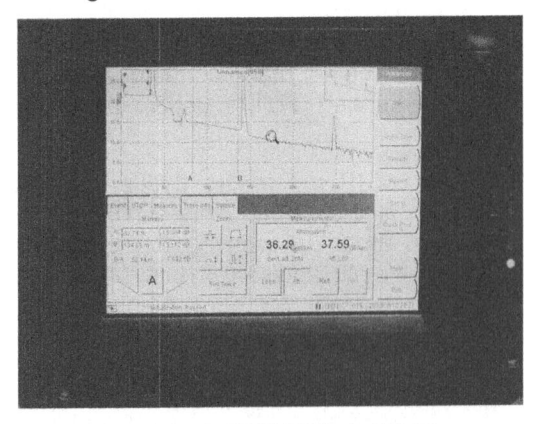

Figure 15-3: EXFO Mini-OTDR

Figure 15-4: USB-Powered OTDR Module

15.3 PRINCIPLES
To understand the method by which an OTDR functions, we must review the basics of attenuation (3.5.1). As light moves through an optical fiber, some of the light is scattered towards the core boundary at an angle greater than the critical angle. Such light escapes from the core, creating attenuation. However, some of the light is scattered backwards towards the input end of the fiber at an angle less than the critical angle. Such light experiences total internal reflection and travels backwards towards the input end (Figure 15-5). Thus, whenever there is light moving in one direction in a fiber, there is a small amount of optical power moving in the opposite direction.

▶ The OTDR measures 'backscattered' power

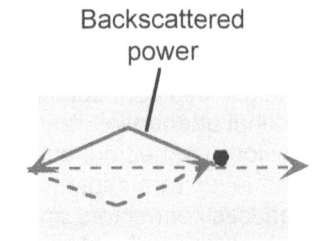

Figure 15-5: Rayleigh Back Scattering

15.4 BLOCK DIAGRAM
In its simplest description, an OTDR consists of four parts (Figure 15-6):

- A high-power laser diode
- A high sensitivity detector
- A bi-directional coupler
- A connector on the front panel

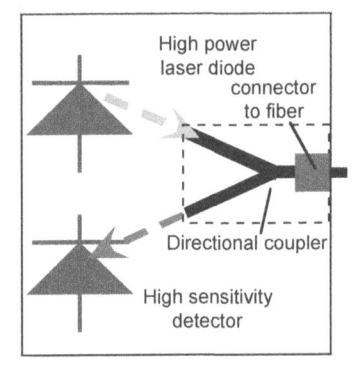

Figure 15-6: Functional Diagram Of OTDR

The laser diode sends a pulse of light into the fiber. At each point along the fiber, some of the light is scattered backwards towards the OTDR. The directional coupler directs this backscattered power to the high sensitivity detector.

15.5 THEORETICAL TRACE

The OTDR measures the power scattered backwards as a function of time. When the RI for the fiber (2.2.3) is entered into the OTDR, it displays optical power level as a function of fiber distance. The OTDR displays this information in the form of a 'backscatter' trace (Figure 15-7).

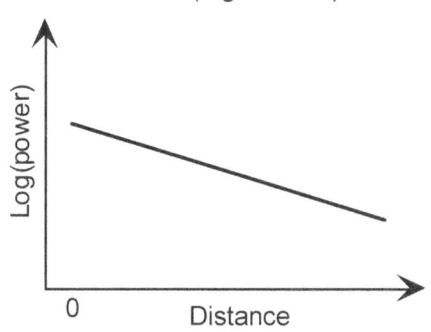

Figure 15-7: Theoretical Backscatter Trace

The backscattered power level is extremely low. For example, the backscatter coefficient for singlemode fibers is approximately 80 dB. With such a value, a 1 mW (10^{-3} W) power level at any point along the fiber will result in a backscattered power level of 10^{-11} W, or one-hundredth of a nanowatt.

Because the backscattered power level is very low, it can contain significant noise when amplified by the OTDR. In order to reduce this noise, the OTDR sends thousands of pulses into the cable. The OTDR averages the measurements from all pulses. With averaging, the noise in the signal is reduced to a level to allow accurate power level measurements.

▶ The OTDR makes multiple measurements

With this capability to measure backscattered power levels at different points along the cable link, we can measure the loss of **almost** every component in a cable link. For example, if we measure the power levels at the beginning and end of a cable segment, we can divide the power loss by the distance between these two locations to determine the attenuation rate, also known as the attenuation coefficient. We can measure the backscattered power levels before and after a connection to determine its power loss.

▶ Power difference is loss

Finally, we can determine cable segment lengths. With these measurements, we can verify acceptably low power loss for almost all components and conformance of the installed cable distances to the design map. In short, the OTDR gives us the ability to verify proper and reliable installation of the link.

15.6 TRACE FEATURES

In this section, we use the symbols shown in Figure 15-8.

- radiussed connector pair
- radiussed connector
- fusion splice
- mechanical splice
- APC connector pair

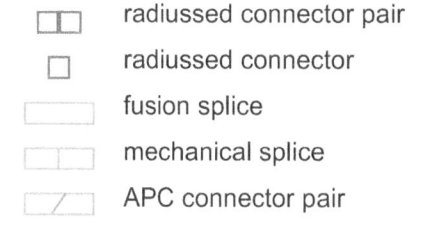

Figure 15-8: Symbols Used

15.6.1 REFLECTANCE

The trace (Figure 15-7) includes only power scattered from atoms in the core. In Chapter 4, you learned that connectors and splices can create reflections. The power from these reflections adds to the backscattered power. When the OTDR adds this reflected power to the backscattered power, the straight-line trace with a negative slope has **peaks** at the locations of reflections. The peaks are also known as **spikes**, but are most accurately called **reflections** (Figure 15-9).

▶ Fiber ends create reflections

The height of a reflectance, or peak, depends on a nature of the connection. The larger the difference in the RI at the connection, the higher will be the reflectance. Reflectance of various types of connections, from highest to lowest, are:

- Air gap connectors
- Flat physical contact connectors
- Radius physical contact connectors
- Mechanical splice

> Fusion splice and APC connectors

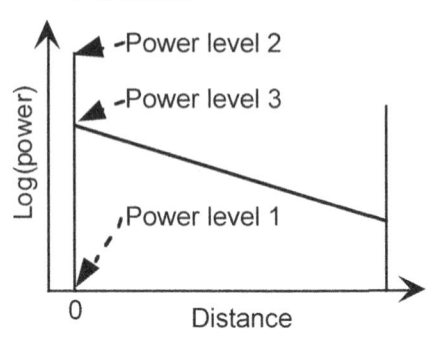

Figure 15-9: Trace With Backscatter And Reflections

15.6.2 DEAD, OR BLIND, ZONES

In Figure 15-9, the peaks are vertical lines. These vertical lines imply an OTDR ability to measure three different power levels in zero time. This implication cannot be correct, since all electronics need a time to respond to changes in power levels. The OTDR cannot display a figure like Figure 15-9 because of this response time.

Instead of vertical lines, as in Figure 15-9, the OTDR creates a trace with peaks that have a forward slant and a finite width (Figure 15-10).

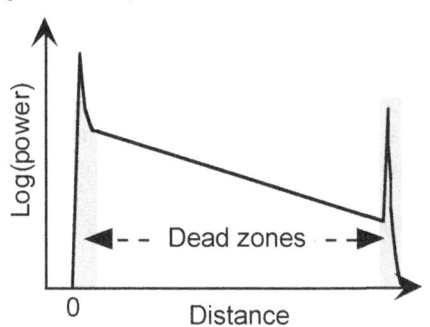

Figure 15-10: Basic Trace With Modified Peaks

Two principles are:

▶ A dead zone exists at any location of a power change.
▶ A power change can occur at all reflective and non-reflective events.

This width can be from a few meters to thousands of meters. The bandwidth limitation of the OTDR detector is one factor that determines the width of these peaks. This peak width obscures the characteristics of the fiber in the peaks. As a result,

▶ Peaks conceal fiber features

Reflectance has two types of tops: sharp and flat. A sharp top indicates that the OTDR detector has not been over-loaded. In this case, the OTDR can be used to measure reflectance. A flat top indicates that the OTDR detector has been overloaded. In this case, the OTDR cannot be used to measure reflectance. When the detector is overloaded, the reflectance value will be an underestimate of the real value. In other words, a reflectance of -35 dB could actually be -25 dB. This relationship is the opposite of what we would wish.

15.6.3 CONCEALED FEATURES

These peaks are called optical **'blind zones'** or **'dead zones'**. These zones conceal features, i.e., reflectance or drops, whenever the features are more closely-spaced than the width of the reflection.

Imagine there are two reflective components that are closely spaced (Figure 15-11). The peak created by the first reflective component can be wider than the distance between the two components. In this case, the peak from the first component can conceal the peak from the second component (Figure 15-11).

Because a reflection from one reflective component can obscure the reflection from another, closely spaced component, we may not be able to measure the loss of every event in a link. Instead, we can measure the total loss of multiple components.

This same principle applies to non-reflective events: a power drop from one component can obscure the power drop from another closely-spaced component.

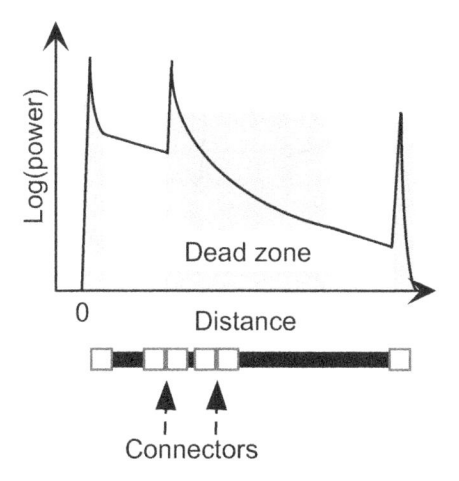

Figure 15-11: Two Closely-Spaced Reflective Events

15.6.4 MAP TO TRACE

Because we do not know what is in the dead zone, we cannot create a map from a trace. Instead, we must have an accurate map to interpret a trace properly. Figure 15-11 provides an example: a single pair of connectors could provide essentially the same trace (Figure 15-12).

Two important principles are:

▶You cannot create an accurate map from a trace.

▶You can only interpret the trace from an accurate map.

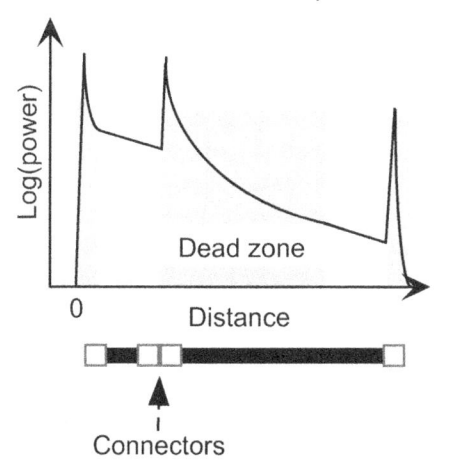

Figure 15-12: Trace Of Connectors, Single Pair

15.7　3 BASIC TRACES

With this basic understanding of the OTDR and its traces, we can reduce thousands of different traces to three basic traces:

➢ A reflective loss
➢ A non-reflective loss
➢ A bad launch

All other traces are combinations of these three basic traces.

15.7.1　REFLECTIVE LOSS

A reflective loss will have the appearance of Figure 15-13. This trace can result from at least five link configurations:

➢ Two segments connected by radius connectors
➢ Two multimode segments connected by mechanical splice
➢ Two singlemode segments connected by a mechanical splice
➢ A broken fiber in a tight tube cable
➢ A single segment with multiple reflections

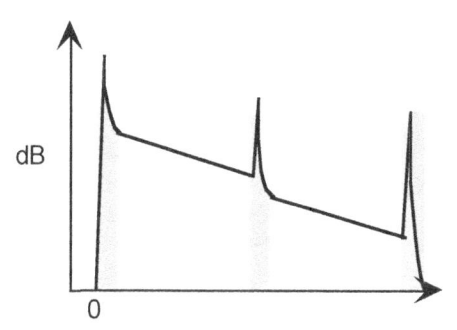

Figure 15-13: Reflective Loss

Radius connectors **always** create reflectance. The OTDR trace of two cables segments connected by radius connectors will exhibit reflectance at the location of the connectors (Figure 15-14). This reflectance results from the imperfectly smooth surfaces of the connectors (5.4.5). This lack of perfect smoothness creates microscopic air gaps, at which the RI changes. The principle is:

▶ Radius connectors always create reflectance.

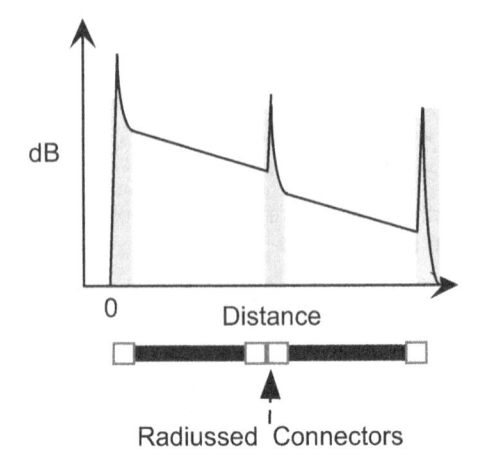

Figure 15-14: Reflective Loss Trace From Radius Connectors

Multimode mechanical splices always create reflectance (Figure 15-15). Reflectance results from the difference between the gel RI and the multiple RIs of the core. The single gel RI will be mismatched to the RIs in most of the core. This principle is:

▶ Multimode mechanical splices always create reflectance.

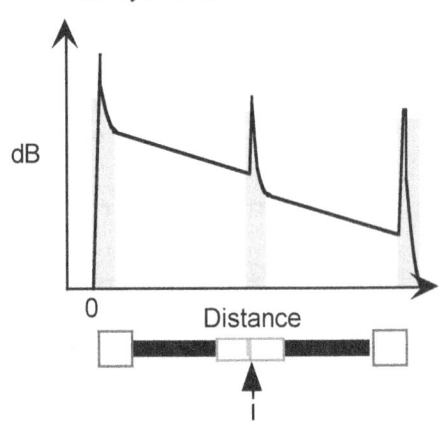

Figure 15-15: Reflective Loss From A Multimode Mechanical Splice

Singlemode mechanical splices may create reflectance (Figure 15-16). Such reflectance occurs if the RIs of the two fibers and that of the gel are not the same.

A broken fiber in a tight tube cable creates reflectance. The OTDR trace of such a break exhibits a reflectance (Figure 15-17). The cause of this reflectance is air between the fiber ends.

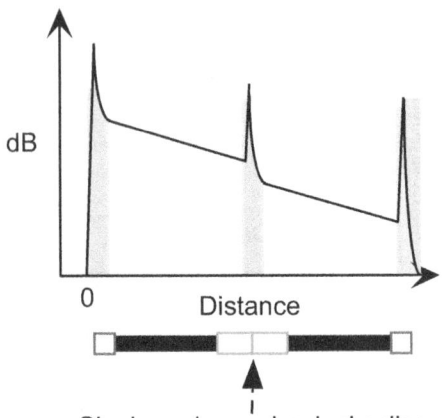

Figure 15-16: Reflective Loss Trace From A Singlemode Mechanical Splice

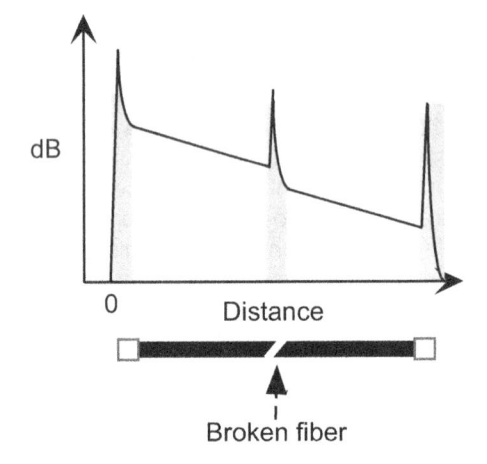

Figure 15-17: Reflective Loss Trace From Broken Tight Tube Cable

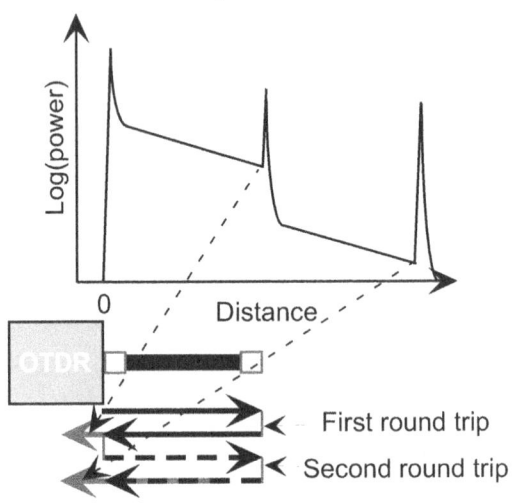

Figure 15-18: Reflective Event From Multiple Reflection

A single segment of cable with a high reflectance connector may create two reflectances (Figure 15-18). Light scattered backwards from the core and reflected by the far end connector can be partially reflected back into the fiber. This light will make a second round trip before entering the OTDR. The light that made this second round trip creates its own peak (Right peak of Figure 15-18). This reflection is a **'multiple reflection'**, or a **'ghost'**. In 15.7.4.4, we discuss ghosts in increased detail.

Figure 15-13 through Figure 15-18 reveal that different link components can create nearly identical traces. This comparison reinforces an important principle:

> ▶ We cannot create a map from a trace. Instead, we must have an accurate map to properly interpret a trace.

15.7.2 NON-REFLECTIVE LOSS

The second basic trace is a non-reflective loss (Figure 15-19). This trace can result from at least four cable configurations:

- Two segments connected with a fusion splice
- Two singlemode segments connected with a mechanical splice
- Two segments connected by APC connectors
- A cable with a violation of any cable performance parameter

Properly made fusion splices of the same fiber type create no reflectance but may exhibit a loss (Figure 15-19). A reflectance may result from a fusion splice if the RIs are different. Because there is little or no change in the RI as light moves from one fiber to another, there is no reflectance. Should the fusion splice exhibit a 0.0 dB power loss, the splice would be invisible to the OTDR (Figure 15-20).

Figure 15-20 reinforces an important rule: because a non-reflective, zero loss component does not appear on a trace,

> ▶ We cannot create a map from a trace.
> ▶ Instead, we must have an accurate link map in order to interpret a trace properly.

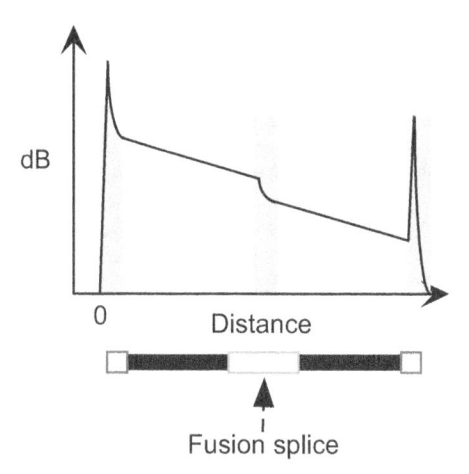

Figure 15-19: Non-Reflective Fusion Splice

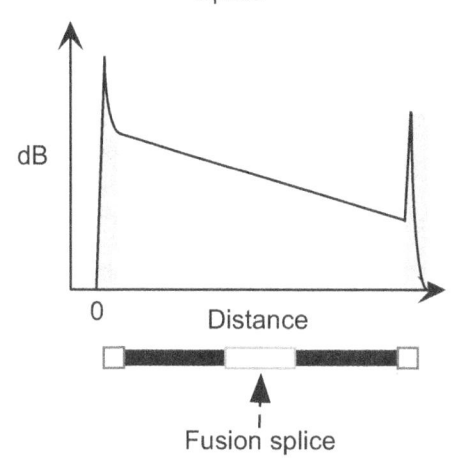

Figure 15-20: Trace For 0 dB Fusion Splice

Singlemode mechanical splices may create no reflectance. The single RI in the core of the singlemode fibers may exactly match the RI of the gel in the splice. Two singlemode segments connected with a mechanical splice may exhibit a loss without a reflectance (Figure 15-21).

Figure 15-16 and Figure 15-21 may create confusion. Both figures could result from the same cable configuration. If the RI of the gel does not exactly match the RIs of the fiber cores, reflectance will result. As stated earlier, the principle is:

> ▶ The trace must be interpreted with an accurate map; a map cannot be created from a trace.
>
> ▶ Two segments connected by APC connectors never create reflectance.

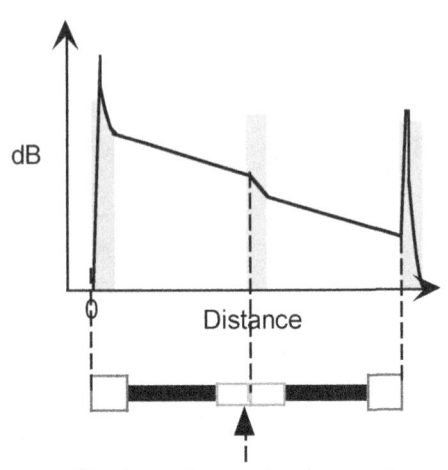

Figure 15-21: Non-Reflective Loss From Singlemode Mechanical Splice

Two segments connected by APC connectors can exhibit a loss without reflectance (Figure 15-22). The APC connectors reflect light backwards at an angle greater than the critical angle of the singlemode fiber. When this light crosses the core boundary, it escapes into the cladding. In this case, there will be no reflectance, and no peak.

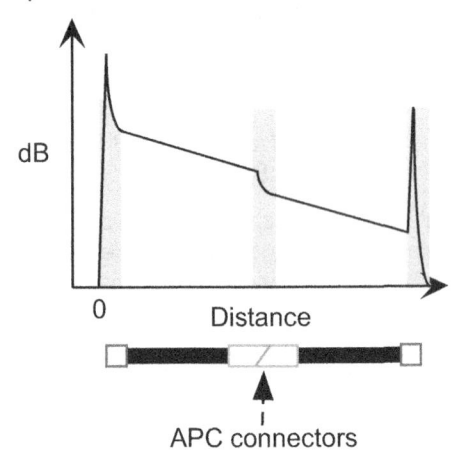

Figure 15-22: Non-Reflective Loss From APC Connectors

A violation of a cable performance parameter can be a violation of the bend radius, the crush load, the operating temperature range, the crush load, or any other cable performance parameter. When this violation occurs, power is lost at that location. We refer to this loss as an 'event', as it is not a connection. This violation results in an event with a non-reflective loss (Figure 15-23). This principle is:

▶ Any violation of a cable performance parameter can create loss without reflectance.

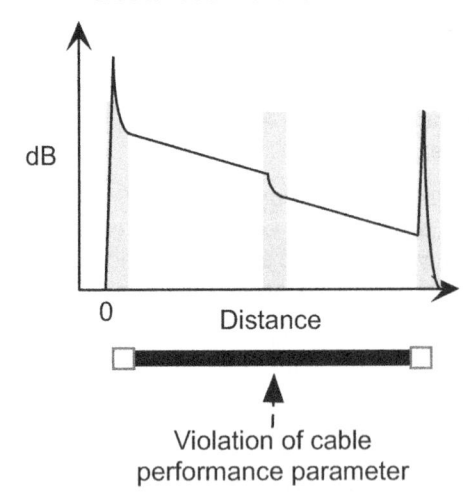

Figure 15-23: Non-Reflective Loss From Cable Parameter Violation

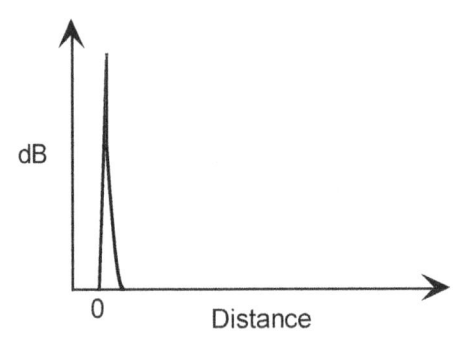

Figure 15-24: Bad Launch

15.7.3 BAD LAUNCH

The third basic trace is a bad launch (Figure 15-24). A bad launch results from at least three situations:

➢ A broken connector at the OTDR
➢ A break in the fiber within the dead zone
➢ A cable shorter than the width of the dead zone

15.7.4 UNUSUAL TRACES

In addition to the three basic traces, there are three unusual traces:

➢ A trace without a far end reflection
➢ A gainer

➤ A multiple, or ghost, reflection

15.7.4.1 NO FAR END REFLECTION

A trace may not have a far end reflection (Figure 15-25). This trace can result from three configurations of the far end:

➤ A cable with high angle cleave
➤ A singlemode fiber with a mechanical splice

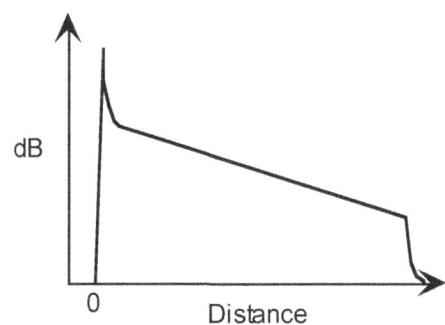

Figure 15-25: Trace Without End Reflection

The APC connector and a high angle cleave reflect light backwards at an angle greater than the critical angle. Thus, no reflection, or peak, is produced.

In the case of the mechanical splice on the far end, exact matching of the index of refraction of the core to the index of refraction of the gel results in no end reflection.

15.7.4.2 GAINERS

Previous measurement sections may create the impression that OTDR connection losses are always negative. Such is not the case. It is possible to measure a positive connection loss, also known as a **'gainer'** (Figure 15-26). This author has observed gainers at high as 1 dB, although most singlemode gainers tend to be less than 0.5 dB.

Gainers are a phenomenon of OTDR testing. A gainer indicates that more optical power is scattered back from after the connection than from before the connection. As such, a gainer is not a real gain in power.

A gainer can result from two different conditions on opposite sides of the connection:

➤ Different attenuation rates

➤ Different core diameters, or mode field diameters

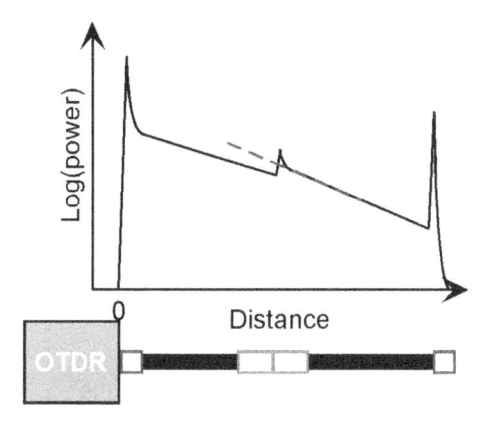

Figure 15-26: Gainer

To understand how different attenuation rates can create a gainer, we return to the cause of attenuation. At each point along its length, an optical fiber scatters power backwards towards the input end.

If we splice two fibers with the same attenuation rate, same core diameter, and same NA with a perfect splice, a 0 dB splice. However, if the attenuation rate of the fiber before the splice is lower than the attenuation rate of the fiber after the splice, the power scattered backwards by the fiber before the splice will be lower than that scattered backwards by the fiber after the splice. A gain results. However, in the opposite direction, the splice will have a power loss, because there will be less power scattered backwards from after than before the splice.

Similarly, differences in multimode core diameters or mode field diameters can result in gainers. If the core diameter of the fiber before the splice is smaller than the core diameter of the fiber after the splice, there will be more atoms scattering power backwards from the fiber after than before, resulting in a gainer. If measured in the opposite direction, the splice will exhibit a loss.

Differences in MFDs can create gainers, but in a manner opposite to that of differences in core diameters. If a large MFD before the splice is smaller than the MFD after the splice, there will be more atoms scattering power backwards from the fiber after than before, resulting in a gainer. In this case, the cladding contributes to the scattering.

Because of these three effects, differences in attenuation rates, in core diameters, and in mode field diameters, we would expect to experience different measurements in different directions.

15.7.4.3 TRUE SPLICE LOSS

Since the fibers in a splice can bias the loss high in one direction and low in the opposite direction, we have another principle:

▶The true splice loss is the average of the losses in both directions.

Averaging the values cancels out the bias.

15.7.4.4 MULTIPLE REFLECTIONS

'**Multiple reflections**', aka '**ghost reflections**', occur when optical power makes more than one round trip through the fiber before entering the OTDR. Understanding multiple reflections is necessary, as they have the appearance of a broken fiber.

Imagine that we have a high reflectance connector on the near end of a cable attached to the OTDR. This connector will reflect light strongly in both directions. The power scattered backwards from the atoms in the core of the fiber and the power reflected back to the OTDR from connectors in the cable under test must travel through this high reflectance connector in order to return to the OTDR. Even if this connector is low loss, this connector will reflect some of the returning power back into the fiber. This power will travel to the far end of the fiber. As it travels on its second round trip through the fiber, some of the power will be scattered backwards by the atoms in the core. In addition, the far end connector will reflect some of this power back to the OTDR.

However, the power from the second round trip arrives after the power that returned after the first round trip. Thus, a single segment may create two apparent segments in a trace (Figure 15-27).

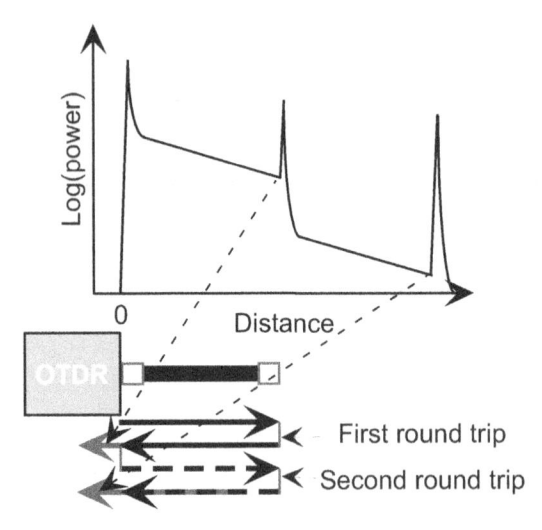

Figure 15-27: Multiple Segments Appear From A Single Segment

The installer must identify multiple reflections as such in order to justify ignoring them. If OTDR reflections are not ghosts, they can be broken fibers in tight tube cables.

The installer can identify ghost reflections by three characteristics:

➢ Length
➢ Height of second peak
➢ Trace from opposite end of cable

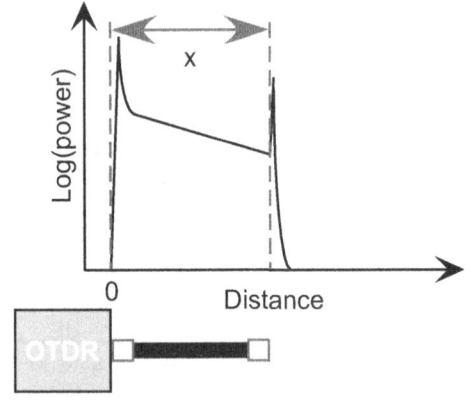

Figure 15-28: Single Segment Trace Expected

Imagine a 100 m segment with a ghost reflection. Without a ghost, we would expect to see a trace like that in Figure 15-28. If we see a second segment (Figure 15-29), we would expect the second segment to have exactly the same length as the first.

As the power entering the OTDR after the second round trip is much lower than that entering after the first, we would expect the height of the second peak to be lower than that of the first. This is the case.

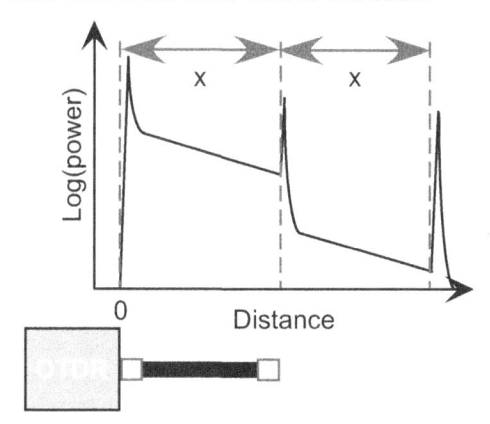

Figure 15-29: Trace With Ghost

Ghosts can appear on a trace from one end of the cable, but not from the other. In this case, the reflection is obviously a ghost, as reflections are from fiber ends. A real fiber end in the cable will create a reflection in both directions.

In this case, the connector on the end from which the ghost appears has a loss higher than the loss of the connector on the opposite end. As the ghost reflections have lower power levels than do real reflections, the closer the signal level is to the OTDR noise floor, the more frequently ghost reflections appear.

The situation in which a ghost reflection is most troubling is a multi-segment cable with a short segment followed by a long segment. In this case, a ghost reflection from the short segment can appear in the middle of the long segment (Figure 15-30). The installer must identify the origin of this reflection at a distance of ~2X, as it could be a broken fiber.

To determine whether the unknown peak is a ghost reflection, the installer tests from both ends. When the cable is tested from the near end, the unknown reflection occurs at about 2X from the near end. If the reflection is a real fiber end, the reflection will occur at a distance of 1x from the far end when the cable is tested from the second end. If the unknown reflection occurs at 2X from the OTDR at both ends of the cable, the reflection is a ghost. If the reflection does not occur in both directions, the reflection is a ghost.

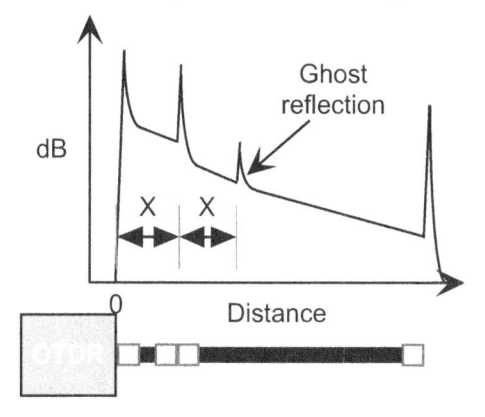

Figure 15-30: Ghost Reflection In Middle

15.7.4.5 HIGH REFLECTANCE CONNECTORS

High reflectance connectors can cause inaccurate measurements of connector loss and attenuation rate. Symptoms are:

- Large dead zone (Figure 15-31)
- Lack of clear demarcation between peak and backscatter (Figure 15-31)
- Overshooting or ringing (Figure 15-32)

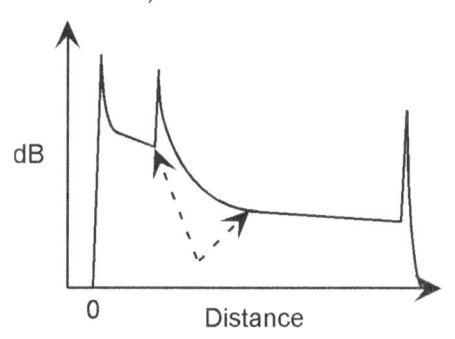

Figure 15-31: High Reflectance Connector

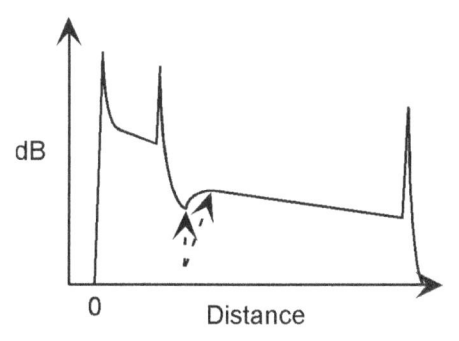

Figure 15-32: Overshooting After High Reflectance Connector

High reflectance can overload the detector, resulting in long recovery time. The horizontal axis is a time axis. With additional power from high reflectance added to the backscatter, one result is increased attenuation rate. This author has observed the increase to result in rejection of link segments because of high attenuation rate. However, when the link is analyzed from the opposite end, the attenuation rate is acceptable. Another method of obtaining accurate rate measurement is cleaning of the connectors. Cleaning can result in a reduced rate.

In order to obtain accurate OTDR measurements, the installer cleans or replaces such connectors. If cleaning or replacement is not practical, application of index matching gel to the connectors can reduce the reflectance and the width of the dead zone. With such reduction, measurements can be accurate. In some cases, overshooting indicates a limitation, or defect, in OTDR circuitry.

A second problem from high reflectance is low connector loss. With additional power from high reflectance added to the backscatter, the power level after the connector is higher than reality. In this case, the risk is acceptance of high loss connectors.

15.7.5 WAVELENGTH EFFECT

As you learned in 14.5, the fiber becomes more sensitive to stress as the wavelength increases. As a result of this relationship, bend radius violations, either of a cable or a fiber in a splice enclosure, may be evident at a long wavelength but not at a short wavelength. Because of this relationship, comparison of OTDR traces at two wavelengths can differentiate between a high loss splice that needs to be remade and a cable or fiber routing problem that can be corrected without splicing.

A splice with the same high loss at both short and long wavelengths indicates a need for replacement. A splice with lower loss at a short wavelength than at a long wavelength indicates the possibility that the fibers and/or buffer tubes need to be rerouted in the splice enclosure to remove stress. Such a splice need not be replaced.

In response to this behavior, OTDR manufacturers developed an 'out of band' OTDR, which operates at a wavelength 1625 nm. This wavelength allows full use of the common transmission wavelengths in DWDM networks and the FTTH/PON CWDM networks. An OTDR operating at this wavelength can provide simultaneous transmission and monitoring of cable integrity.

15.8 USING THE OTDR

15.8.1 SET UP

The purpose of using an OTDR is to make power loss measurements of all components in a link. In order to create valid measurements, the installer must provide the OTDR with six types of information:

- Wavelength
- Pulse width
- Maximum length of cable to be tested
- Either the maximum time allowed for the test or the number of pulses to be analyzed
- Index of refraction
- Backscatter coefficient

The wavelength is the same as that at which the transmitter operates. The pulse width determines the amount of power launched into the fiber in each pulse. As the cable length increases, the pulse width must increase to ensure adequate power from the far end of the cable. As the pulse width increases, the dead zone wide increases. As the dead zone becomes wider, the spacing between components must increase if the loss of each component is to be measured separately.

The maximum length of cable to be tested limits the time the OTDR takes to complete the test. The installer sets this value above the length of the cable, but as close to that length as possible.

The OTDR launches a pulse into the fiber, waits until the pulse arrives back from the far end, and then launches the next pulse. By setting the maximum length grater than but close to the length of the cable, the installer minimizes test time.

The time of the test or the number of pulses to be analyzed determines the noise level in the final backscatter trace. From experience, the installer learns the values that produce straight line cable segments. If the time or number is low, each cable segment will have noise that destroys the needed straight line trace.

The RI calibrates the OTDR to the fiber being tested. With calibration, the OTDR provides accurate fiber length measurements. However, the fiber length is greater than the cable length.

Some current generation OTDRs allow an input for 'excess fiber length' or 'helix factor'. This input corrects the calculation of length measurement so that the displayed length is cable length, not fiber length. For loose tube cables, this value is close to 2%.

The backscatter coefficient calibrates the OTDR to the fiber under test. With this calibration, the attenuation rates are accurate.

15.8.2 LAUNCH CABLE

In order to make an OTDR trace, the installer connects a launch cable between the OTDR and the cable under test. With a launch cable, the trace will exhibit at least two segments (Figure 15-33).

The launch cable is also known as a 'pulse suppressor'. This label is misleading, in that the cable does not suppress the pulse. Rather, the cable allows measurement of the near end connector by creating a straight-line trace before the near end connector of the cable under test.

▶▶ Use launch cable

The launch cable has a fiber with the same core diameter, or mode field diameter, the same NA, and the same connector as on the cable under test. An additional requirement is a low loss connector on both ends.

With a length longer than the dead zone, the launch cable serves two functions:

➢ Protection of the OTDR port
➢ Measurement of the loss of the near end connector

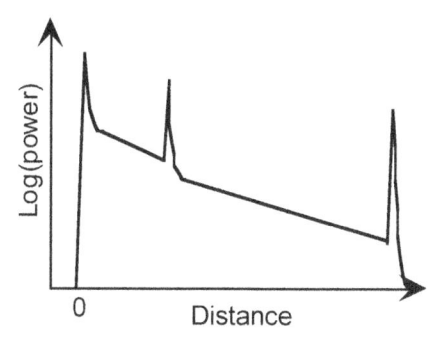

Figure 15-33: Trace With Launch Cable

When testing a large number of fibers, the installer connects the launch cable to the cable under test. Occasionally, and eventually, the connector on the far end of the launch cable becomes damaged. The repair of the launch cable connector is much less expensive than repair of an OTDR port.

If the length of the launch cable is greater than the width of the dead zone, the installer will be able to measure the loss of the connector at the near end of the cable. Without this launch cable, the loss of the near end connector may not be measurable.

If the launch cable is longer than any cable the installer may test, multiple reflections will appear beyond the end of the cables under test. Thus, multiple reflections will not appear in the trace. Of course, this method of avoiding ghosts may not be practical if the cables are many miles long.

15.8.3 MEASUREMENTS

Once the installer sets up the OTDR, he obtains a trace of the cable. From this trace, the installer makes loss and length measurements. To make these measurements, the installer can allow the OTDR software to determine all measurements. As we all well know, software is not perfect!

Because of this lack of perfection, the installer needs to know how to place cursors on the trace in order to make accurate measurements.

He will do so in two situations. First, he places cursors to verify the accuracy of the software numbers. Second, he places cursors when he prefers manual measurements. In this section, we present the rules for proper cursor placement…so that you need not curse at cursors!

One final note that applies to all measurements made by moving cursors to the proper location:

> ➤ The installer makes all measurements with both axes expanded enough to see fine details.

If either axis is not sufficiently expanded, the installer may place the cursors on the trace incorrectly.

The installer can make three types of OTDR measurements:

> ➤ Segment length
> ➤ Splice and connector loss
> ➤ Attenuation rate

15.8.3.1 LENGTH MEASUREMENTS

Segment length measurements require placement of either one or two cursors. Cursors are placed at the beginning and at the end of the segment.

For the first segment attached to an OTDR, the beginning of the segment is at the OTDR front panel, which is at 0 m. For this measurement, the installer need position only one cursor.

The end of any segment is defined by either:

> ▶A peak for a reflective connection
>
> ▶A change in slope or drop off, for a non-reflective connection or event

The method for length measurement of the first segment with a reflective connection is:

> ▶▶The curser is at the lowest point of the trace before the peak (Figure 15-34).

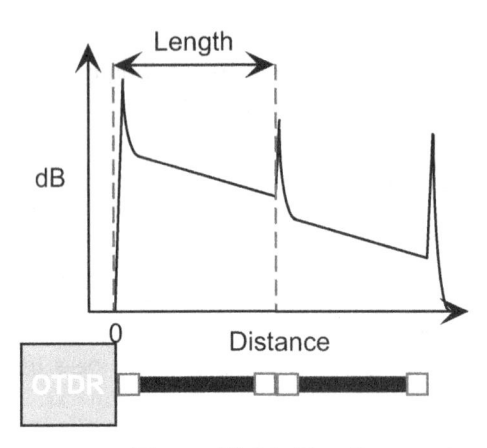

Figure 15-34: First Segment Length Measurement, Reflective Connection

The first segment length measurement method with a non-reflective connection is:

> ▶▶The curser is at the lowest point of the backscatter trace before the change in slope, or drop off, that marks the end of the first segment (Figure 15-35).

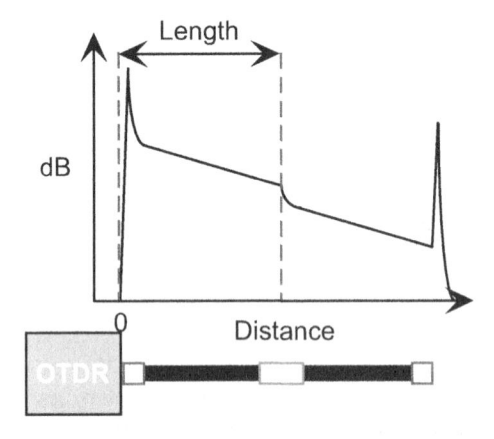

Figure 15-35: First Segment Length Measurement, Non-Reflective Event

For length measurements of other than the first segment, the installer places two cursors on the trace. The installer places the first cursor at the end of the previous segment and the second cursor at the end of the segment of interest.

The method for length measurement with reflective connections is:

> ▶▶The first cursor is at the lowest point on the trace before the peak at the end of the previous segment (Figure 15-36).

▶▶The second cursor is at the lowest point of a trace of the segment before the peak that marks the end of that segment (Figure 15-36)

The method for length measurement with non-reflective connections is:

▶▶The first cursor is at the lowest point of the straight-line trace before the drop that marks the end of the previous segment
▶▶The second cursor at the lowest point of a straight-line trace of the segment being measured before the drop that marks the end of that segment (Figure 15-37 and Figure 15-38).

In summary:

▶A length measurement includes a peak or a drop off at the beginning and excludes a peak or drop off at the end.

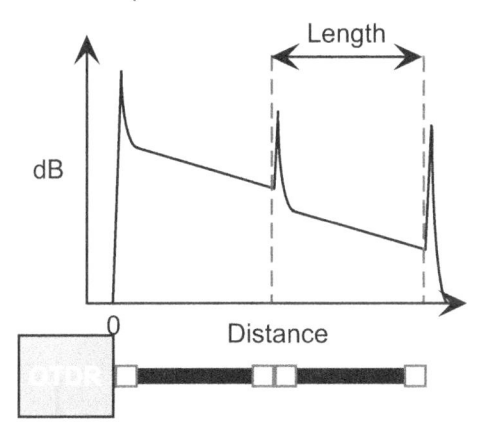

Figure 15-36: Segment Length With Reflective Ends

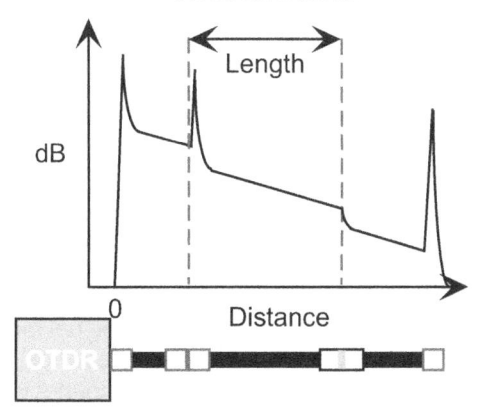

Figure 15-37: Segment With Reflective and Non-Reflective Ends

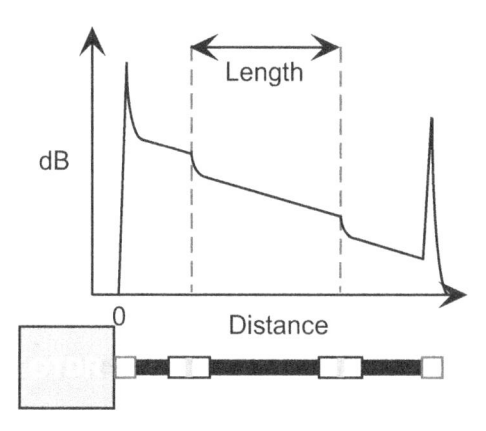

Figure 15-38: Segment Length With Non-Reflective Connections At Both Ends

15.8.3.2 CONNECTION LOSS MEASUREMENTS

There two types of connection loss measurements: estimated and accurate. The installer makes estimated measurements; the computer in an OTDR makes accurate measurements.

15.8.3.2.1 ESTIMATED CONNECTION LOSSES

This estimated loss method is also known as the two-point method. The principle for estimated connection loss measurements is:

▶Placement of two cursors that straddle the connection, but are in the backscatter, or straight lines, on both sides of the peak or drop.

These measurements overestimate the actual loss. This overestimate results from the fiber attenuation that occurs between the two cursors. As an overestimate, this method is conservative.

The principle for estimated reflective connection loss measurement is:

▶Two cursors placed in the backscatter trace on both sides of the peak, as close to the peak as possible without being in the peak (Figure 15-39).

If the installer places cursors away from the peak (Figure 15-40), the measured loss increases. If the installer places cursors in the peak, the measured loss increases (Figure 15-41) or decreases (Figure 15-42).

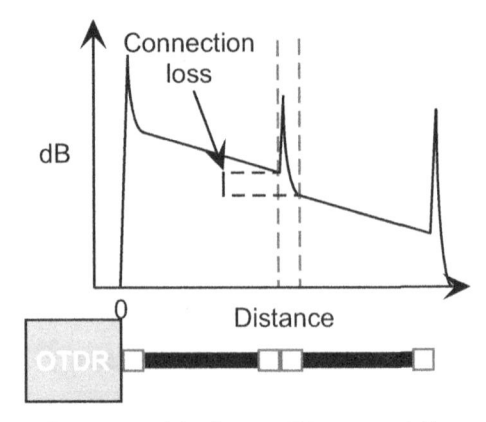

Figure 15-39: Cursor Placement For Estimated Reflective Connection Loss

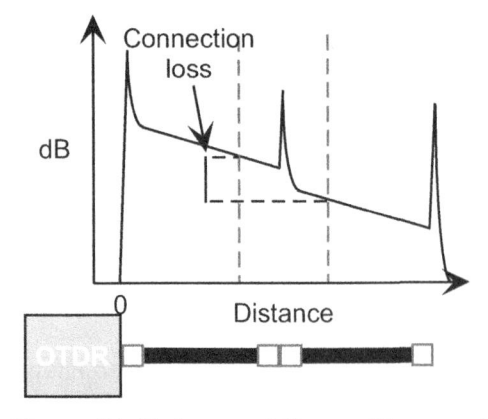

Figure 15-40: Incorrect Cursor Placement

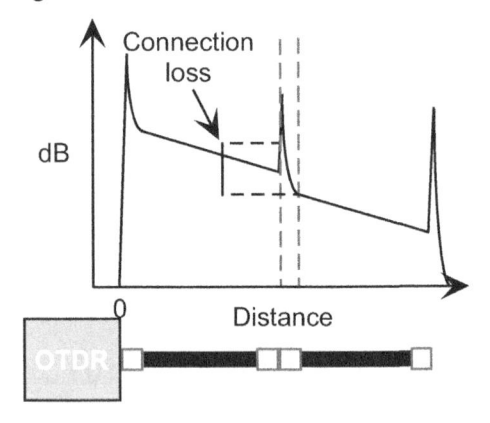

Figure 15-41: Incorrect Cursor Placement

The principle for estimated loss measurement of non-reflective connections is:

▶ Placement of cursors in the backscatter on both sides of the drop off, but as close as possible to the drop off without being in the drop off (Figure 15-43).

If the installer places the cursors away from the drop off (Figure 15-44), the measured loss increases. If the installer places the cursors in the drop-off, the measured loss decreases (Figure 15-45).

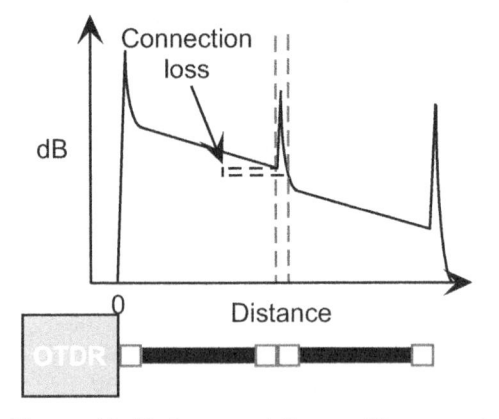

Figure 15-42: Incorrect Cursor Placement

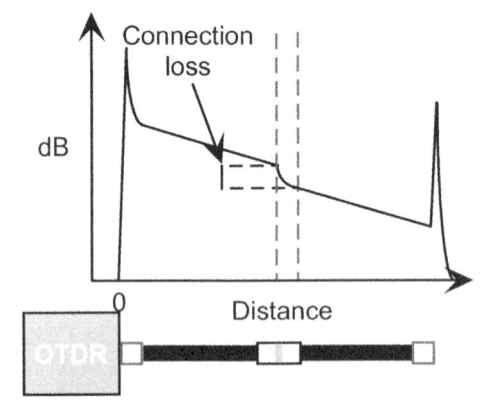

Figure 15-43: Cursor Placement, Non-Reflective Connection Loss

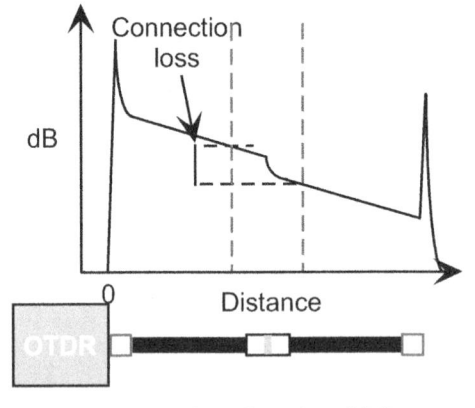

Figure 15-44: Wide Spacing Of Cursors

As indicated, connector loss measurement requires placement of two cursers in backscatter on both sides of the connection. This requirement cannot be met for the connector on the far end of a cable. Measurement of far end connector loss requires one of two approaches:

➢ The installer moves the OTDR to the far end. With this movement, the far end becomes the near end.
➢ The installer installs a second 'launch' cable on the far end.

This second cable is a 'receive' cable, similar to the 'receive' test cable in insertion loss testing. With this installation, the trace meets the requirement of backscatter on both sides of the connector. Thus, it is possible, though not convenient, to measure the loss of the far end connector from the near end.

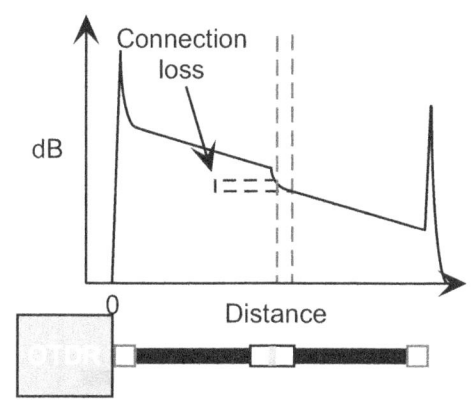

Figure 15-45: Cursors In Drop Off

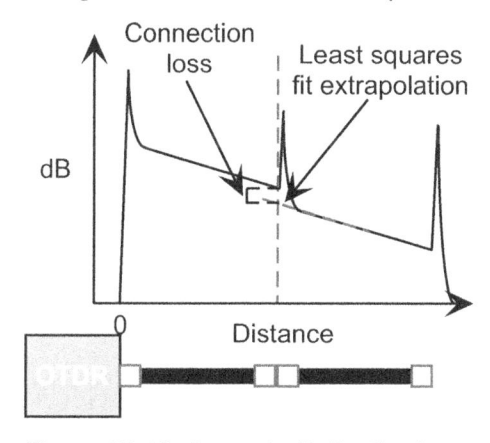

Figure 15-46: Accurate Reflective Loss Measurement

15.8.3.2.2 ACCURATE CONNECTION LOSSES

The OTDR can make accurate loss measurements, also called 'splice loss' measurements, automatically. When this measurement method is active, the computer in the OTDR performs a 'least squares analysis' (LSA) of the straight-line trace that follows the connection. The LSA is also known as a 'least squares fit', The LSA analysis results in a linear equation for the backscatter line after the connection.

The computer extrapolates the line defined by the equation back to the end of the previous segment. The computer calculates the power drop from the end of the previous segment to the extrapolation of the subsequent segment (Figure 15-46 and Figure 15-47). As this method includes no fiber attenuation in the loss value, it provides accurate loss values.

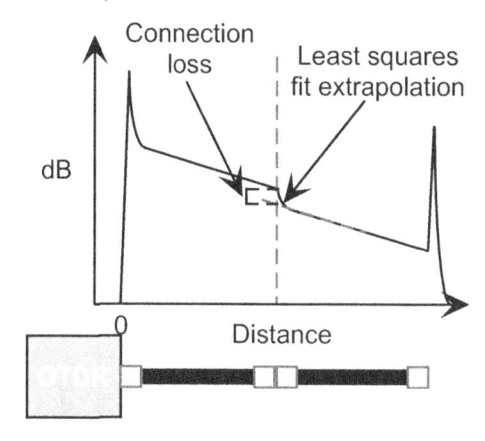

Figure 15-47: Accurate Non-Reflective Loss Measurement

This method requires that the installer place the curser near the connection. The computer in the OTDR performs the calculation, without precise curser placement. Note that this method is the same for both reflective and not reflective connections.

15.8.3.3 ATTENUATION RATE MEASUREMENTS

Attenuation rate measurements require two cursors: one each at the beginning and end of the segment. The principles for attenuation rate measurement are:

▶ Placement of two cursers as far apart as possible in the same straight-line segment (Figure 15-48)

▶ Cursers must not enclose any features such as a drop (Figure 15-49), or a peak (Figure 15-50).

The straight line trace is an important feature of an attenuation rate measurement. If a single segment does not have a straight-line trace, there is stress on the cable at the location that is not straight.

Moreover, a high attenuation rate in a segment that is not straight indicates an installation error. Finally, if the high attenuation rate occurs in a cable segment that is straight, the cause is defective cable.

> Thus, it is possible to tell the difference between installation error and defective cable.

Wide placement creates a value that is most representative of the fiber. Exclusion of drops and peaks results in accurate values.

If the backscatter trace is not straight, the attenuation rate measurement is biased higher than the actual value. In this case, the installer can make attenuation rate measurements on each side of the non-uniform loss (Figure 15-51).

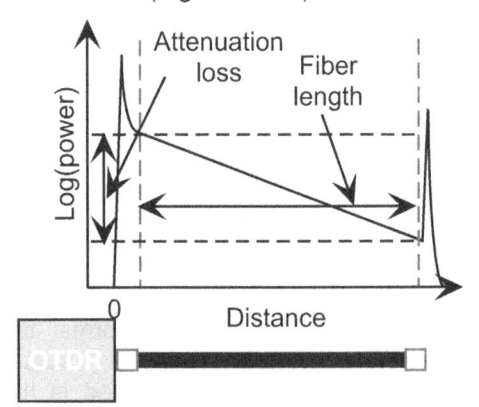

Figure 15-48: Attenuation Rate Measurement

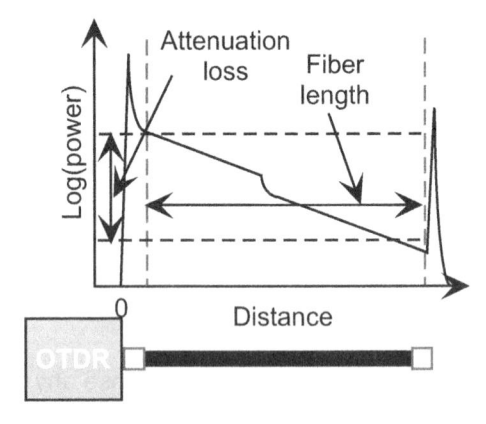

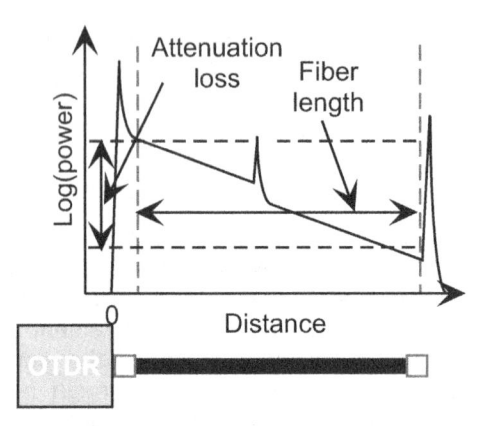

Figure 15-49: Improper Cursor Placement For Attenuation Rate Measurement

Figure 15-50: Improper Cursor Placement For Attenuation Rate Measurement

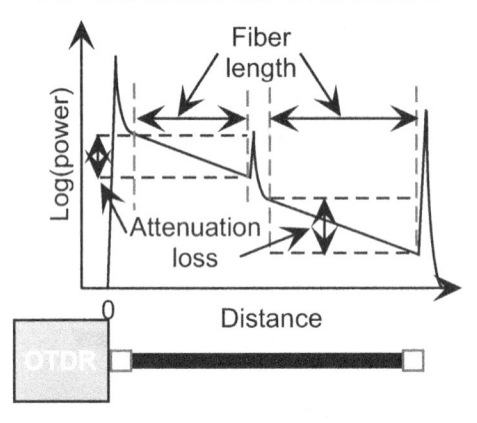

Figure 15-51: Correct Attenuation Rate Measurement With Non-Uniformity

A common error is placement of cursers in the peaks or drops that define the ends of the segment. To avoid making this error, the installer moves the two cursors slightly away from peaks or drop offs to ensure that the cursors are properly placed in the straight backscatter trace (Figure 15-52).

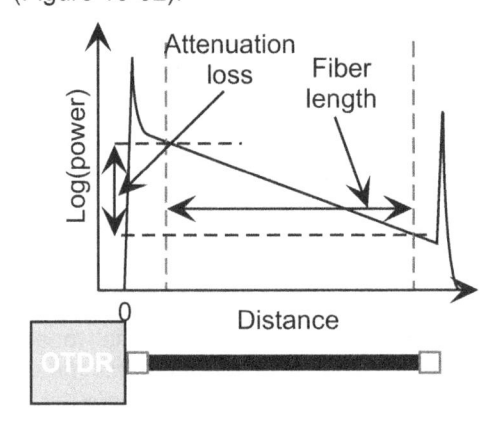

Figure 15-52: Attenuation Rate Measurement With Reduced Cursor Separation

Moving the cursors slightly away from peaks or drop offs changes the attenuation rate value slightly. However, the interpretation of the attenuation rate measurement will not change. That is, measurements made with both cursor placements will indicate the same condition, i.e., an acceptable rate or an unacceptable rate.

With the information provided thus far, you can determine whether a high attenuation rate is due to defective product or defective installation. The principles for such determination are:

> ▶ If the attenuation rate is high and the backscatter trace is straight, i.e., uniform, the problem is due to defective cable.
> ▶ If the attenuation rate is high and the backscatter trace is not straight (Figure 15-49), the problem is due to defective installation.

To understand this, remember that a properly designed, properly manufactured, and properly installed cable segment will have a uniform trace. If the trace is uniform and high, the problem is defective cable. If there is defective installation, the location of the defect is indicated by a non-uniform trace (Figure 15-23).

15.8.3.4 DIRECTIONAL DIFFERENCES

In Chapter 14, you learned that insertion loss measurements will be different in opposite directions. In 15.7.4.2, you learned that gainers result from differences in the fibers at a splice. From these two sections, it is reasonable to expect that OTDR loss measurements in opposite directions will result in different values. Such is the case. Such is the case.

15.9 FTTH/PON LINK TRACES

Thus far, all of the traces presented have point-to-point topologies. The FTTH/PON link differs, in that the FTTH/PON link is point to multi-point. This difference creates complication in interpretation of the OTDR trace in one direction.

From the central office (CO) optical network terminal (ONT) to the subscriber optical line terminal (OLT), the 'downstream' direction, the link is point to multipoint. In the reverse direction, the 'upstream' direction, the link is point to point.

In the downstream direction, light travels through the fiber, through a splitter and into multiple output fibers (Figure 15-53). The backscattered power travels from each fiber, through the splitter back to the OTDR. This travel path produces a trace like that in Figure 15-54. Note that the first section of the backscatter after the splitter includes backscatter from two fibers, while that from the second section includes backscatter from one fiber.

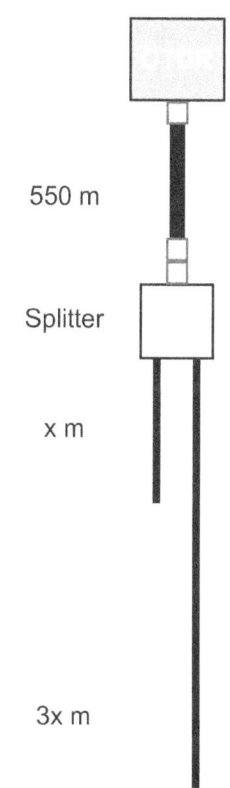

Figure 15-53: FTTH/PON Link With 1x2 Splitter

The trace in Figure 15-54 results from a simple configuration of a 1 x 2 splitter. In an FTTH/PON, the link could be a single splitter with up to 64 output fibers or multiple splitters with a total split of 64. The features of the trace are the sum of reflections and backscatter from multiple link components. The interpretation of traces from such realistic link configurations can be difficult and complex. This difficulty makes troubleshooting a subscriber problem from the CO more difficult than troubleshooting a point-to-point link.

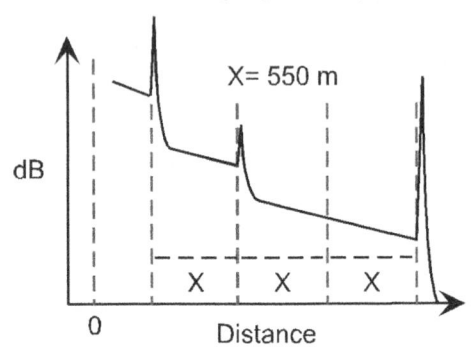

Figure 15-54: Downstream Trace Through 1x2 Splitter

In the upstream direction, the features of the trace are created by components of a single link. As such, it is possible to measure attenuation rates, connection and splitter losses. In general, it is relatively easy to troubleshoot from the subscriber OLT to the CO ONT.

15.10 OTDR INSERTION LOSS COMPARISON

It is not possible to compare or justify measurement made by these two methods. Three differences invalidate such a comparison. First, the insertion loss test simulates transmission from a transmitter to a receiver: the light travels in one direction. An OTDR test does not simulate, in that the light travels in two directions.

Second, an OTDR trace can have gainers. Gainers reduce the total loss indicated on a trace. Gainers do not occur in an insertion loss test. Thus, OTDR gainers can result in an 'insertion loss' that is less than that obtained from an insertion loss test.

Third, unless there is a launch cable on the far end of the cable under test, a trace cannot include the loss of the far end connector. Requiring a launch cable on the far end requires a second person for OTDR testing. The second person eliminates one advantage of the OTDR trace: that it can be performed for one end.

15.11 THE REAL WORLD

In the real world, traces do not always appear as they would in a training text. Figure 15-55 and Figure 15-56 provide an example of this fact and the importance of testing at both wavelengths.

Figure 15-55 shows a fiber that was carrying SONET OC-48 traffic at 1310 nm. The link length was 12,893 m. The link was operating properly. The client planned to upgrade the link to SONET OC-192 at 1550 nm. To do so, the client requested 1550 nm testing.

Figure 15-56 shows the trace for 1550 nm. The scales of the two traces are different: the reflectance at 0 m is at the OTDR. The second reflectance to the right of the first is the 101 m launch cable. The next drop is at 499 m. Note that the OTDR did not receive a reflectance from the far end at 1550 nm. Note also that there is a drop at about 200 m. This drop does not appear in the 1310 nm trace (Figure 15-55).

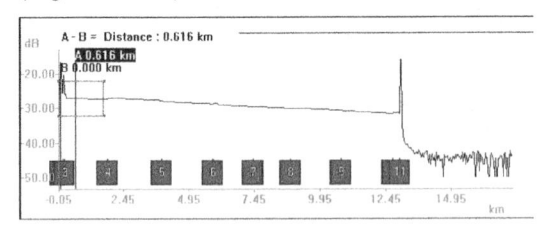

Figure 15-55: Singlemode Link At 1310 nm

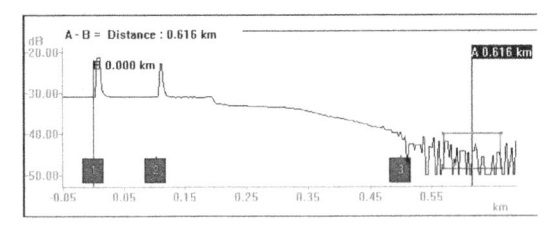

Figure 15-56: The Same Singlemode Link At 1550 nm

Stress on the fiber was low enough that it did not appear on the 1310 nm trace. However, it was high enough to prevent transmission at 1550 nm.

15.12 REVIEW QUESTIONS

1. The attenuation rate in segment A is: __ higher / __ lower than that in Segment B.

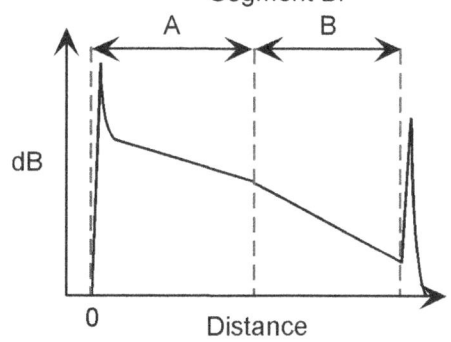

2. Why it is not possible to perform simultaneous, bi-directional communication at the same wavelength on the same fiber?

3. Select the best answer. The reflection in the circle of the figure in Question 7 could have:

   ___ One connector pair
   ___ Two connector pairs
   ___ Any number of connector pairs

4. What fact must be true to answer Question 3?

5. Can gainers occur at connectors? Justify your answer.

6. After the first test of a link, a splice exhibits high loss. Your splice installer sees the value and tells you he will go to correct this condition. What should you say to him? Justify your answer.

For Questions 7-13, multiple answers are possible. In the context of these questions, a bad cleave is a high angle cleave that reflects power outside the critical angle.

---

7. The feature(s) in the circle below could be:

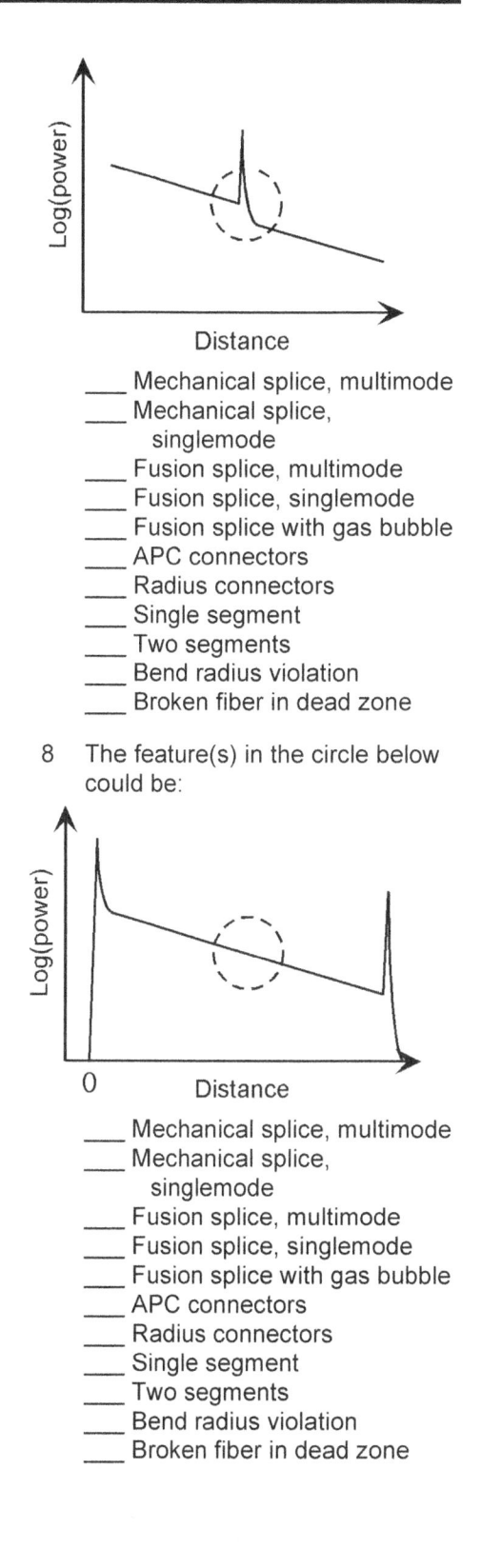

   ___ Mechanical splice, multimode
   ___ Mechanical splice, singlemode
   ___ Fusion splice, multimode
   ___ Fusion splice, singlemode
   ___ Fusion splice with gas bubble
   ___ APC connectors
   ___ Radius connectors
   ___ Single segment
   ___ Two segments
   ___ Bend radius violation
   ___ Broken fiber in dead zone

8. The feature(s) in the circle below could be:

   ___ Mechanical splice, multimode
   ___ Mechanical splice, singlemode
   ___ Fusion splice, multimode
   ___ Fusion splice, singlemode
   ___ Fusion splice with gas bubble
   ___ APC connectors
   ___ Radius connectors
   ___ Single segment
   ___ Two segments
   ___ Bend radius violation
   ___ Broken fiber in dead zone

9 The feature(s) in the circle below could be:

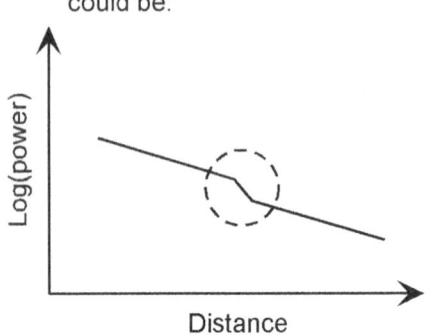

[graph: gradually declining trace with small step down in circled region]

___ Mechanical splice, multimode
___ Mechanical splice, singlemode
___ Fusion splice, multimode
___ Fusion splice, singlemode
___ Fusion splice with gas bubble
___ APC connectors
___ Radius connectors
___ Single segment
___ Two segments
___ Bend radius violation
___ Broken fiber in dead zone

10 The feature(s) in the circle below could be:

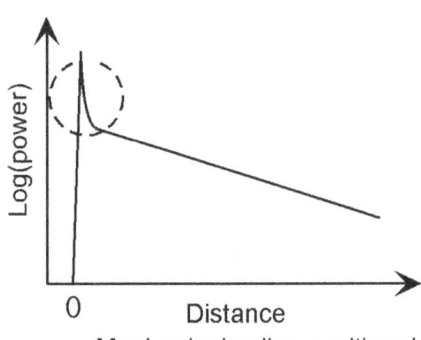

[graph: initial tall spike circled, then declining trace]

___ Mechanical splice, multimode
___ Mechanical splice, singlemode
___ Fusion splice, multimode
___ Fusion splice, singlemode
___ Fusion splice with gas bubble
___ APC connectors
___ Radius connectors
___ Single segment
___ Two segments
___ Bend radius violation
___ Broken fiber at OTDR
___ Broken fiber in dead zone

11 The feature(s) in the circle below could be:

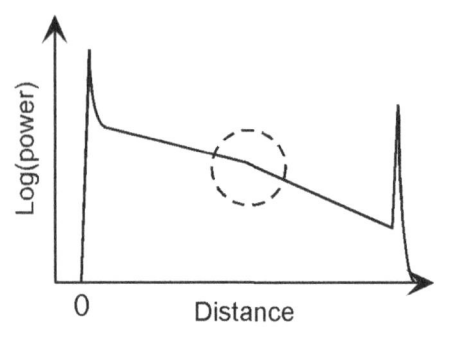

[graph: initial spike, declining trace with small feature circled mid-trace, ending with tall spike]

___ Mechanical splice, multimode
___ Mechanical splice, singlemode
___ Fusion splice, multimode
___ Fusion splice, singlemode
___ Fusion splice with gas bubble
___ APC connectors
___ Radius connectors
___ Single segment
___ Two segments
___ Mechanical splice on cable end
___ Bend radius violation
___ Broken fiber in dead zone

12 The feature(s) in the circle below could be:

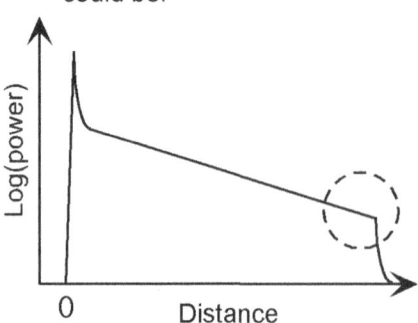

[graph: initial spike, declining trace, ending with drop at circled region at cable end]

___ Mechanical splice, multimode
___ Mechanical splice, singlemode
___ Fusion splice, multimode
___ Fusion splice, singlemode
___ Fusion splice with gas bubble
___ APC connectors
___ Radius connectors
___ Single segment
___ Two segments
___ APC connector on cable end
___ Bad cleave on cable end
___ Mechanical splice on cable end
___ Bend radius violation
___ Broken fiber in dead zone

PROFESSIONAL FIBER OPTIC INSTALLATION 15-23

13 The feature(s) in the circle below could be:

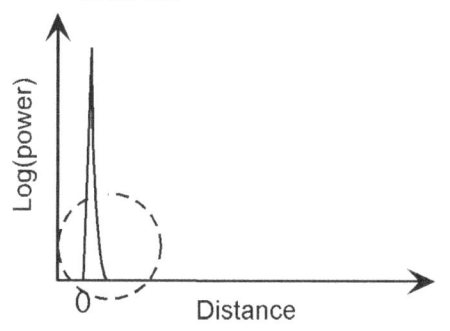

[graph: Log(power) vs Distance, showing a narrow peak near 0 with a dashed circle around the base]

___ Mechanical splice, multimode
___ Mechanical splice, singlemode
___ Fusion splice, multimode
___ Fusion splice, singlemode
___ Fusion splice with gas bubble
___ APC connectors
___ Radius connectors
___ Single segment
___ Two segments
___ APC connector on cable end
___ Bad cleave on cable end
___ Mechanical splice on cable end
___ Bend radius violation
___ Broken fiber at OTDR
___ Broken fiber in dead zone

14 State the two rules of cursor placement for attenuation rate measurement.

15 State the rule of cursor placement for estimated connection loss measurement.

16 State the rule of cursor placement for accurate connection loss measurement.

17 State the rule of cursor placement for segment length measurement.

18 Is the estimated connection loss greater than or less than the accurate connection loss?

19 Explain your answer to the previous question.

For each of the traces in Questions 20-27, identify the type of measurement being made.

Use the following codes:

X= invalid cursor placement

AR= attenuation rate measurement
AC= accurate connection loss
EC= estimated connection loss measurement
FS= first segment length measurement
SS= subsequent segment length measurement

20 The type of measurement is ____

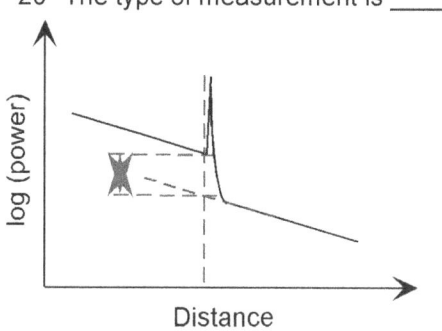

[graph: log(power) vs Distance with invalid cursor X marked]

21 The type of measurement is ____

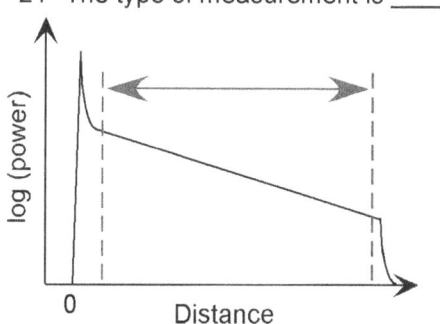

[graph: log(power) vs Distance with double-arrow across segment]

22 The type of measurement is ____

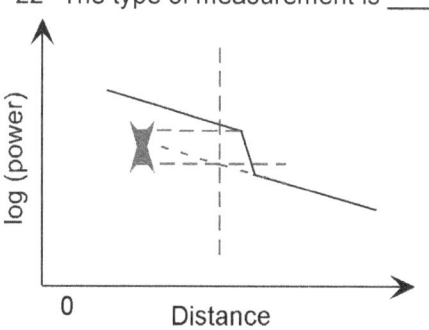

[graph: log(power) vs Distance with invalid cursor X marked at a step]

© PEARSON TECHNOLOGIES INC.

23 The type of measurement is ____

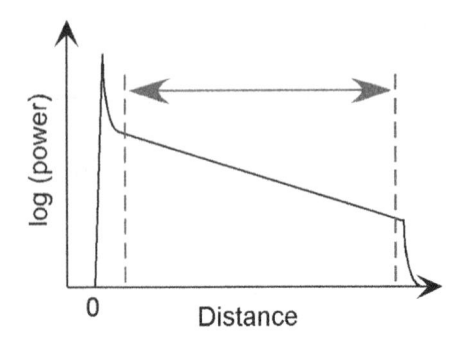

24 The type of measurement is ____

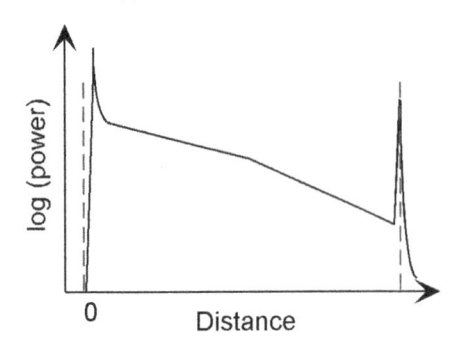

25 The type of measurement is ____

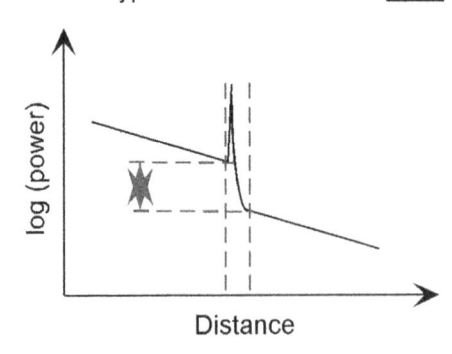

26 The type of measurement is ____

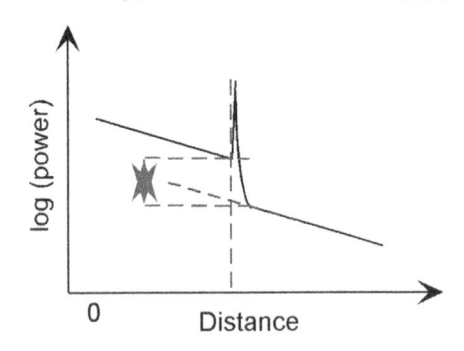

27 The type of measurement is ____

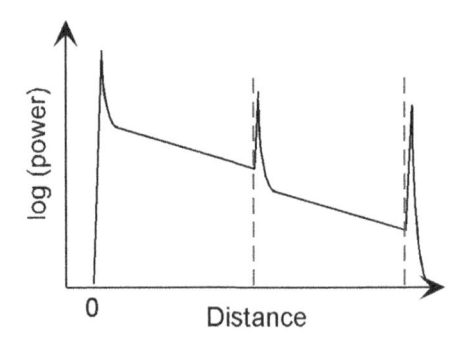

Instructions For Questions 28-32: identify the features that could be in each of the following traces. If there are specific conditions (singlemode, multimode, APC connector, fusion splice, etc.) that must be met, or locations in which the feature must be located, identify those conditions or locations. Multiple answers are possible.

28 The trace below *could* contain:

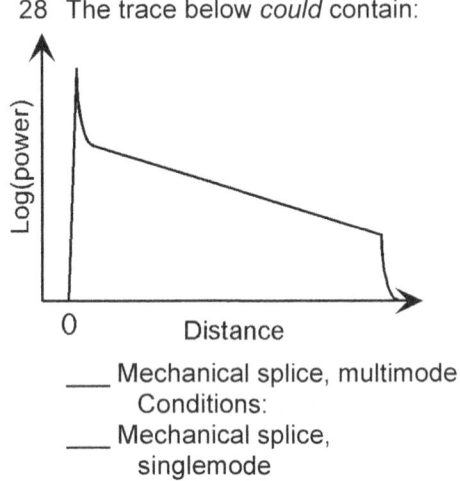

___ Mechanical splice, multimode
Conditions:
___ Mechanical splice, singlemode
Conditions:
___ Fusion splice, multimode
Conditions:
___ Fusion splice, singlemode
Conditions:
___ Fusion splice with gas bubble
Conditions:
___ APC connectors
Conditions:
___ Radius connectors
Conditions:
___ Single segment
Conditions:
___ Two segments
Conditions:
___ APC connector on cable end

Conditions:
___ Bad cleave on cable end
    Conditions:
___ Mechanical splice on cable end
    Conditions:
___ Bend radius violation
    Conditions:
___ Broken fiber at OTDR
    Conditions:
___ Broken fiber in dead zone
    Conditions:

29 The trace below *could* contain:

[Graph: Log(power) vs Distance, showing OTDR trace with two intermediate peaks]

___ Mechanical splice, multimode
    Conditions:
___ Mechanical splice, singlemode
    Conditions:
___ Fusion splice, multimode
    Conditions:
___ Fusion splice, singlemode
    Conditions:
___ Fusion splice with gas bubble
    Conditions:
___ APC connectors
    Conditions:
___ Radius connectors
    Conditions:
___ Single segment
    Conditions:
___ Two segments
    Conditions:
___ APC connector on cable end
    Conditions:
___ Bad cleave on cable end
    Conditions:
___ Mechanical splice on cable end
    Conditions:
___ Bend radius violation
    Conditions:
___ Broken fiber at OTDR
    Conditions:
___ Broken fiber in dead zone

    Conditions:
    Conditions:

30 The trace below *could* contain:

[Graph: Log(power) vs Distance, showing OTDR trace with end peak]

___ Mechanical splice, multimode
    Conditions:
___ Mechanical splice, singlemode
    Conditions:
___ Fusion splice, multimode
    Conditions:
___ Fusion splice, singlemode
    Conditions:
___ Fusion splice with gas bubble
    Conditions:
___ APC connectors
    Conditions:
___ Radius connectors
    Conditions:
___ Single segment
    Conditions:
___ Two segments
    Conditions:
___ APC connector on cable end
    Conditions:
___ Bad cleave on cable end
    Conditions:
___ Mechanical splice on cable end
    Conditions:
___ Bend radius violation
    Conditions:
___ Broken fiber at OTDR
    Conditions:
___ Broken fiber in dead zone

31 The trace below *could* contain:

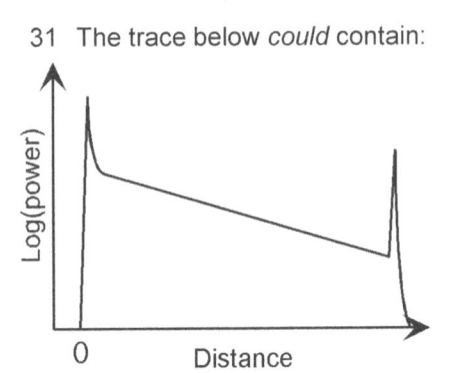

[Graph: Log(power) vs Distance, showing a typical OTDR trace with initial reflection peak, sloping backscatter, and end reflection peak]

___ Mechanical splice, multimode
    Conditions:
___ Mechanical splice, singlemode
    Conditions:
___ Fusion splice, multimode
    Conditions:
___ Fusion splice, singlemode
    Conditions:
___ Fusion splice with gas bubble
    Conditions:
___ APC connectors
    Conditions:
___ Radius connectors
    Conditions:
___ Single segment
    Conditions:
___ Two segments
    Conditions:
___ APC connector on cable end
    Conditions:
___ Bad cleave on cable end
    Conditions:
___ Mechanical splice on cable end
    Conditions:
___ Bend radius violation
    Conditions:
___ Broken fiber at OTDR
    Conditions:
___ Broken fiber in dead zone

32 Explain two ways you can use an OTDR to measure the loss of a connector on the far end of a link.

33 Will the directional difference in a FTTH/PON link be larger or smaller than that in a data communication link? Explain your answer.

34 What wavelength is not used for communication? Explain your answer.

35 Testing of a FTTH/PON link is performed at ___ wavelengths. List the wavelengths.

36 An FTTH/PON link feeds a splitter with four output ports. The output ports are fused to four output fibers (Figure 13-88). The four output fibers have lengths longer than the dead zone of the OTDR, which is approximately 50 m. The four output fibers have lengths of 550 mm, 1100 m, 1650 m and 2200 m. The four output fibers have reflective connectors. Assume that there is a launch cable of 550 m connected to the input of the splitter. The launch cable has a APC connector on the end connected to the splitter. Assume that the cable has been installed correctly. Using Figure 15-58 to Figure 15-61, draw the OTDR traces you would expect to see.

Hint: draw the trace with the 2200 m leg. Then draw the trace with the 2200 and 1650 m legs, etc.

37 What four features or characteristics does the OTDR identify that no other piece of test or inspection equipment can identify?

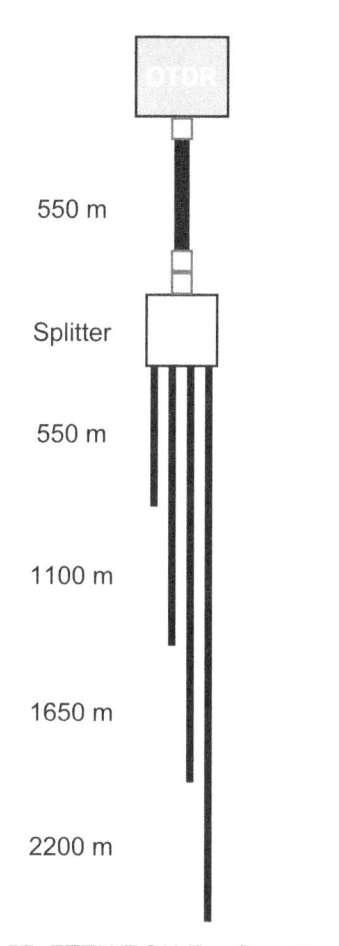

Figure 15-57: FTTH/PON For Question 37

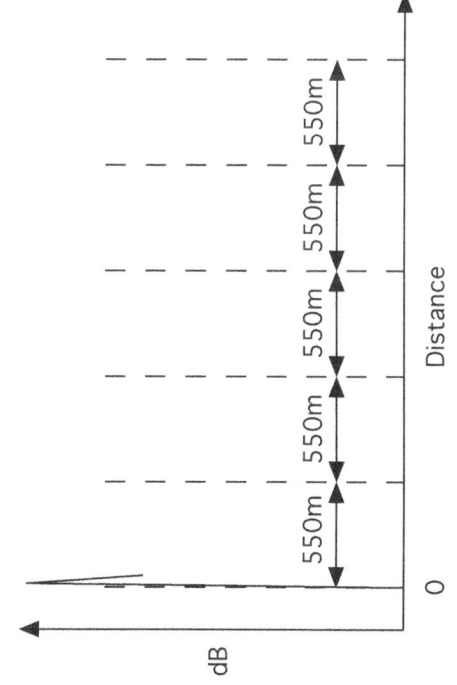

Figure 15-58: Blank Trace For Question 38, 2200 m Leg

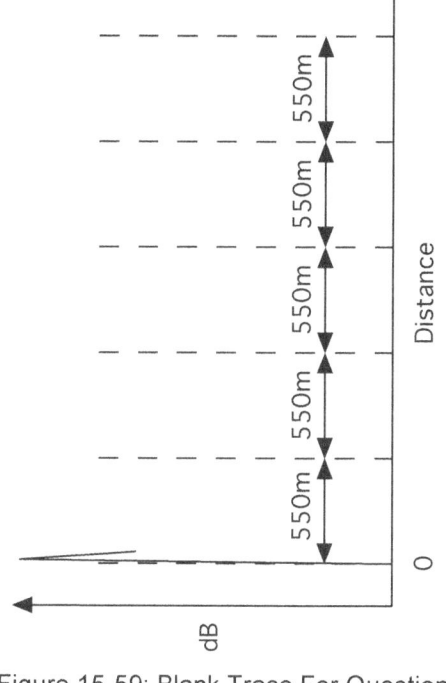

Figure 15-59: Blank Trace For Question 38, 2200 m and 1650 m Legs

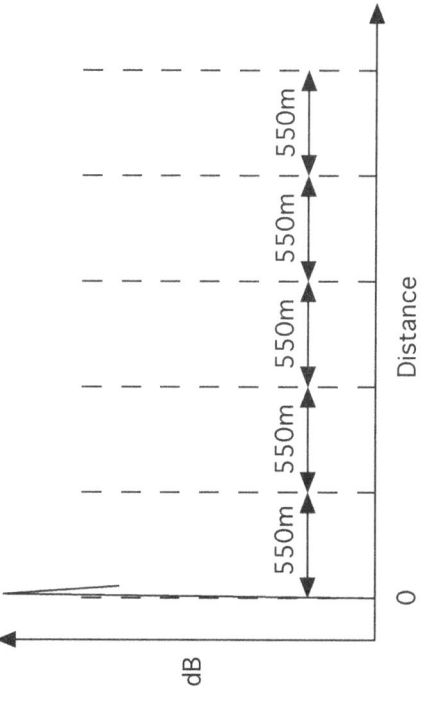

Figure 15-60: Blank Trace For Question 38, 2200 m, 1650 m and 1100 m Legs

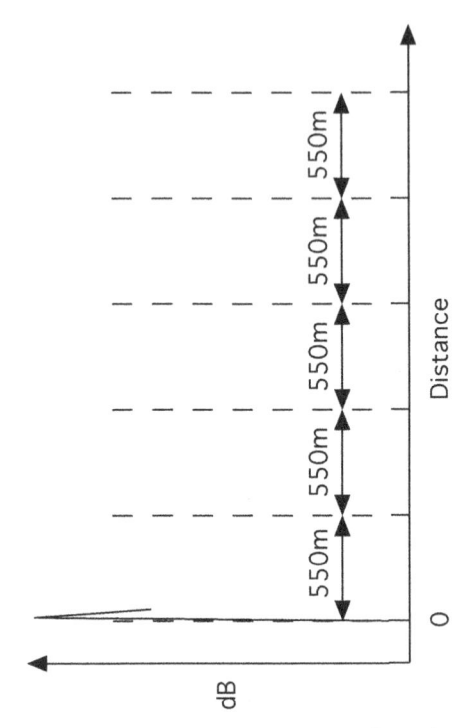

Figure 15-61: Blank Trace For Question 38, 2200 m, 1650 m, 1100 m and 550 m Legs

16 REFLECTANCE AND ORL- PRINCIPLES AND METHODS

Chapter Objectives: you learn how to make and interpret reflectance and optical return loss (ORL) tests.

16.1 INTRODUCTION

16.1.1 REFLECTANCE

The installer tests reflectance to ensure signal accuracy (5.4.5). Excessive connector reflectance can create multiple pulses at the receiver for a single transmitter pulse.

The installer tests reflectance on pigtails and short patch cords. The installer performs reflectance testing on pigtails prior to their installation on the ends of the main cable. The installer does not perform reflectance testing of connectors installed on a long length of cable because the backscatter from the fiber between the cable ends will result in a high reflectance measurement.

As a practical matter, installers perform reflectance tests on singlemode connectors but not on multimode connectors. There is limited field test equipment for such multimode testing. Singlemode field test equipment (Figure 16-1) is common.

The approved reflectance test method is FOTP-107. The installer uses FOTP-107 to test the reflectance of connectors on patch cords or pigtails.

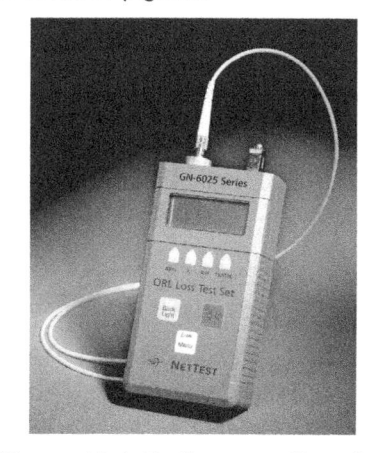

Figure 16-1: Reflectance Test Set (Courtesy of NetTest)

16.1.2 OPTICAL RETURN LOSS

Optical return loss (ORL) is a measurement of the total optical power returning to the input end of the fiber. excessive, this power can interfere with accurate transmission.

ORL is the sum of power backscattered from the core (15.3) and reflected backwards from connections (5.4.5). In other words, it is the sum of reflectance and backscatter. Fortunately, both reflectance and ORL are measured with the same test unit.

16.2 REFLECTANCE PRINCIPLES

By definition (Equation 5-1), reflection is a measurement of the ratio of incident power to reflected power. By this definition, a reflectance test requires two power measurements. In addition, the reflectance test equipment may have internal reflections that need to be removed from the reflected power calculation. Reflectance values are negative.

16.3 REFLECTANCE TESTING

16.3.1 EQUIPMENT REQUIRED

The FOTP-107 reflectance test requires:

- A reflectance test set
- A the least two low reflectance reference leads
- Low loss, clean barrels
- A mandrel with a diameter appropriate to the wavelength of the test
- Connector cleaning supplies

The reflectance test sets consists of a high-power, stable, laser diode, a high sensitivity detector, and three directional couplers (Figure 16-2). The test set must have a high-power laser in order to create measurable reflected signal strength. For example, a 200-microwatt laser diode and a -60 dB reflectance connector will result in a reflected signal strength of 200 nanowatts, an extremely low power level. This low power level requires a high sensitivity detector in the test set.

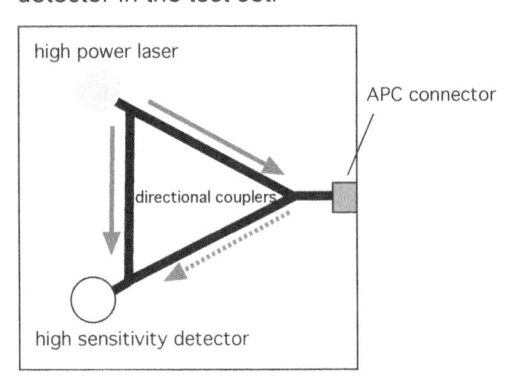

Figure 16-2: Structure Of Reflectance Test Set

The reference lead needs an APC connector on the end at the test set and a low reflectance connector on the other end. The APC connector reflects no power into the test set.

The low reflectance connector is of the type to be tested, e.g., ST-compatible, SC, LC. The second connector of the reference lead should have reflectance that is 5 to 10 dB below the acceptance value. With a high reflectance reference connector, the installer will reject low reflectance connectors.

16.3.2 PROCEDURE

The reflectance test has six steps:

- Clean all connectors
- Set a reference power level
- Null out power from internal reflections in the test set
- Check the reference lead against a known low reflectance reference lead
- Connect the reference lead to the connector to be tested
- If required, wrap the opposite end of the jumper around a mandrel

16.3.2.1 CONNECTOR CLEANING

The installer cleans all connectors to be tested and both connectors on the reference lead. This cleaning needs to be done without leaving residue. Isopropyl alcohol can leave such residue, which will result in high reflectance measurements.

16.3.3 SET UP

The installer turns on the test set and allows it to stabilize. With this set up, the test equipment determines the power delivered to the low reflectance reference connector. The installer connects a reference lead to the test set and wraps the opposite end of the reference lead around a mandrel (Figure 16-3). The mandrel is not required for APC connectors.

The mandrel removes light that is reflected from the opposite end. The mandrel has a diameter of 8 mm for 1310 nm testing and 10 mm for 1550 nm testing With this set up, the reflectance test set 'nulls out' any power from internal reflections.

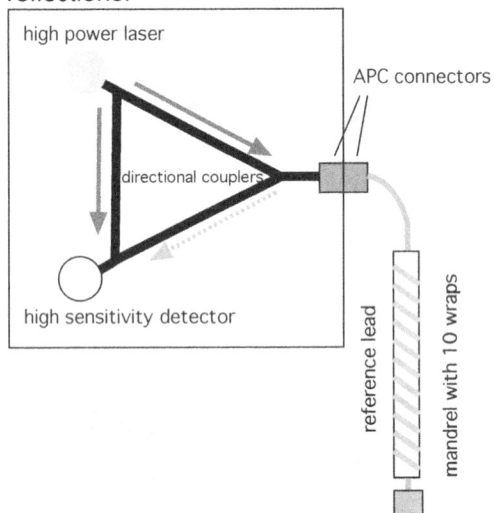

Figure 16-3: Reflectance Reference Configuration

16.3.4 VERIFY LOW REFLECTANCE

With use, reference leads develop damage and high reflectance. Before performing any tests, the installer tests the low reflectance reference lead with a known, low reflectance 'master' reference lead. This test verifies the low reflectance of the reference lead.

To verify low reflectance, the installer connects two reference leads together and wraps the far end of the second reference lead on a mandrel (Figure 16-5). The reflectance value should be lower than the acceptance value. For example, if the acceptance value is -50 dB, the reflectance value should be below -55 dB. The test shown in Figure 16-4 results in qualification of both reference leads.

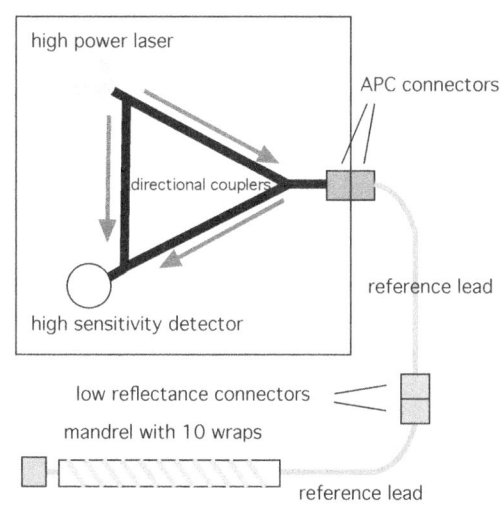

Figurte 16-4: Verification of Low Reflectance Reference Leads

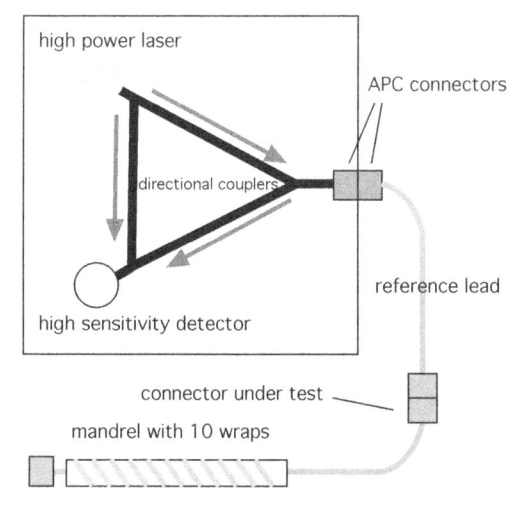

Figure16-5: Reflectance Test

16.3.5   TEST REFLECTANCE

To test reflectance, the installer connects the connector under test to the reference connector. If the connector on the opposite end of the cable under test is not an APC connector, the installer wraps that opposite end 10 times around a mandrel (Figure 16-5). If the connector on the opposite end of the cable is an APC connector, the installer need not wrap the opposite end.

16.4   ORL TESTING

ORL is defined in a manner similar to reflectance, except that the power ratio is the reciprocal of the reflectance ratio. As a result, ORL values are positive. The larger the ORL value is, the more accurate the transmission will be.

10 Log (incident power/ reflected power)

Equation 16-1

ORL testing is performed in a manner similar to both reflectance testing and one lead insertion loss testing: ORL testing requires a reference lead on both ends of the cable under test.

ORL is performed with a reflectance test set and two low reflectance test leads. The leads are qualified as in Figure 16-4. The test is performed as in Figure 16-6.

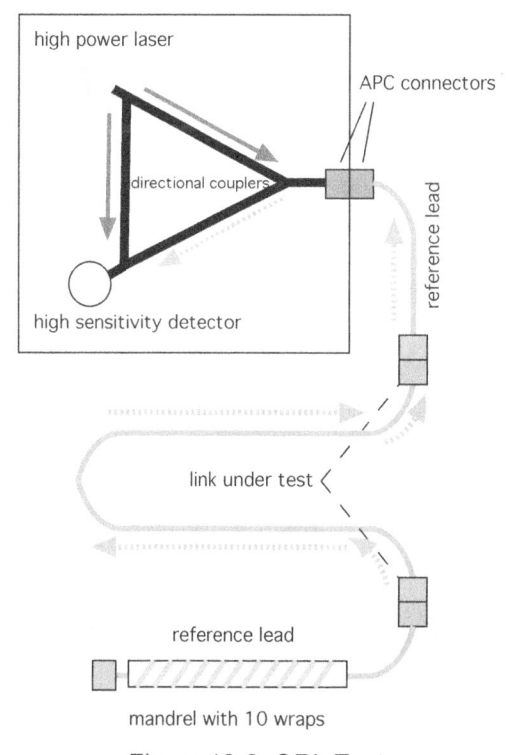

Figure 16-6: ORL Test

The reference lead on the far end of the cable allows removal of the high reflectance that occurs from an 'open' connector. Of course, the far end reference lead is unnecessary when the far end connector is APC.

16.5 INTERPETATION

Interpretation of both reflectance and ORL tests is simple: the measured power must be less than a maximum value. The difference results from the difference in the sign of the value. Reflectance maximums are stated in – dB. The measured value must be less, or more negative, than the requirement.

ORL minimums are positive. Thus, the measured value must be greater, or more positive, than the requirement.

A high reflectance connector, either due to damage or dirt, often creates a high reflectance value and high ORL value from that end. The test results from the opposite end of the link can be low and acceptable.

16.6 FOTP-8

The FOTP-8 standard describes a method of measuring reflectance with an OTDR. It describes an equation that can be used with measurements made from the trace. If this method is automated in software and the software properly complies with the requirements of the peak, this method should work.

However, there is a key requirement, that the peak not be flat, which is not met with the usual OTDR set up. Paragraph 5.6.2 states:

> "Ensure that the receiver is not saturating here; this may be done by attenuating the signal with an OTDR control, by an in-line variable attenuator, or by introducing a loop prior to the splice or connector.".

Such a test requires setting up the OTDR specifically for this measurement.

16.7 STANDARDS

The Building Wiring Standard, TIA/EIA-568-C, requires measurement of reflectance by FOTP-107.

The TIA/EIA-568-C reflectance requirements are:

- Multimode connectors: ≤ -20db
- Singlemode connectors: ≤ -26 dB
- Analog CATV applications: ≤ -55 dB

16.8 REVIEW QUESTIONS

1. True or false: a -45 dB connector has a lower reflectance than a -60 dB connector.

2. True or false: a -45 dB connector has better reflectance than a -60 dB connector.

3. True or false: dirt on a connector results in a reduced reflectance.

4. True or false: reflectance testing is performed on multimode connectors only.

5. Is the diameter of the mandrel for testing at 1550 nm larger than that for 1310 nm?

6. Explain your answer to Question 4. Hint: see 14.5

7. A singlemode link is 215,000 feet (65.5 km) long. It has radius connectors. You and your assistant perform ORL tests. The test results are all greater than 30 dB, which is your maximum allowed value. These results were typical of links with lengths between 10 and 40 km. The day after testing, your assistant states: "You know, I forgot to install the mandrel during the tests we made yesterday." You have two choices: use the results from yesterday or retest. What do you do? Justify your answer.

8. You have completed an OTDR test on a link. It is a 10 km, singlemode link like that in Figure 16-6 that has an APC connector

on the far end. To make an ORL test, do you need to send a technician to the far end?

9. The connectors on your link have radiussed ends. If reference lead with the mandrel in Figure 16-6 were removed, what would happen to the ORL value?

10. The connectors on your link in Figure 16-6 are APCs. If the mandrel wrap in Figure 16-6 were removed, what would happen to the ORL value?

11. You measure the ORL of a link from both ends, end A and B. The ORL value from B is much higher than that from end A. What one conclusion might you draw?

17 DISPERSION TESTING PRINCIPLES AND METHODS

Chapter Objectives: you learn the principles and methods of testing of chromatic dispersion and polarization mode dispersion.

17.1 INTRODUCTION

Dispersion testing is testing of chromatic dispersion (CD) and polarization mode dispersion (PMD). Dispersion testing is performed in two situations:

- Singlemode links that are to be upgraded from OC-48, (2.488 Gbps), to OC-192 (9.953 Gbps)
- Singlemode data links upgraded from 1 Gbps to Ethernet 10Gb Ethernet (10GBASE-LX, 12.5 Gbps), 40 Gb or 100 Gb Ethernet.

There are three reasons for this testing. First, such testing was rarely performed on singlemode links after installation from 1083-1999. Such testing was not necessary for the initial bit rates of 2.5 Gbps and below. Such testing was not deemed necessary, as PMD did not interfere with transmission at these rates.

Second, fibers made in the 1980s and early 1990s were not made with a goal of minimal PMD. As a result, some of these fibers had relatively poor PMD performance.

A third reason for testing is to qualify links for future upgrades to bit rates of 40 Gbps and 100 Gbps. Such links will have significantly reduced allowable values for both CD and PMD.

For the purpose of this chapter, we redefine the term 'chromatic dispersion' to be the total of material, waveguide, and chromatic dispersion, as presented in 2.3.3. Separate measurement of these three types of dispersion is unnecessary and impractical. Thus, the first principle of CD measurement is:

▶ Measured CD is the sum of chromatic, waveguide and material dispersions

17.2 BACK TO BASICS

As stated (2.3.3), dispersion is one of the two characteristics that limit transmission distance by reducing signal accuracy. The limiting parameter is the bit rate: the width of the time interval at the receiver limits the total pulse width of the optical pulse at the receiver.

The following presentation does not reflect the actual principles that determine the design of receiving optoelectronics. Rather, this presentation is an easy-to-understand explanation of the significance of dispersion.

OC-48 transmission is an example. At this transmission rate, the width of the time interval is 400 ps. If the designer places a guard band on this interval, the actual interval is smaller. For the purpose of simplifying this presentation, we set the actual interval at 90% of the total interval. Thus, the width of the pulse at the receiver is 360 ps.

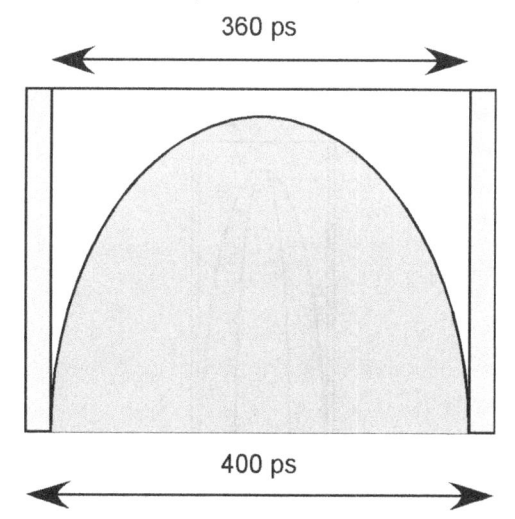

Figure 17-1: Time Interval For OC-48

But this width results from three factors:

- Initial pulse width
- PMD

➤ Chromatic dispersion, as redefined in 17.1

We will ignore the initial pulse width, as it can be below 1 ps. Chromatic, material and waveguide dispersions are deterministic, in that they are constant with time. Thus, PMD is the total dispersion less this third factor (Figure 17-2).

As Figure 17-2 Indicates, PMD is a small part of the total dispersion allowable at OC-48 (~2.5 Gbps). When we upgrade a link to OC-192 (~10 Gbps), the time interval is reduced by 75 % (Figure 17-3). At this rate, PMD becomes a significant part of the dispersion allowed. Another consequence is a significant reduction in the chromatic dis-persion (Figure 17-3).

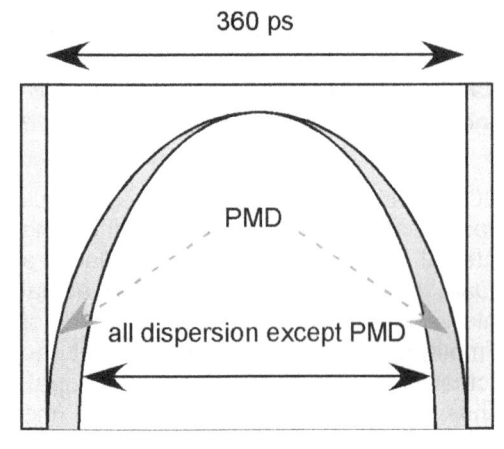

Figure 17-2: Total Dispersion, (CD And PMD) For OC-48

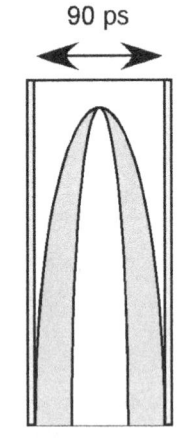

Figure 17-3: Total Dispersion, (CD And PMD) At OC-192

17.3 CHROMATIC DISPERSION

17.3.1 REASON FOR TEST

There are two reasons for this test. The first is that an upgrade to a higher bit rate requires a reduction in the allowable amount of chromatic dispersion (Figure 17-2 and Figure 17-3). The second is that the calculation of link CD is tedious.

Such a calculation would require knowledge of the actual chromatic dispersion slopes (Figure 3-13) and segment lengths for all fibers in a link. For example, a link of 215,000 feet could contain 32 different fibers. These fibers could be from different manufacturers. Each of these fibers could have a slightly different dispersion rate. In short, the actual dispersion of a link could be significantly different from the calculated value. As the calculation is tedious, a five-minute measurement is preferred.

17.3.2 TEST METHODS

Chromatic dispersion is measured over a wavelength range of 100 nm. This leads to the principle:

▶ CD test methods require testing at multiple wavelengths.

The data produced are fitted to an equation appropriate to the fiber under test. From this equation, the CD is determined. This equation creates the graph shown in Figure 3-13 through Figure 3-15.

CD testing can be performed with equipment at one end or at both ends. Measurement with equipment at one end is based on OTDR technology. The OTDR measures the round trip travel time of pulses at different wavelengths. These values are fitted to an appropriate equation. This equation defines the dispersion slope and the zero dispersion wavelength. With this information, the dispersion at any wavelength is calculated. This method is a 'spectral group delay in time domain' method.

This method requires end reflectance. This requirement limits testing to links without APC connectors. Links with APC connectors can be tested by adding a cable without an APC connector to the far end.

Measurement with equipment at both ends is based on one of three methods

- Phase shift method defined by ITU-T G650.1 and IEC 60793-1-42
- Spectral group delay in time domain method
- Differential phase shift defined by ITU-T G650.1 and IEC 60793-1-42

The details of these methods are beyond the scope of this text. For additional information, see the excellent presentation in the reference in 17.5.

17.3.3 TEST RESULTS

Table 17-1 presents the measured CDs at 1550 nm. The test equipment was an EXFO Model 5700. This equipment uses OTDR technology. The test method was FOTP-175B.

The fibers were G.652 fibers installed between approximately 1985 and 1995. In this table, we include the 1550 nm calculated maximum dispersion (CCD). We performed this calculation with a value of 17.405 ps/nm/km. This value results from the dispersion slope of 0.092 ps/nm^2/km. This slope is the current maximum slope for G.652 fiber. These fibers were from at least three manufacturers.

In this table:

- CD is the measured chromatic dispersion, in ps/nm.
- L is the fiber length, in km
- CCD is the calculated maximum chromatic dispersion, in ps/nm
- Ratio is the ratio of measured to calculated chromatic dispersion

Note that most fibers have measured CD that is less than, but close to, the maximum calculated value. However, six, or 15%, of the fibers have CD values appreciably greater than the maximum.

	CD ps/nm	L, km	CCD ps/nm	Ratio
1	218.99	13.141	228.72	95.7%
2	216.71	13.141	228.72	94.7%
3	218.79	13.152	228.90	95.6%
4	215.91	13.154	228.94	94.3%
5	382.86	22.730	395.61	96.8%
6	385.67	22.820	397.19	97.1%
7	385.88	22.834	397.43	97.1%
8	385.57	22.891	398.42	96.8%
9	642.43	23.545	409.80	156.8%
10	640.16	23.550	409.88	156.2%
11	408.24	24.350	423.82	96.3%
12	407.54	24.352	423.84	96.2%
13	411.74	24.362	424.02	97.1%
14	410.90	24.366	424.09	96.9%
15	640.16	26.953	469.11	136.5%
16	642.43	26.954	469.13	136.9%
17	533.24	31.632	550.55	96.9%
18	532.60	31.632	550.55	96.7%
19	522.08	31.637	550.64	94.8%
20	523.05	31.640	550.69	95.0%
21	596.22	35.411	616.33	96.7%
22	596.22	35.416	616.42	96.7%
23	596.22	35.426	616.59	96.7%
24	624.79	37.463	652.05	95.8%
25	624.59	37.468	652.13	95.8%
26	624.35	37.473	652.22	95.7%
27	596.22	37.473	652.22	91.4%
28	596.22	37.473	652.22	91.4%
29	640.16	38.082	662.81	96.6%
30	642.43	38.087	662.90	96.9%
31	642.43	43.279	753.27	85.3%
32	640.16	43.289	753.45	85.0%

Table 17-1: Chromatic Dispersion Test Results

17.4 PMD

17.4.1 CAUSE

In an ideal fiber, the stress state will be uniform with respect to the x- and y-axes (Figure 17-4). In this state, the material density and the RI will be the same in both axes.

However, the stress state in all fibers is not uniform. Non-uniformity results from three sources:

- Fiber imperfections from the manufacturing process

- Non-uniform stress imposed on the fiber by the cable structure (i.e., twisting)
- Non-uniform stress imposed on the cable by the environment

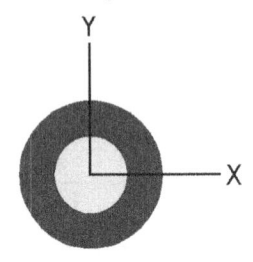

Figure 17-4: Orthogonal Stress State in Ideal Fiber

Fiber manufacturing results in core offset and core non-circularity (3.3.3). Both conditions produce non-uniform stresses.

Cable manufacturing imposes stress on the fiber. In MFPT cables, the excess fiber length results from the fiber being coiled in the buffer tube. In both loose tube and tight tube cable designs, the buffer tube spirals around the cable core. This spiraling imposes additional stress on the fiber.

Environmental stresses include temperature-induced changes in dimensions, vibration, and shock. These stresses can, and do, change with time. For example, an aerial cable will have different stresses on windy and windless days. As a second example, a cable buried near a railroad track may have no vibrational stresses unless a train is running. Thus, PMD is not constant. Finally, note that these three factors increase with increasing transmission distance.

Since this uniform stress state does not exist, light traveling in the fiber will exhibit different IRs in the orthogonal axes. This behavior is known as 'bi-refringence'.

When light travels in a singlemode fiber, the light behaves as an energy field. Remember the mode field diameter? This field is described as having a vector. This vector can be resolved into two perpendicular components. These components are known as the 'fast axis' (X) and 'slow axis' (Y) (Figure 17-5) and the principle states of polarization (PSP).

If the fiber is ideal (Figure 17-4), the fast and slow axes have the same IR. However, no fiber is ideal. The fast and slow axes have unequal IRs. Since the IRs are not the same in both axes, the portions of the energy field in the two axes will travel at different speeds and arrive at the fiber end at different times. This is PMD dispersion.

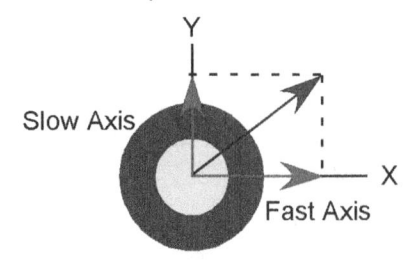

Figure 17-5: Orthogonal Stress State In Real Fiber

Since PMD varies with time, the measured value must be a time average. Thus,

▶ PMD is a time-averaged value

Since links must be designed to work with a maximum dispersion, the average must be converted to a statistical maximum. Thus, we have another PMD measurement principle:

▶ The measured average PMD can be converted to a maximum dispersion

The distribution of PMD has been shown to have conform to the mathematical equation for a 'Maxwellian' distribution. The maximum dispersion depends on the statistics of the Maxwellian distribution.

To define a PMD that is meaningful, we need define a probability. The probability of concern is the percentage of time that the link has a PMD in excess of the maximum. We shall call this the 'outage' probability. This probability depends on the data format.

17.4.2 LIMITS

For SONET transmission, the outage probability is 10^{-5}. This value corresponds to 99.999% availability. With this value, and a maximum possible PMD of 30 ps, we can back-calculate an average PMD. This average is 10 ps. This is the limit applied to links that are to transmit SONET OC-192 signals.

For 10 Gb Ethernet transmission, the outage probability is 10^{-7}. This value corresponds to 99.99999% availability. With this value, and a maximum possible PMD of 19 ps, we can back-calculate an average PMD. This average is 5 ps. This is the limit applied to links that are to transmit 10 Gb Ethernet signals. Table 17-2 presents these values.

	PMD Ave.	PMD Max.	Outage Prob.	CD ps/nm
OC-192	10 ps	30 ps	10^{-5}	1176
10 GBE	5 ps	19 ps	10^{-7}	738

Table 17-2: Summary Of PMD Limits For OC-192 And 10 Gb Ethernet

17.4.3 TEST METHOD

▶▶All PMD test methods require testing at multiple wavelengths

When PMD is added to CD on long links carrying 10 Gbps signals, the PMD can cause erratic accuracy problems. Since the PMD is statistical in nature, intermittent low PMD will enable accurate transmission, while Intermittent excessive PMD will result in signal inaccuracy.

This author has observed this statistical nature. During testing of telephone links, this author made PMD tests on a link that went under a railroad track. One test was with no activity of the track; the second was with a train moving over the cable. The result was predictable: the PMD value was higher with the train than without the train.

17.4.4 TEST RESULTS

Table 17-3 presents the results of field tests of PMD at 1550 nm The test equipment was an EXFO Model 5700. This equipment uses OTDR technology. The test method was FOTP-243. The fibers were G.652 fibers installed between approximately 1985 and 1995. We calculated the 1550 nm maximum PMD with a value of 0.1 ps/(/km)$^{0.5}$. This value is the current maximum value for G.652 fiber.

In this table:

- PMD is the measured dispersion
- L is the fiber length
- CPMD is the calculated PMD
- Ratio is the ratio of measured to calculated PMD

Note that 19 of the 33 fibers (57.5%) had PMD in excess of 100% of the maximum value for current fibers. This result is not surprising, in that these fibers were manufactured before PMD limits were established.

	PMD ps	L, km	CPMD ps	Ratio
1	0.11	13.141	0.363	30%
2	0.06	13.141	0.363	17%
3	0.13	13.152	0.363	36%
4	0.08	13.154	0.363	22%
5	0.08	22.730	0.477	17%
6	0.18	22.820	0.478	38%
7	0.06	22.834	0.478	13%
8	0.19	22.891	0.478	40%
9	3.29	23.545	0.485	678%
10	5.47	23.550	0.485	1127%
11	0.11	24.350	0.493	22%
12	0.13	24.352	0.493	26%
13	0.24	24.362	0.494	49%
14	0.23	24.366	0.494	47%
15	5.47	26.953	0.519	1054%
16	3.29	26.954	0.519	634%
17	0.50	31.632	0.562	89%
18	0.48	31.632	0.562	85%
19	2.17	31.637	0.562	386%
20	2.66	31.640	0.562	473%
21	4.36	35.411	0.595	733%
22	4.36	35.416	0.595	733%
23	4.36	35.426	0.595	733%
24	6.49	37.463	0.612	1060%
25	6.72	37.468	0.612	1098%
26	6.80	37.473	0.612	1111%
27	4.36	37.473	0.612	712%
28	6.80	37.473	0.612	1111%
29	4.36	37.473	0.612	712%
30	5.47	38.082	0.617	886%
31	3.29	38.087	0.617	533%
32	3.29	43.279	0.658	500%
33	5.47	43.289	0.658	831%

Table 17-3: PMD Test Results

17.5 REFERENCE

Reference Guide To Fiber Optic Testing, Volume 2, JDSU, B. Collings et al, 2010.

17.6 REVIEW QUESTIONS

1. What four factors determine the amount of chromatic dispersion? Hint: review 2.3.3, 2.3.3.2, and 3.4.3.3.

2. What four factors determine PMD?

3. True or false: PMD is constant with time.

4. True or false: CD is constant with time.

5. True or false: the amount of allowable CD is higher at 10 Gbps or OC-192 than at 2.5 Gbps or OC-48.

6. True or false: the amount of allowable PMD is higher at 10 Gbps or OC-192 than at 2.5 Gbps or OC-48.

7. True or false: PMD is more important at OC-48 than at OC-192.

8. An OC-192 link has a cable in a conduit under the runways. The airport is busy, with landings every few minutes. The network has monitoring equipment. This equipment monitors the real throughput on this link. This equipment indicates that the real throughput is much lower than the OC-192 rate. However, from midnight to 5 am, this equipment indicates OC-192 throughput. What is the most likely cause of this problem?

18 OTHER TESTS AND EQUIPMENT

Chapter Objectives: you learn of other tests and equipment.

18.1 INTRODUCTION

Installers perform tests and inspections and use equipment other than those in Chapters 14-18 and 20. The tests are:

- Protocol tests
- Optical return loss

In addition, the installer may use equipment other than that previously mentioned. Such equipment includes:

- Fiber identifier
- Visual fault locator
- Attenuator
- Traffic detector
- Talk set
- Microscope

18.2 OTHER TESTS

18.2.1 PROTOCOL TESTING

In data networks, testing of loss, OTDR, and reflectance is sufficient to indicate proper performance and installation with high reliability. On the other hand, FTTH networks, which carry voice, data, and video information streams, may exhibit reduced performance due to effects of the three wavelengths present.

For example, the combined reflectance of the downstream wavelengths of 1490 nm and 1550 nm may create reflectance sufficient to result in inaccurate transmission, even though the reflectance of each wavelength is sufficiently low. Because of this potential, FTTH networks can require testing at the protocol level. Protocol testing is of the electrical signal. The installer uses a protocol analyzer, such as those from EXFO, for such testing.

18.2.2 BACK REFLECTION

In the past, back reflection was used to indicate reflectance of a component other than a connector. At the time of this of this writing, the term back reflection is not commonly used. The test method is the same as that for reflectance (16).

18.3 OTHER EQUIPMENT

18.3.1 FIBER IDENTIFIER

The fiber identifier is light source that the installer places on the end of a fiber. This tool enables installers to identify fibers from opposite ends and to verify continuity. This source can be white light or red light. This tool is a low power device.

18.3.2 VFL

A visual fault locator or visual fault finder (VFL), or is a high power visible red laser source. In general, the fault locator is used to find faults in short lengths of fiber or cable. This tool is most useful when the OTDR indicates a dead zone problem.

When the installer places the VFL on a fiber, any light that escapes from the core will be visible. The installer uses this locator to find breaks in a tight tube cable, in a splice tray or splice enclosure, loss in a mechanical splice, loss in some cleave and crimp connectors, and bend radius violations in splice enclosures.

While most fault locators are not useful on a cable with a dark or thick jacket, they are useful for finding breaks and excess loss in; light colored indoor cables, 900 µ buffer tubes, and near cable ends.

This tool is essential for determining high loss in cleave and crimp connectors. When a link with such connectors exhibits high loss, the question becomes: which end has high loss? When placed on the far end of a link terminated with such connectors, the near end connector will glow if the fiber and the preinstalled fiber are not in full contact or if the cleave on the cable fiber has dirt or a high angle.

When a VFL is installed onto a cleave and crimp connector, it will usually glow, as the VFL launces light into the cladding of the fiber. Such light can be lost inside the connector. For this reason, the VFL test is valid when the VFL is on the end opposite from the end with the connector being examined (Figure 18-1). It may be possible to test the near end connector if a test lead is inserted between the VFL and the connector.

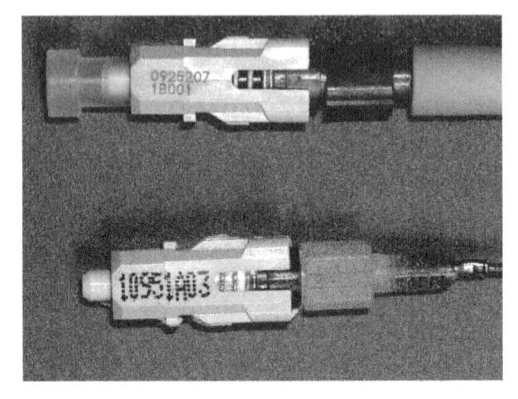

Figure 18-1: VFL Test On Cleave And Crimp Connector (Bottom, Glowing)

18.3.3 ATTENUATOR

An attenuator is a device that produces a deliberate loss in an optical link. Attenuators can be used to verify the optical power budget available from a transmitter, and to verify the sensitivity of a receiver. Attenuators can be fixed or variable, calibrated or uncalibrated.

Attenuators can be devices plugged into a barrel at a patch panel. Alternatively, the barrel can include an attenuator. Such barrels can be disassembled for insertion of the attenuator, a partially darkened thin film. Such barrels have a color different from that of the standard barrel. As a matter of practice, attenuators are placed on the cable end nearest the receiver.

18.3.4 TRAFFIC DETECTOR

The traffic identifier is a receiver that is placed on a fiber to indicate the presence of a signal. The identifier bends the fiber slightly, allowing light to escape from the core. Some traffic identifiers will indicate the direction of the traffic. These devices are most useful in identifying fibers with traffic to avoid disrupting traffic.

18.3.5 TALK SET

A fiber talk set enables voice communication over fiber. Talk sets allow either full or half duplex communication. Half duplex communication requires a single fiber. The talk set can connect to the fiber via a connector or via a mechanism that bends the fiber. By bending the fiber, the talk set does not require a terminated fiber. Some meter-source sets include an integrated talk set.

18.3.6 MICROSCOPE

The microscope enables inspection of connector ends (20). Such inspection reveals defects on polished connectors. Microscopes have an adapter sized to the ferrule diameter. Current generation of microscopes use LED light sources. Such sources have long battery life.

Microscopes can have two illumination modes: on axis, or coaxial, and off axis. Off axis illumination can reveal surface contamination not visible with on axis illumination.

Microscope magnification ranges from 100-400. Industry opinion on the ideal magnification is split. Some professionals, including this author, believe that 400x is best, as it reveals all features that can influence loss and reflectance. Others believe that 100 or 200x is best. Their argument is that 400x reveals detail that is confusing and in excess of what is needed. Their argument is valid, in that the installer must learn to ignore certain features that are visible at 400x but not at lower magnifications.

18.4 SUMMARY

18.5 REVIEW QUESTIONS

1. Identify two locations at which a VFL might be useful.

2. What three characteristics does a VFL identify?

3. An installer tells you that every cleave and crimp connector he installs glows when the VLF is placed on the connector. What do you suggest that he change?

19 CERTIFICATION PRINCIPLES AND METHODS

Chapter Objectives: you will learn alternative certification strategies. In addition, you learn how to certify an installed link for high reliability.

19.1 INTRODUCTION

Certification is the process of interpreting the test results. This process has the two objectives of verification:

- Of sufficiently low loss through the link
- That each component in a link has been reliably installed

With these two verifications, the link will work properly, i.e., optically accurately, and provide high reliability.

Certification of network links requires two steps: certification of insertion loss tests and certification of OTDR tests. Certification of reflectance tests is not required, since the reflection test is a maximum value allowed.

In order to certify a link, the installer must choose a certification strategy to interpret the test results. In this chapter, we present the process of interpretation and the advantages of three certification strategies. The three strategies are:

- Use of maximum loss values
- Use of typical loss values
- Use of mid-point loss values

The process of certification has four parts:

- Obtaining required information
- Performing insertion loss calculations
- Calculating insertion loss acceptance values
- Calculating OTDR acceptance values

19.2 REQUIRED INFORMATION

In order to certify a network, the installer calculates the link power loss. These calculations require seven datum:

- Accurate map
- Attenuation rate, maximum
- Attenuation rate, typical
- Connector loss, maximum
- Connector loss, typical
- Splice loss, maximum
- Splice loss, typical

The product data are available from data sheets or from web pages.

19.3 INSERTION LOSS CALCULATIONS

The insertion loss calculation depends upon the test method. The installer chooses from the three insertion loss methods, the one lead, the two lead, and the three lead methods (14.9.3).

For most installers, this choice is determined by the requirement for compliance with TIA/EIA-568-C. With such compliance, the installer uses the one lead method. Otherwise, the installer uses one of the other two methods.

These three methods treat the end connectors differently:

- The one lead method treats two end connectors as two pairs
- The two lead method treats two end connectors as one pair

Thus, an insertion loss calculation for one lead method will be higher than a calculation for two lead method by the loss of one connector pair. As this text has a focus on compliance with TIA/EIA-568-C, we present certification with the one lead method.

We present the optical power loss calculation in Table 19-1. The optical power loss is the sum of the losses in all components. This loss includes cable loss, connector loss, splice loss, and passive device loss. For reasons that will become apparent, the installer performs this calculation with both the maximum and typical values.

Loss in	dB/km		#km		dB
Cable		*		=	
	dB/pair		#pairs		
Connector		*		=	
	dB/splice		#splices		
Splice		*		=	
			Total	=	

Table 19-1: The Optical Power Loss Calculation

19.4 DEVELOPMENT OF A STRATEGY

We perform a one lead method calculation of the link in Figure 19-1. This link consists of two segments, with segment lengths of 1500 m and 500 m. There is a mid span patch panel with single pair at the panel. We present the component specifications in Table 19-2.

Connectors

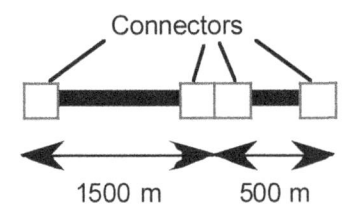

1500 m   500 m

Figure 19-1: Map 1

	Maximum	Typical
	dB/km	dB/km
Attenuation rate	3.5	3.0
	dB/pair	dB/pair
Connector loss	0.75	0.30
	dB/splice	dB/splice
Splice loss	0.15	0.10

Table 19-2: 850 nm, 62.5μ Multimode Values[54]

With these specifications, the maximum and typical losses are 9.25 dB (Table 19-3) and 6.9 dB (Table 19-4), respectively

What happens if the installer were to use the maximum loss as an acceptance value? He would allow up to 2.35 dB of excess loss. Excess loss occurs only through installation errors. Installation errors reduce network reliability.

➤ The risk of using the maximum loss as an acceptance value is reduced reliability.[55]

Loss in	dB/km		#km		dB
Cable	3.5	*	2.0	=	7.00
	dB/pair		#pairs		
Connector	.75	*	3	=	2.25
	dB/splice		#splices		
Splice	0.15	*	0	=	0.0
			Total	=	9.25

Table 19-3: Calculation, Maximum Loss

Loss in	dB/km		#km		dB
Cable	3.0	*	2.0	=	6.0
	dB/pair		#pairs		
Connector	0.3	*	3	=	0.9
	dB/splice		#splices		
Splice	0.1	*	0	=	0.0
			Total	=	6.9

Table 19-4: Calculation, Typical Loss

If use of the maximum loss as an acceptance value can result in reduced reliability, the installer might consider using the typical value. However, properly installed cables, connectors and splices can have loss values slightly higher than the typical values. For example, the 3M multimode ST® compatible connector has a typical loss of 0.3 dB/pair but can exhibit loss as high as 0.4 dB/pair when installed properly.[56]

Were the installer to use the typical value as a maximum acceptance value, he would reject properly installed link components. Such a rejection would increase the cost of the network without any increase in reliability.

Since the installer does not want to use either the maximum loss or the typical loss as an acceptance value, he can consider an acceptance value mid-way between these two losses. We call this value the 'mid point acceptance value'. For the link in Figure 19-1, the mid-point acceptance value is 8.075 dB.

---

[54] At the time of this writing, 50μ attenuation rate values are less than those in this table.

[55] TIA/EIA-568-C allows acceptance of the maximum loss.

[56] Pearson Technologies Inc. has installed or supervised more than 10,000 such connectors. The value of 0.4 dB is the result of work with this product.

▶▶The high reliability insertion loss acceptance value is halfway between calculated maximum and calculated typical values.

There is nothing magic about this strategy. Use of this mid-point value as an acceptance value assumes that the installer will either make minor mistakes or major mistakes. Minor mistakes can increase the loss from the typical value to just below the acceptance value. Major mistakes can, and usually do, increase the loss to above the acceptance value.[57]

19.4.1 INSERTION LOSS CERTIFICATION

The basic strategy (19.4) is designed to avoid acceptance of conditions of reduced reliability. The basic strategy is based on the expectation of typical performance most of the time. With this expectation, the insertion loss strategy reduces to three steps:

▶▶Accept test values less than or equal to the acceptance value

▶▶Investigate test values greater than the acceptance value and less than the calculated maximum value

▶▶Reject test values greater than the calculated maximum value

The first and third of these steps should be obvious. The second step requires additional examination. It is possible that properly installed cables, connectors and splices will have loss values higher than the acceptance value and less than the maximum value. This author's experience is that this is highly unlikely. Thus, products with values between these two values must be investigated to verify proper installation. If installation is proper, the installer accepts these components.

19.4.2 OTDR CERTIFICATION

The installer investigates an insertion loss value between the acceptance and the maximum values with OTDR measurements and with microscopic inspection of the connectors. In addition, he may inspect splices and cleave and crimp connectors with a visual fault locator (VFL).

To interpret these measurements, we follow the same strategy we used with the insertion loss test.

19.4.2.1 OTDR ACCEPTANCE VALUES

The installer expects that each component will test less than or equal to midway between the typical and the maximum values. With this expectation, the OTDR acceptance values are:

▶▶Attenuation rate acceptance value= (maximum rate + typical rate)/2
▶▶Connector pair acceptance value= (maximum loss + typical loss)/2
▶▶Splice loss acceptance value= (maximum loss + typical loss)/2

In addition to meeting these three acceptance values, the trace must meet a requirement for uniformity.

19.4.2.2 UNIFORMITY PRINCIPLE

The OTDR trace for every cable segment must be a straight line. A straight line indicates uniform loss through the segment. The principles are:

▶A properly designed, properly manufactured, properly installed cable segment always exhibits a straight-line trace.

▶Any deviation from a straight trace indicates an installation error.

It is essentially impossible to violate a cable performance parameter so that the attenuation rate is uniform and excessive. Thus, a high attenuation rate cable segment with a straight-line trace indicates a defective cable, not improper installation.

---

[57] These two statements reflect the author's 36 years of experience and are an opinion.

19.5 AN ALTERNATIVE STRATEGY

The mid point strategy allows the installer to identify conditions of reduced reliability. However, this strategy does not guarantee identification of all such conditions.

Another, more expensive, strategy enables identification of conditions of reduced reliability with loss values that would be acceptable by the mid point strategy. This strategy has three steps:

- Comparison of OTDR attenuation rates of each segment prior to installation to those rates after installation
- Visual inspection of all connectors with acceptance requirements for the core, cladding, and ferrule surface (20.4)
- Use of a maximum connector loss value less than the mid-point value

With comparison of the attenuation rate of each segment prior to installation to that after installation, the installer can determine proper installation. With proper installation, there will be no increase in attenuation rates.

With inspection of all connectors prior to testing, the installer can identify conditions of reduced reliability. Acceptance requirements for a connector that requires polishing would be:

- A featureless core
- A clean cladding
- A clean ferrule (20.4)

The acceptance requirement for a connector that does not require polishing would be:

- A connector that does not glow when a VFL is placed on the opposite end of the fiber

With these three requirements, the connectors would have the maximum reliability possible.

It is possible, and highly likely, that polish connectors will be improperly installed at the mid-point value. Our experience indicates that the mid point value, 0.525 dB/pair, is never reached except when there are installation errors. For example, From installation of approximately 10,000 connectors, this author has observed that multimode, 3M Hot Melt, ST™-compatible and SC connectors never exceed 0.40 dB/pair unless there is damage visible on the core.

Non-polish connectors have performance different from polish connectors. His author's experience indicates that properly installed cleave and crimp connectors may exceed the mid-point value.

Since some reduced reliability can result from use of the mid-point value for connector loss, the installer could consider use of a reduced value. To determine an appropriate reduced value, the installer would need to know the statistics of connector loss for the products he intends to use. Such statistics may not be readily available. However, were these statistics available, the installer could set a statistical upper limit based on a statistical knowledge of connector loss.

19.6 SUMMARY

In order to certify fiber optic links, the installer must choose a strategy for determining acceptance values. Use of the maximum loss value can result in acceptance of reduced reliability. Use of the typical loss value results in increased cost with no benefit. Use of a mid-point acceptance value results in detection of most conditions of reduced reliability with no major disadvantage. When the installer chooses the mid-point strategy, the installer calculates acceptance values according to these simple requirements:

▶▶Insertion loss= (maximum loss + typical loss)/2

▶▶Attenuation rate= (maximum rate + typical rate)/2

▶▶Connector loss= (maximum loss/pair + typical loss/pair)/2

▶▶Splice loss= (maximum loss/splice + typical loss/ splice)/2

▶▶ Each cable segment has a uniform OTDR trace.

19.7 REVIEW QUESTIONS

1. Using one lead insertion loss method, the multimode map (Figure 19-2) and specifications (Table 19-5), calculate the maximum and typical insertion loss, the insertion loss acceptance value and the OTDR acceptance values.

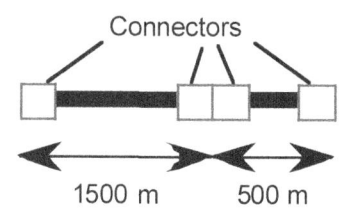

Figure 19-2: Map For Question 1

	Maximum	Typical
	dB/km	dB/km
Attenuation rate	1.5	0.7
	dB/pair	dB/pair
Connector loss	0.75	0.30
	dB/splice	dB/splice
Splice loss	0.15	0.10

Table 19-5: 1300 nm Multimode Values

2. Using one lead insertion loss method, the singlemode map (Figure 19-3) and specifications (Table 19-6), calculate the maximum insertion loss, the typical insertion loss, the insertion loss acceptance value and the OTDR acceptance values. You may assume that the patch cord between the two segments is short enough to ignore its length.

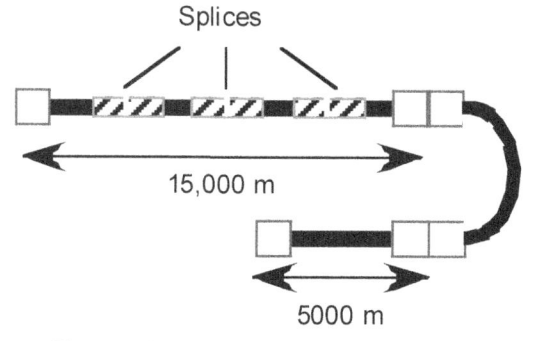

Figure 19-3: Map For Question 2

	Maximum	Typical
	dB/km	dB/km
Attenuation rate	0.5	0.35
	dB/pair	dB/pair
Connector loss	0.75	0.30
	dB/splice	dB/splice
Splice loss	0.15	0.10

Table 19-6: 1310 nm Singlemode Values

3. Using one lead insertion loss method, the map in Figure 19-4, and multimode specifications in Table 19-5, calculate the maximum insertion loss, the typical insertion loss, the insertion loss acceptance value and the OTDR acceptance values for this collapsed backbone.

4. Using one lead insertion loss method, the map in Figure 19-4, and multimode specifications in Table 19-2, calculate the maximum insertion loss, the typical insertion loss, the insertion loss acceptance value and the OTDR acceptance values for this collapsed backbone.

5. Using one lead insertion loss method, the map in Figure 19-5 and specifications in Table 19-5, calculate the maximum insertion loss, the typical insertion loss, the insertion loss acceptance value and the OTDR acceptance values for this building riser backbone.

6. Using one lead insertion loss method, the singlemode map in Figure 19-6, and specifications in Table 19-6, calculate the maximum insertion loss, the typical insertion loss, the insertion loss acceptance value and the OTDR acceptance values.

19-6　　CERTIFICATION PRINCIPLES AND METHODS

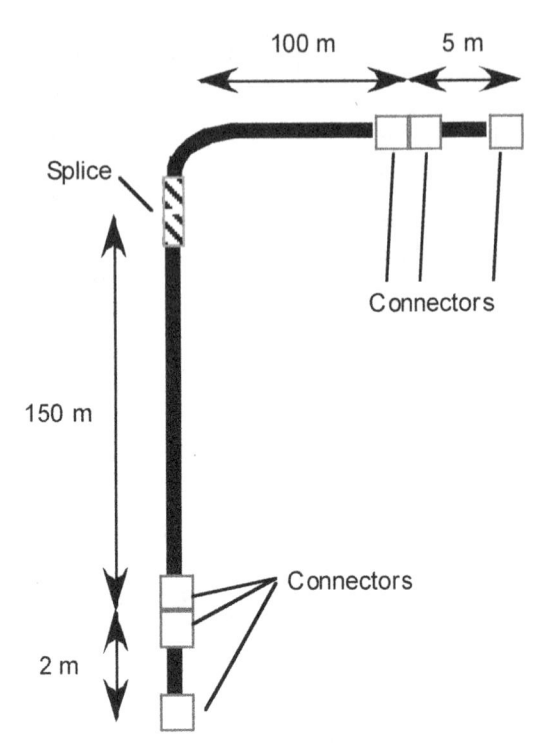

Figure 19-4: Map For Question 3

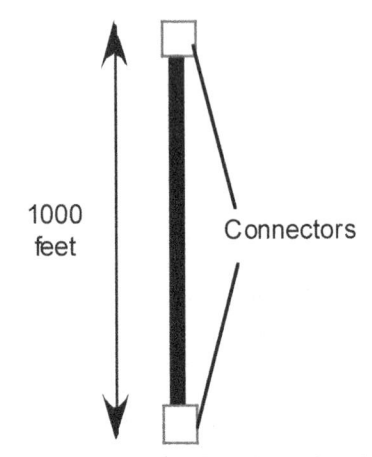

Figure 19-5: Map For Question 4

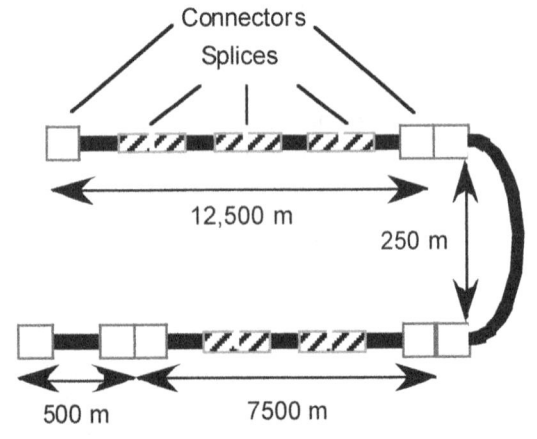

Figure 19-6: Map For Question 5

20 CONNECTOR INSPECTION

Chapter Objectives: you learn how to inspect and rate the appearance of a connector. You learn how to interpret the appearance of 'bad' connectors in order to identify the appropriate corrective actions.

20.1 APPLICABILITY

This chapter applies to the inspection of all fiber optic connectors that require polishing. For all connectors except the 'cleave and crimp' products, the condition of the end face of the connector correlates highly with the power loss of the connector.

That is, a 'good' appearance of the end face correlates highly with low loss. Because the 'cleave and crimp' connectors include a mechanical splice in the back shell, the correlation between end face appearance and power loss is not as high as in connectors that require polishing.

20.2 EQUIPMENT REQUIRED

A 400-magnification connector inspection microscope with an IR filter[58]

- Lens grade tissues (Kim wipes or equivalent)
- 98% isopropyl alcohol
- Electro-Wash® Px

Some fiber optic professionals recommend magnifications of 100 and 200. These reduced magnification microscopes allow the installer to miss features that increase loss and reflectance. The 400-magnification microscope enables installers to see all features of concern. Unfortunately, this magnification has two disadvantages. The first disadvantage is that it will reveal features that are irrelevant. Use of 400x microscopes requires the installer to learn those features that can be ignored.

The second disadvantage of 400x is inability to view the entire ferrule surface. An inspection at 400x may not reveal dirt on the ferrule outside of the field of view. 100x inspection will. In this case, loss test results will not be consistent with microscopic evaluation. Cleaning of the ferrule is the solution to this inconsistency.

Note: photographs in the text of this chapter are at a magnification of slightly less than 200; photographs in the Review Questions are at various magnifications. To determine the actual magnification, multiply the fiber diameter, in inches, by 200.

20.3 PROCEDURE

20.3.1 CLEANING

Prior to inspection, all connectors require cleaning. The installer may perform this cleaning as part of the installation activity or just prior to inspection.

The installer performs cleaning with tissues and a cleaning solution. The tissues are lens grade and lint free, such as Kim Wipes or woven tissues.

The liquid can be 98% isopropyl alcohol or a connector cleaning solution such as Electro-Wash® Px (Chemtronics, Kennesaw GA). A third alternative is lens grade tissues pre-moistened with isopropyl alcohol.

This author's opinion is that isopropyl alcohol is less expensive than Electro-Wash. ElectroWash is more effective, more expensive, and easier to use than is isopropyl alcohol. This author uses isopropyl alcohol during training and ElectroWash during field inspection.

Rubbing alcohol and medical wipes are not appropriate, as both may contain oil or water, both of which leave a residue. Residue can complicate interpretation and increase loss and reflectance.

The installer wipes the connector twice: with a moistened tissue area and with a dry tissue area. With isopropyl alcohol, the installer wipes the connector with a moistened area and with a dry area.

---

[58] www.Westoverscientific.com

Failure to wipe immediately in a dry area can result in watermarks on the connector. Watermarks complicate interpretation of the condition of the connector and can result in increased reflectance.

With ElectroWash Px, the installer creates a cleaning pad of tissue. He wipes from a moistened area to a dry area three times. With this method, ElectroWash leaves no residue.

20.3.2 GENERAL INSTRUCTIONS

The installer inspects with a microscope that has an adapter that will accept the connector ferrule. He performs four steps on each connector :

- Remove the cap from the connector
- Install the connector into the microscope
- Turn on and focus the microscope
- View and evaluate the connector (Figure 20-1)

Figure 20-1: Use Of Inspection Microscope

20.3.3 BACK LIGHT

Whenever possible, the installer performs connector inspection both with (Figure 20-2) and without (Figure 20-3) back light. Backlight can both reveal and conceal features in the core. The principle is:

▶▶ The installer rates the connector 'good' if it is 'good' both with and without backlight.

To inspect with back lighting, another person holds a white light to the connector on the opposite end of the fiber. The core should light up. If the core does not light up, the fiber is broken. The installer inspects and evaluates the connector as in 20.4.

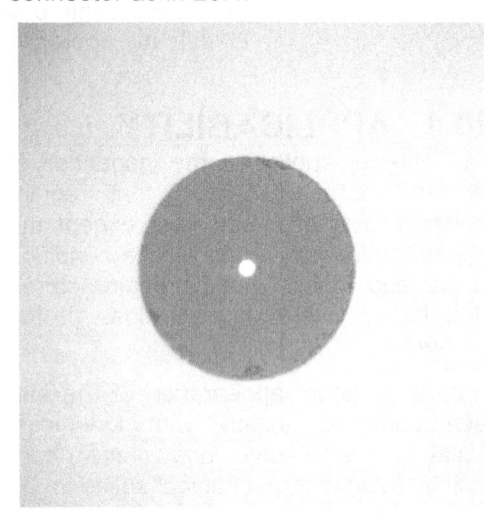

Figure 20-2: Connector With Back Light

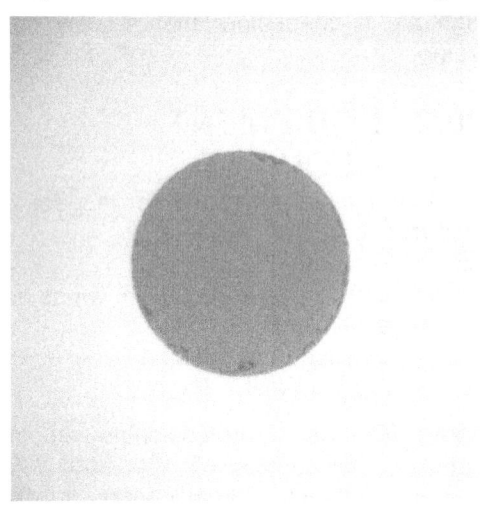

Figure 20-3: Same Connector No Back Light

If the cladding lights up during back lighting, the fiber is broken in the connector, usually in the ferrule (Figure 20-4 and Figure 20-5). While infrequent with adhesive connectors, backlight in the cladding is not unusual in 'cleave and crimp' connectors (24) installed by novices.

During back lighting, multimode cores can exhibit concentric rings. These rings result from the multiple layers in the core (3.4.2).

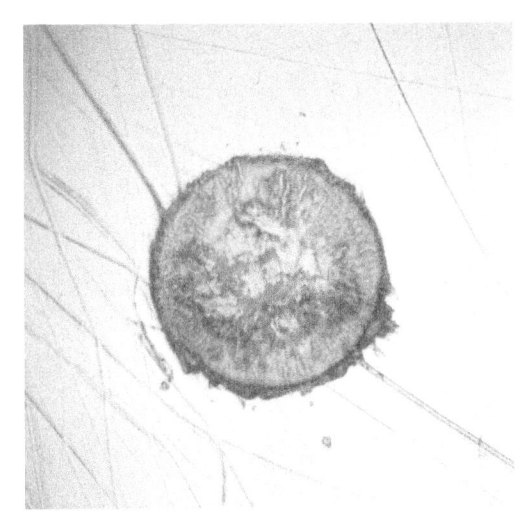

Figure 20-4: Shattered End With Back Light

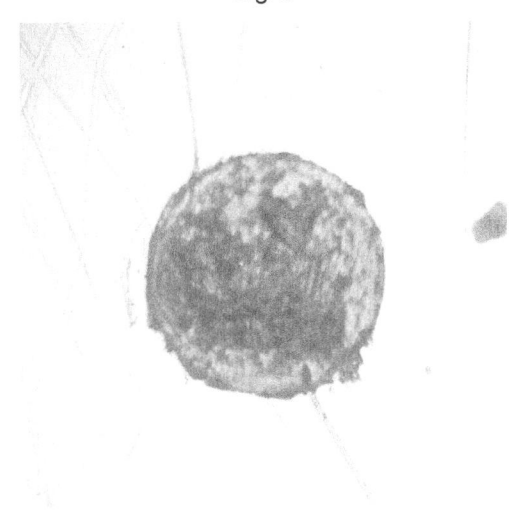

Figure 20-5: Shattered End

20.4 EVALUATION CRITERIA

In order to rate the condition of a connector, the installer need remember two words:

- Core
- Contact

The light travels within the core, or mode field. Therefore the condition of the core determines the loss of the connector. The connectors must make full contact in order to avoid excess loss from non-contact.

With these two words, the installer can understand the criteria for rating a connector as 'good'.

A good connector has a

▶ 'Good' core
▶ Clean cladding
▶ Clean ferrule

20.4.1 CORE

Any condition that can divert or block the light from its normal path will increase the loss of the connector. Thus, a 'good' connector must have a core that is (Figure 20-6):

▶ Round
▶ Clear
▶ Featureless
▶ Flush

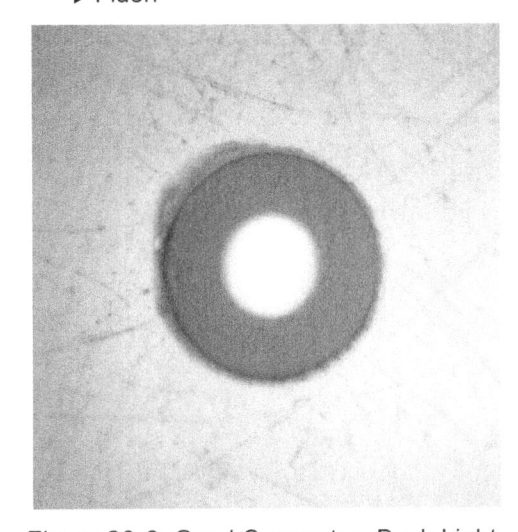

Figure 20-6: Good Connector- Back Light

A round core means that none of the core is below the surface of the ferrule (Figure 12-18). If the core is below this surface, it cannot be featureless, because it cannot be polished.

A clear core has no oil or grease that can make the image unclear or fuzzy. A featureless **core has no features, or faults,** that can block or divert the light from its normal path. **Features** are:

- Cleaning residue (Figure 20-7)
- Scratches (Figure 20-8)
- Cracks (Figure 20-9 and Figure 20-10)
- Pits
- Dirt (Figure 20-11)
- Dust

One final concept on features: if a feature, such as a speck of dirt, is large enough to be visible, it is too large to accept. If repeated cleaning does not remove it, it is not dirt, but a pit in the surface fiber.

Finally, the core must be flush with the ferrule surface. When the core and the ferrule surface are both in focus, the fiber is flush with the surface. If it is above the surface, the fiber can be damaged from use. If it is below the surface, there will be an air gap between this fiber and the mated fiber. As you learned in (5), air gaps increase loss.

Fibers can be above or below the ferrule surface. There are two causes of this. The first is uncured adhesive or epoxy The second is incomplete polishing. Of course, incomplete polishing usually occurs with adhesive on the ferrule surface.

Figure 20-7: Cleaning Residue

20.4.2 EVERYWHERE ELSE

To achieve low loss by full core contact, the cladding and ferrule surface must be:

▶Clean

Dirt on ferrule can prevent full ferrule contact (Figure 20-12).

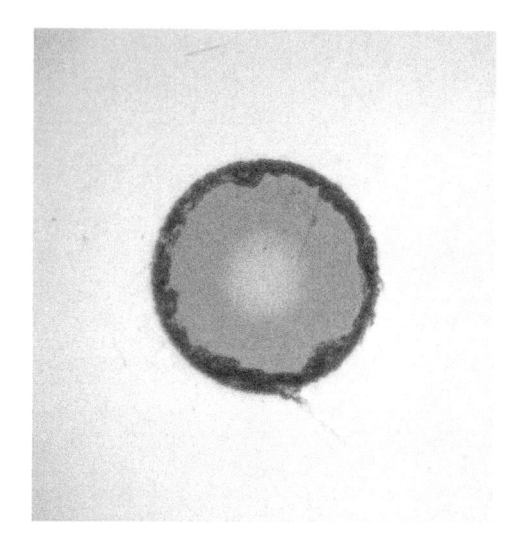

Figure 20-8: Scratch In Core

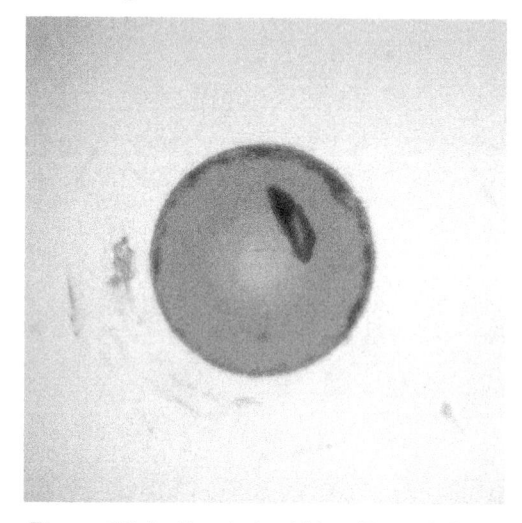

Figure 20-9: Crack And Non-Round Core

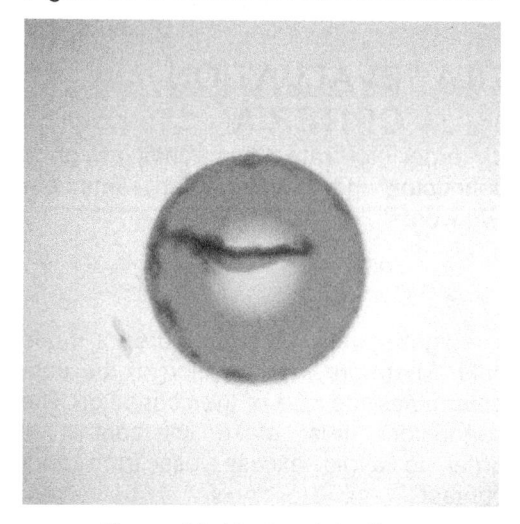

Figure 20-10: Crack In Core

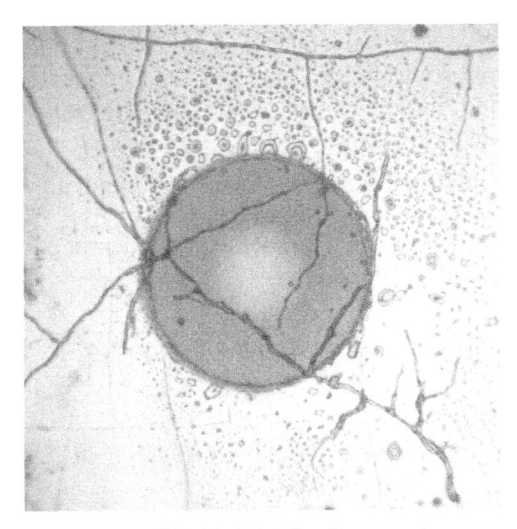

Figure 20-11: Dirt On Connector

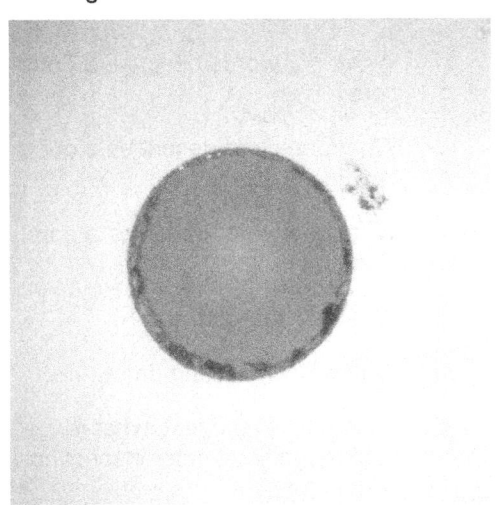

Figure 20-12: Dirt On Ferrule

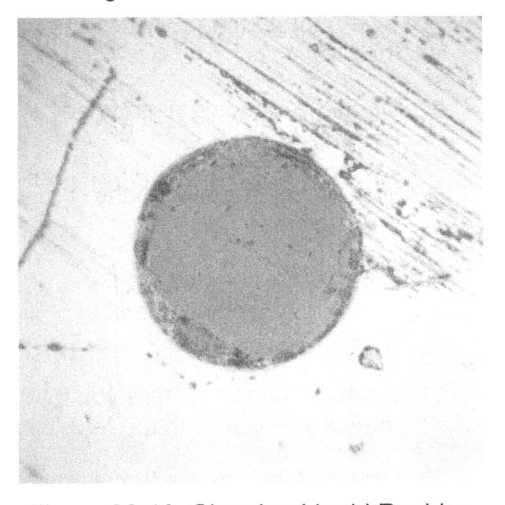

Figure 20-13: Cleaning Liquid Residue

Figure 20-14: Acceptable, Imperfect Clad

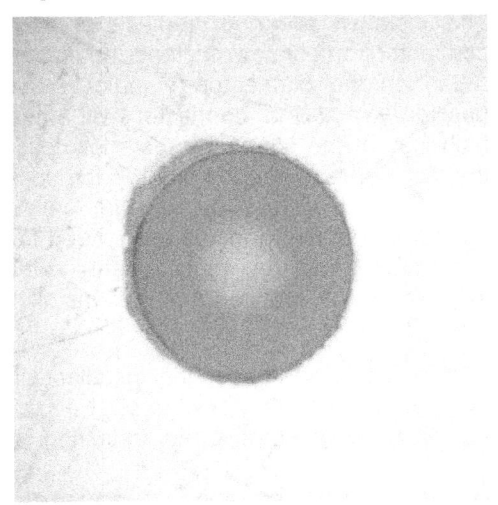

Figure 20-15: Ferrule Features

20.4.3 BAD CLAD? BE GLAD!

As light travels in the core, the cladding need not be perfect. An imperfect cladding means polishing was not perfect. The polishing procedure may need improvement. Thus, the cladding need not be round. It need not be completely present. It need not meet the criteria for a good core (20.4.1). However, it must be clean.

20.4.4 FERRULE FEATURES

The only requirement for the ferrule is that it be clean. Features in the ferrule surface will not interfere with proper transfer of light. While this is a dangerous statement, this author states:

▶ With respect to the ferrule surface (almost) anything goes.

20.4.5 CONNECTOR DISPOSITION

In reality, connectors will have one of three ratings:

- Good
- Bad
- Terrible (Figure 20-5)

The definition of a 'good' connector is that of a *perfect* connector. Small scratches, cracks or contamination in the core may not divert enough light to cause the connector to exhibit high loss.

The installer does not replace a 'bad' connector immediately. Instead, he places a tag on the connector to indicate the possibility that the connector will have high loss. If the connector tests low loss, the installer removes the label. If the connector tests high loss, the installer leaves the label in place to indicate the need for connector repair or replacement. After inspecting the connector, the installer replaces the cap.

With a little experience, the installer will recognize terrible connectors. He replaces these connectors prior to testing.

20.5 TROUBLESHOOTING

20.5.1 DIRT ON CONNECTOR

Potential cause: dirt on microscope lens

Action: clean lens; or, rotate connector in microscope; if dirt does not move, dirt is on lens. Clean lens if possible.

20.5.2 FAINT STAINS ON CONNECTOR

Potential cause: moisture stains from alcohol
Action: dry wipe immediately after cleaning with alcohol; or clean with Electro-Wash Px

20.5.3 NO FIBER FOUND

Potential cause: out of focus

Action: move focusing ring through entire range of motion; or, back light connector and move focusing ring through entire range of motion

Potential cause: connector not completely inserted into microscope
Action: insert connector into microscope adapter completely

20.5.4 CORE DOES NOT BACKLIGHT

Potential cause: broken fiber
- Action: locate break with fault indicator or VFL

20.6 REVIEW QUESTIONS

1. What four words describe a good core?
2. What one word describes a good ferrule surface?
3. What one word describes a good cladding?
4. What is backlighting?
5. Why is back lighting done?
6. According to the text, what is the best magnification for inspecting connectors?
7. Why do some cores test good even though they may have a 'bad' rating?
8. Does the cladding need to be perfectly round?
9. Explain your answer to the previous question.
10. What is the ring outside the cladding and inside the fiber hole?
11. What characteristic tells you that the core is flush with the ferrule?
12. Provide two reasons that a fiber is not flush with the ferrule.

PROFESSIONAL FIBER OPTIC INSTALLATION 20-7

For each of the photographs in Review Questions 15 to 32: rate the connector good or bad and identify the reason for the connector being bad. Multiple reasons are possible. Use:
- NR: Not round
- FIC: Feature in core
- DOC: Dirt on core
- DOCl: Dirt on cladding
- DOF: Dirt on ferrule
- NF: Not flush

13. Rate this connector.

14. Rate this connector.

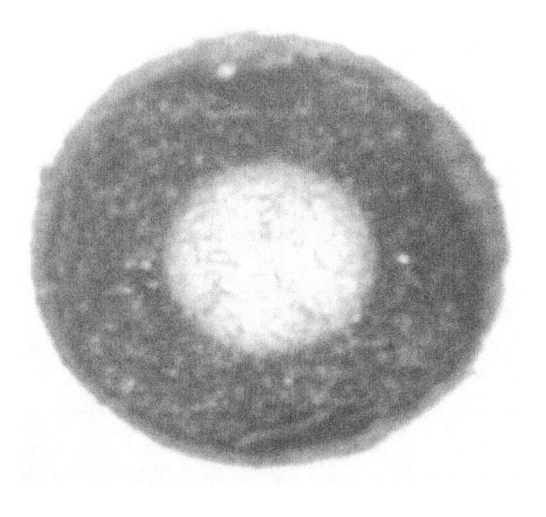

15. Rate this connector.

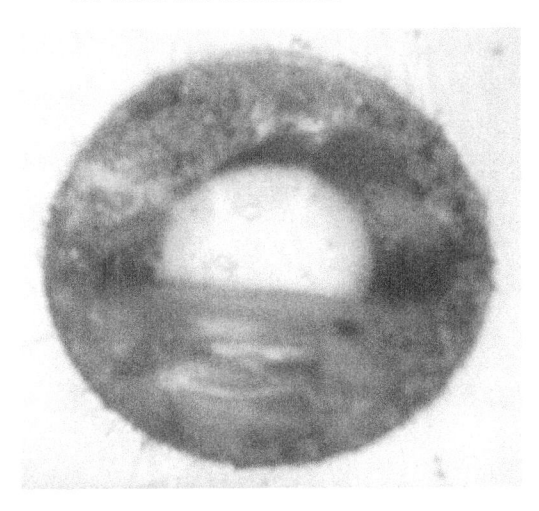

16. Rate this connector.

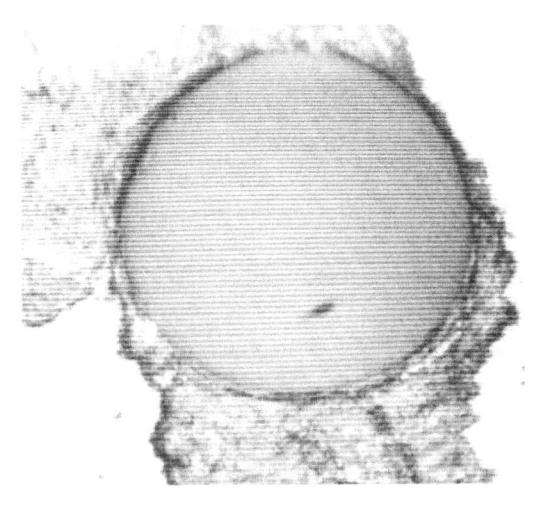

17. Rate this connector.

© PEARSON TECHNOLOGIES INC.

18. Rate this connector.

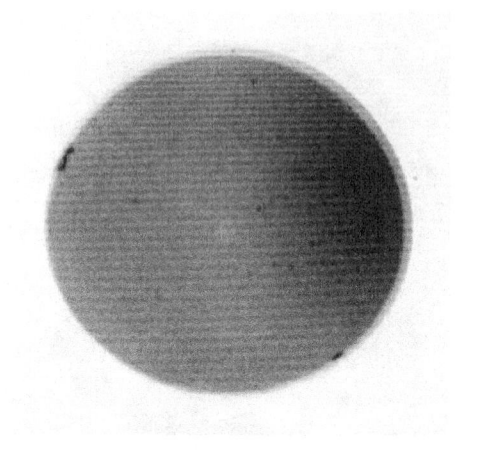

19. Rate this connector.

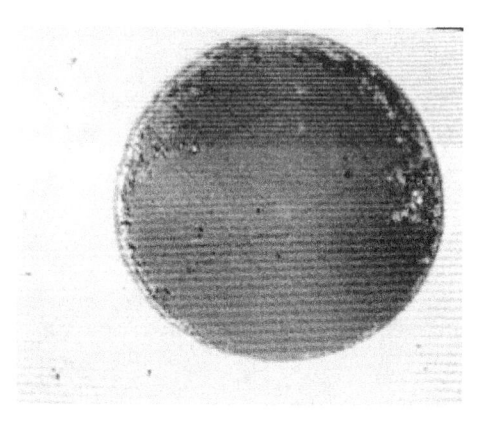

20. Rate this connector.

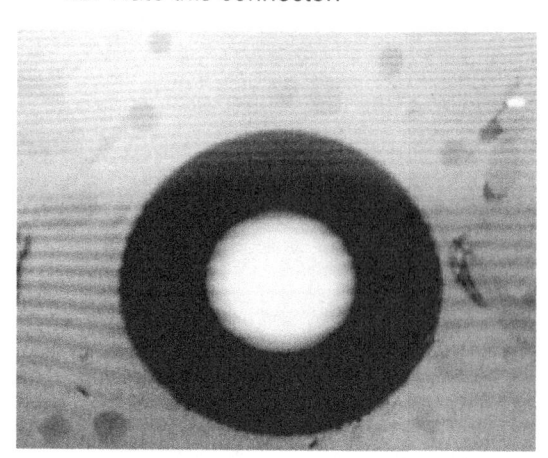

21. Rate this connector.

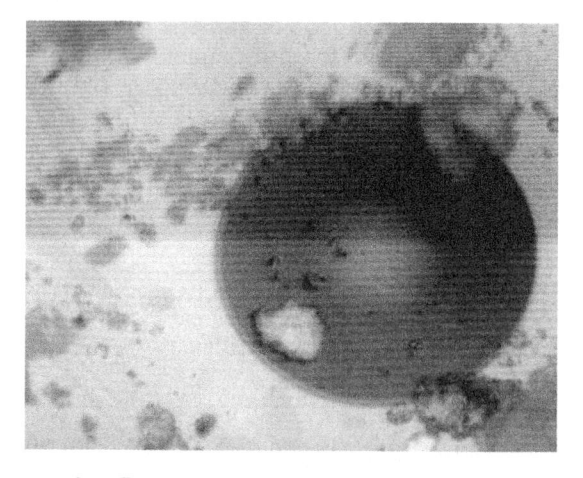

22. Rate this connector.

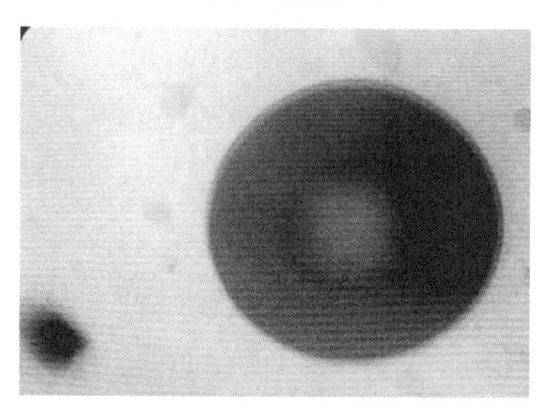

23. Rate this connector.

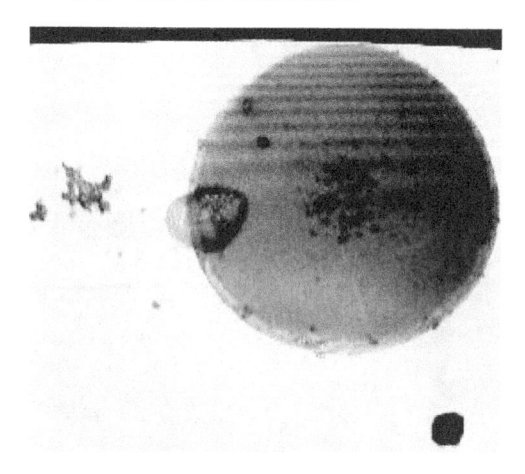

Professional Fiber Optic Installation

24. Rate this connector.

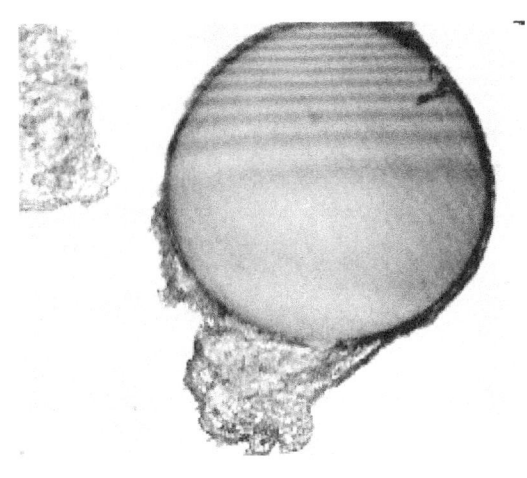

25. Rate this connector.

26. Rate this connector.

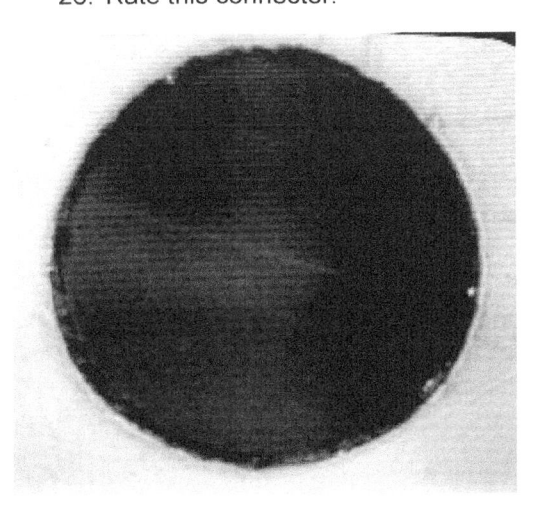

27. Rate this connector.

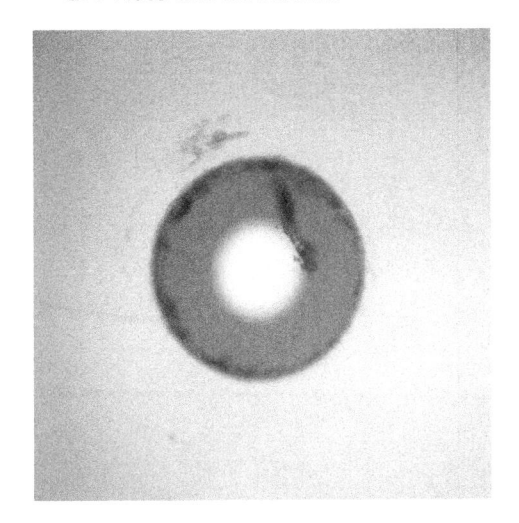

28. Rate this connector.

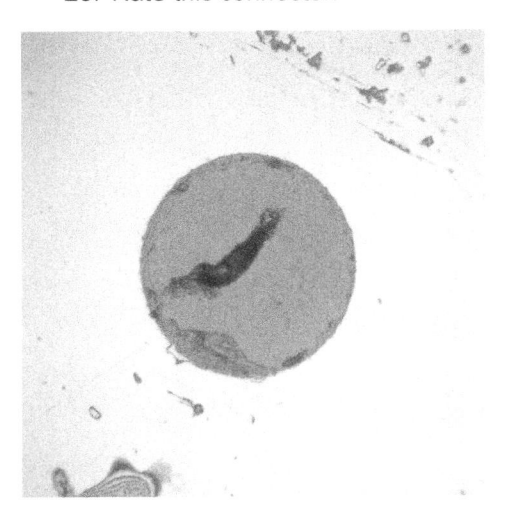

29. Rate this connector.

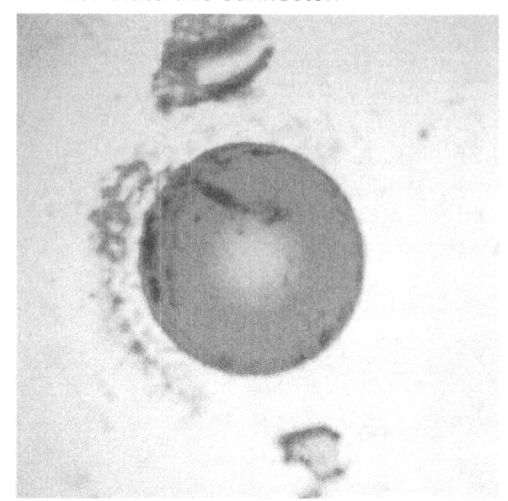

PART FOUR
INSTALLATION PROCEDURES

HOW TO USE THESE PROCEDURES
To develop an overview of the entire process covered by this document, the installer reads the chapter through several times. He uses the One Page Summary to perform the activity.

SAFETY PRECAUTIONS
For all installation activities that involve the creation and handling of bare fibers, the installer follows the safety rules and guidelines of 10.12.

21 CABLE END PREPARATION

Chapter Objectives: by following these procedures, you prepare the ends of loose tube and tight tube cables for pulling and termination. In addition, you learn how to handle, without damage, the 250µ fiber exposed at the ends of loose tube cables. Finally, you perform these activities without damaging the fibers.

21.1 INTRODUCTION

The installer handles the fiber in two forms: cable and fiber. During splicing and connector installation, he handles the fiber, both with and without the primary coating. He prepares the cable end for two operations: pulling and end termination.

In this chapter, we present five procedures:

- Fiber handling (21.3)
- End preparation of loose tube cables for pulling (21.4.1)
- End preparation of tight tube cables for pulling (21.4.2)
- End preparation of loose tube cables for termination (21.5.1)
- End preparation of tight tube cables for termination (21.5.2)

As these procedures are generic, they apply to many, but not all, cables. For exact instructions, we recommend that the installer review data sheets and installation instructions for the cables he is to install.

21.2 TOOLS AND SUPPLIES

- Pull rope, installed in conduit or inner duct
- Swivel with shear pin rated below installation load of cable, General Machine Products part number 71406 or equivalent[59]
- Electrician's tape
- Jacket removal tool, Ideal 45-162 or Clauss RCS-20
- Break off blade knife
- Kevlar© cutter
- Tubing cutter, (Ideal 45-162)[60]
- Fiber stripper, Clauss No Nik, 203 or Miller jacket stripper
- Fiber optic cable gel/grease remover, D'Gel™ (Polymicro Technologies) or Electrosol™ (American Polywater Corporation)
- 98 % isopropyl alcohol
- Clean rags or heavy duty paper towels
- Unscented talc (Aka 'dry fiber lubricant')
- Fiber, with primary coating of 245-250µ, 3' long
- 12' of 3 mm single fiber cable
- Matches
- Bare fiber collection bottle
- Pliers
- Disposable gloves

21.3 FIBER HANDLING

The activities in this section sensitize the installer to what the fiber will and will not withstand. From these activities, the installer will recognize those steps at which he needs to be careful and those steps at which he needs to be very careful.

21.3.1 COATED FIBER

Coated fiber has properties that are different from those of uncoated fiber. These activities sensitize the installer to the differences.

21.3.1.1 TENSILE BEHAVIOR

The installer wraps a fiber around the fingers of both hands. He pulls as hard as he dares. Drawing blood is not a requirement of this activity. The installer answers the following questions.

---

[59] www.GMtools.com, GMP swivel part numbers are 71399-71406, 70188, and 71418.

[60] This cutter is for loose buffer tubes with a diameter of 2-4 mm. For larger buffer tubes, use a different cutter.

21.3.1.2 REVIEW QUESTIONS

1. Is fiber stronger or weaker than expected?
2. What region of the fiber allows the fiber to retain its intrinsic high strength?
3. How well does this region function?

21.3.1.3 BENDING BEHAVIOR

The installer makes a large loop near the end of the fiber. Slowly, he pulls the loop smaller until the fiber breaks. He estimates the diameter at which the fiber breaks. He repeats this step.

With a 3 mm diameter cable, the installer makes a loop at the minimum long term bend radius (4.6.2). He compares this cable diameter to the diameter at which the fiber broke.

21.3.1.4 REVIEW QUESTIONS

1. If the installer respects the minimum bend radius, will he ever experience breakage in bending?
2. Is the minimum bend radius conservative or realistic?
3. Was it easier to break the fiber in bending or in tension?
4. When installing fiber cables, will the installer need to be more careful with tension or with bending?

21.3.2 UNCOATED FIBER 1

These activities sensitize the installer to the properties of bare fiber after significant exposure to the air or after damage to the cladding.

21.3.2.1 BENDING BEHAVIOR

With a match, the installer burns off 2-3" of the primary coating. He wipes the burnt coating with his fingers. He bends the bare fiber until it breaks. He repeats this step. He places the bare fiber in the bare fiber collection bottle.

Note that this is not the recommended method for removing primary coating.

21.3.2.2 REVIEW QUESTIONS

1. What happens to the bend strength of the fiber when the primary coating is removed?
2. How well does the primary coating function?
3. Consider fitting a 125 ± 0.5µ fiber into a connector or splice with a 126µ or 127µ hole. What might happen?
4. How can the installer avoid this?
5. When does the installer need to be careful?
6. When does he need to be very careful?

21.3.3 UNCOATED FIBER 2

These activities sensitize the installer to the properties of the fiber after proper removal of the primary coating.

The installer wraps the fiber around one finger 5 times at 12" from the end. Any finger will do. Following the procedure in 22.3.6.2.2, the installer strips 3" of primary coating from the fiber.

Important Note

It is possible to remove the inner layer of primary coating, but leave the outer layer. For tight tubes, it is possible to remove the buffer tube, but leave the inner primary coating layer. In both cases, the fiber appears larger than it should be.

The installer examines the fiber to ensure that no primary coating remains. Bare fiber will have a glass smooth appearance. Primary coating will have a translucent, but not glass smooth, appearance.

21.3.3.1 BENDING BEHAVIOR

By holding the fiber in both the bare and coated areas, the installer bends the bare fiber until it breaks. He estimates the diameter at which it broke. He repeats this step. He places the fiber in the fiber collection bottle.

21.3.3.2 REVIEW QUESTIONS

1. What happens to the bend strength of the fiber when the primary coating is removed correctly?

7. How well does the primary coating function?

8. The strength the installer observes with flame stripping is the strength he would expect after exposure of the cladding for 12 hours. Should he remove the last layer of protection from the fiber one day and install the connector or splice the next? Why or why not?

21.4 PULLING

21.4.1 LOOSE TUBE

21.4.1.1 ATTACH SWIVEL

The installer chooses a break away swivel rated at a load less than the maximum installation load rating of the cable. The swivel must fit into the conduit into which the cable is to be installed. He follows the directions for installing the shear pin into the swivel (Figure 21-1). With two half hitches, he ties a pull rope to the swivel. With electricians tape, he tapes the loose end of the pull rope back along the rope. He wraps the last layer of tape towards the swivel.

21.4.1.2 SET SLITTER DEPTH

While holding the jacket slitter against the end of the cable, the installer sets the depth of the slitter blade to slightly less than the thickness of the cable jacket (Figure 21-2).

Figure 21-1: Break Away Swivel With Pull Rope

21.4.1.3 MAKE TEST CUT

The installer places the slitter on the jacket 2" from end. He rotates the slitter around the jacket several times. He removes the slitter. He pulls on the jacket to slide the jacket off the end of the cable (Figure 21-3). The jacket should not slide off. If it does, the slitter was set excessively deep. In addition, the strength members should not be visible (Figure 21-3 and Figure 21-4).

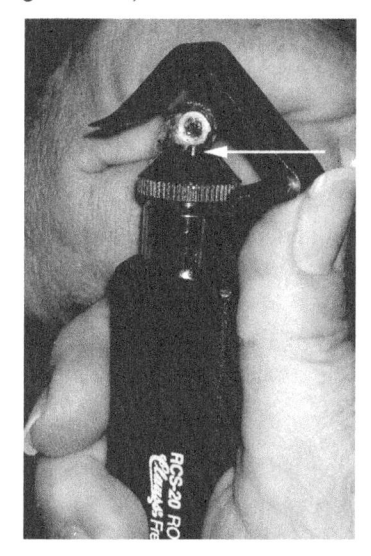

Figure 21-2: Setting Slitter Depth

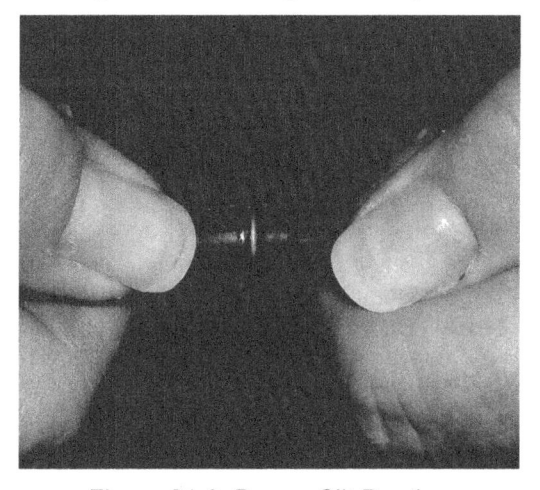

Figure 21-3: Proper Slit Depth

If the jacket slides off, the installer sets the blade to a shallower depth and cuts off the cable with the slit. He makes another test cut on the jacket at a distance of about 2" from the jacket end. He tries to pull the jacket from the cable.

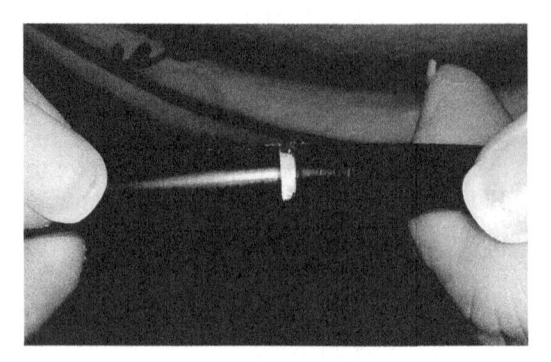

Figure 21-4: Excessive Slit Depth

21.4.1.4 REMOVE TEST CUT

The installer bends and rotates the cable at the slit. If the slitter depth is correct, the jacket becomes fully separated (Figure 21-4). He pulls the jacket from the cable. He examines the strength members and buffer tubes for damage. There should be none.

If there is damage, he reduces the blade depth. He makes another test cut at a distance of about 2" from the previous cut.

21.4.1.5 REMOVE JACKET

The installer repeats Steps 21.4.1.4 and 21.4.1.5 at a distance of 18" from the end of the cable. He removes the jacket. 18" is the length necessary to have enough strength member to tie to the swivel. If necessary, the installer can remove more than this length.

21.4.1.6 REMOVE CORE MATERIALS

The installer separates the buffer tubes from the strength members. If a binder tape is present, he cuts it off. He cuts the buffer tubes flush with the end of the jacket. He cuts and removes the central strength member at the end of the jacket. He cuts the ripcord 8" from the end of the jacket.

21.4.1.7 GEL/GREASE FILLED

If there is no grease, the installer proceeds to 21.4.1.10.

21.4.1.8 REMOVE GREASE

He dons plastic gloves. The installer moistens paper towels or clean rags with gel/grease remover. He wipes the strength members with the towels until he has removed all the grease.

21.4.1.9 REMOVE GREASE REMOVER

The installer moistens paper towels or clean rags with isopropyl alcohol. He wipes the strength members with the paper towels until he has removed all the gel/grease remover.

21.4.1.10 ATTACH SWIVEL

The installer twists and feeds the strength members through the lower housing of the swivel. He ties the strength members to the swivel in a double half hitch (Figure 21-5). He tapes the strength members so the tape ends on the cable.

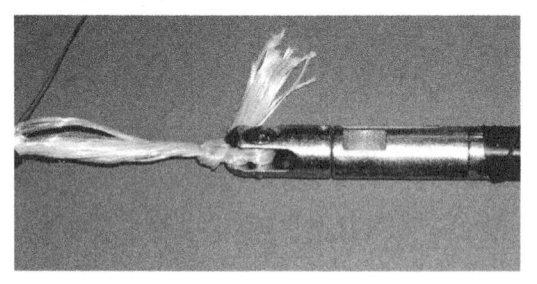

Figure 21-5: Strength Members Attached To Swivel

21.4.1.11 INSTALL CABLE

Following the appropriate procedure, the installer pulls the cable into location. He cuts off 3' from the end of the cable. Some installers cut off more than 3'.

21.4.2 PREMISES CABLE

This procedure is for 2-24 fiber distribution or premises cables.

21.4.2.1 ATTACH SWIVEL TO PULL ROPE

As in 21.4.1.1, the installer attaches a pull rope to the swivel.

21.4.2.2 SLIT JACKET

The installer bends the cable at 12" from the end of the cable. He places a sharp knife blade against the outside of the bend (Figure 21-6). Gently, he saws the blade until the blade cuts through to the strength members. He rotates the bend so that jacket that is not cut is on the outside of the bend. He repeats the sawing action on the outside of the bend until the blade cuts through the jacket. He repeats this step until he has cut the jacket around its circumference. He removes the jacket.

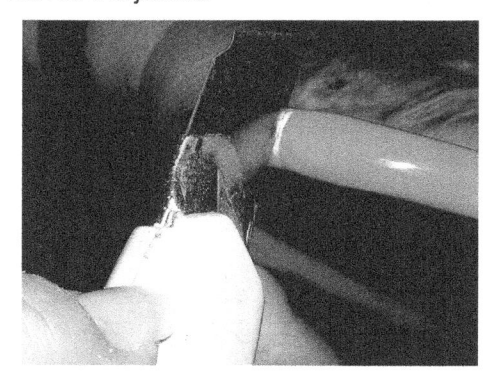

Figure 21-6: Cutting Of Premises Cable Jacket

21.4.2.3 REMOVE FIBERS

The installer separates the fibers from the strength members. If a binder tape is present, he cuts it off. He cuts the fibers flush with the jacket.

21.4.2.4 REMOVE CENTRAL STRENGTH MEMBER

Some cables may require retention of the central strength member. The installer reviews and complies with the manufacturer installation procedure for the cable. If appropriate, the installer cuts and removes the central strength member at the jacket.

21.4.2.5 INSTALL STRENGTH MEMBERS INTO SWIVEL

The installer twists the strength members together and feeds them through the swivel. He ties two half hitches into the strength members (Figure 21-7). He covers the strength members with 1-3 layers of electricians tape. The tape covers the end of the jacket. The last layer of tape ends on the jacket of the cable.

21.4.2.6 PULL CABLE

Following the appropriate procedure, the installer pulls the cable into location. He cuts off 3-6' from the end of the cable.

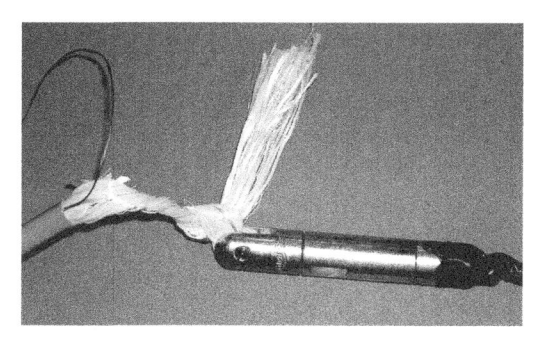

Figure 21-7: Premises Cable Attached To Pull Rope

21.5 TERMINATION

21.5.1 LOOSE TUBE

21.5.1.1 PULL CABLE

Following the appropriate procedure, the installer pulls the cable into location. He cuts off 3-6' from the end of the cable.

For a central loose tube design, the installer untwists for fiber bundles, removes the binding thread or yarn from each bundle, and cleans each bundle, as described below in 21.5.1.14 to 21.5.1.16.

Strength member length

Central strength member length

Buffer tube length

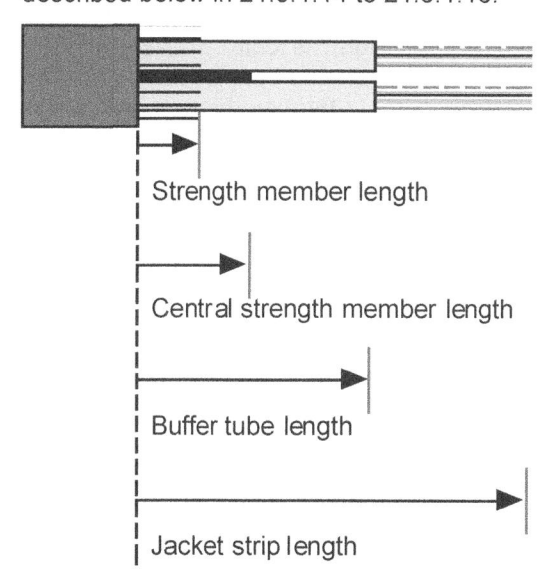

Jacket strip length

Figure 21-8: End Preparation Dimensions

21.5.1.2 DETERMINE DIMENSIONS

The installer determines the following four dimensions for the enclosure into which the cable is to be installed (Figure 21-8):

➢ Length of jacket to remove

- Length of strength member to leave
- Length of central strength member to leave
- Length of buffer tube to leave

The enclosure instruction sheet specifies these lengths.

21.5.1.3 PREPARE WORK AREA

If the cable has gel and/or grease, the installer sets up a work surface covered with a disposable plastic sheet or newspapers.

21.5.1.4 PRE INSTALL CABLE

The installer can install the cable into the enclosure either before or after end preparation. This author recommends that novice installers install the cable prior to end preparation. Experienced installers can install the cable after end preparation, as long as they can install buffer tubes into the enclosure without violating their bend radius. Such violation can result in broken fibers.

With the assumption of installation prior to end preparation, the installer feeds the cable through the entrance hole and out the front of the enclosure. He pulls enough cable through the hole so that the cable reaches a work surface that makes the cable relatively easy with which to work. After the end preparation, the installer pushes back the excess cable through the hole. This excess cable becomes a service loop.

21.5.1.5 ACCESS RIPCORD

If at least 8" of the ripcord is available, the installer proceeds to 21.5.1.7.

21.5.1.6 REMOVE JACKET

To remove the jacket, the installer performs 21.4.1.2-21.4.1.4. He repeats these steps at a distance of at least 8" from the end of the cable. He removes the jacket.

If the cable does not slide off, the installer rotates the blade in the slitter by 90°. While holding the blade in this position, he places the blade in the cut at 8" but not on top of the ripcord. While forcing the blade to stay in the jacket, he pulls the slitter to the end of the jacket. Alternatively, he uses a knife to make a longitudinal cut from the ring cut to the end of the cable. He removes the jacket to access the ripcord.

21.5.1.7 MAKE RING CUT

The installer repeats 21.4.1.3 at a distance from the end of the cable equal to the length of jacket removal length (Figure 21-8). This is the 'ring cut'.

The installer wraps the exposed ripcord around the plastic covered handle of a pair of pliers. He wraps the ripcord on top of itself. He pulls the ripcord along the cable to the ring cut. He removes the jacket by pulling the cable core through the slit.

Note that some cables have jackets that allow their removal without using the ripcord.

21.5.1.8 PREPARE CORE

The installer separates the strength members from the cable core. He cuts the external strength members under the jacket to the strength member length (Figure 21-8). He cuts the ripcord 8" from the end of the jacket. Leave this ripcord for potential future use. He cuts the binding tape at the jacket. He removes the tape from the core.

Optional step: the installer places 4-6" of heat shrink tubing over the end of the jacket. This tubing provides a cosmetic function of concealing the end of the jacket.

He untwists and separates the buffer tubes from the central strength member. He cuts the central strength member to the central strength member length (Figure 21-8).

21.5.1.9 CLEAN CORE

If there is no grease, the installer proceeds to 21.5.1.12.

21.5.1.10 REMOVE GREASE

The installer dons plastic gloves. The installer moistens paper towels or clean rags with gel/grease remover. He wipes the buffer tubes and strength members with the towels until the buffer tubes begin to squeak. After the squeaking begins, he wipes the buffer tubes and strength members with gel/grease remover one more time.

21.5.1.11 REMOVE GREASE REMOVER

The installer moistens paper towels or clean rags with isopropyl alcohol. He wipes the buffer tubes and strength members with the paper towels until the buffer tubes begin to squeak. After the squeaking begins, he wipes the buffer tubes and strength members with isopropyl alcohol two more times.

21.5.1.12 PREPARE FIBERS

One buffer tube at a time, the installer performs steps 21.5.1.13 to 21.5.1.16.

21.5.1.13 REMOVE BUFFER TUBE

The installer places the tubing cutter on one buffer tube at the buffer tube length (Figure 21-8). Without applying pressure to the tubing cutter, he rotates the cutter 30° in one direction and 30° back (Figure 21-9). This action will score, but not cut through the buffer tube.

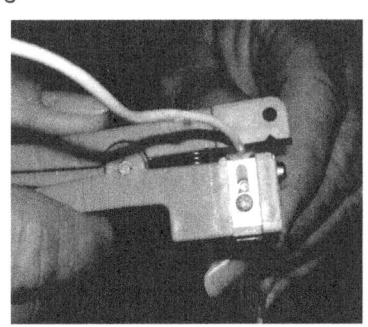

Figure 21-9: Cutter On Buffer Tube

Some gels are thick enough to prevent removal of the buffer tube in a single length. For cables with such gel, the installer removes the buffer tube by the method in the following paragraph in several short segments.

With his thumb on the buffer tube opposite to the score, he rapidly bends ('snaps') the buffer tube until it breaks. While holding the buffer tube straight, he pulls the buffer tube from the fibers slowly. While pulling the buffer tube from the fibers, he drags a paper towel or clean rag along the fibers to remove some of the water-blocking gel. The installer removes whatever gel comes off with this one wipe. At this time, he does not attempt to remove all the gel.

21.5.1.14 REMOVE GEL

If there is no gel, the installer proceeds to 21.5.1.17.

With a paper towel or clean rag moistened with water blocking compound/gel remover, the installer wipes the fibers from the end of the buffer tube to the end of the fibers. When wiping, he squeezes the fibers lightly. He repeats this wiping until the fibers begin to squeak. He repeats this wiping two more times.

21.5.1.15 REMOVE GEL REMOVER

The installer moistens paper towels or clean rags with isopropyl alcohol. He wipes the fibers with the paper towels until the fibers begin to squeak. After the squeaking begins, he wipes the fibers with isopropyl alcohol two more times.

21.5.1.16 LUBRICATE AND SEPARATE FIBERS

If the installer is preparing this cable for splicing, this step is optional. If he plans to install these fibers into a furcation kit, this step is necessary.

The installer places approximately 1/2 teaspoon of 'dry fiber lubricant' (unscented talc) in the palm of one hand. With the thumb of his other hand, he forces the fibers into the talc and draws the entire length through the talc. He repeats this step two more times. The fibers will automatically untwist and separate.

He counts the fibers. The number of fibers should be 6 or 12. If the installer broke a fiber, he cuts all the fibers at the end of the buffer tube. He removes a length of jacket equal to the length of buffer tube he cut. He repeats the Steps 21.5.1.7 to 21.5.1.16.

21.5.1.17 ATTACH CABLE

If the installer has fed the cable through an access hole in the enclosure, he feeds the excess cable back through the hole and arranges the extra cable as a service loop. He attaches the cable strength members to the enclosure in the manner specified in the enclosure instructions.

21.5.2 PREMISES

These instructions are for 2-24 fiber premises or distribution cables.

21.5.2.1 DETERMINE DIMENSIONS

The installer determines the following four dimensions for the enclosure into which the cable is to be installed (Figure 21-8):

- Length of jacket to remove
- Length of strength member to leave
- Length of central strength member to leave
- Length of buffer tube to leave

21.5.2.2 INSTALL CABLE INTO ENCLOSURE

As in 21.5.1.4, the installer pre-installs the cable.

21.5.2.3 SLIT JACKET

The installer bends the cable at the jacket strip length (Figure 21-8). As in 21.4.2.2, he removes the jacket.

21.5.2.4 SEPARATE FIBERS

If a binder tape is present, the installer cuts it off at the jacket. He separates the fibers from the central strength member.

21.5.2.5 TRIM FLEXIBLE STRENGTH MEMBERS

The installer separates the fiber(s) from the aramid yarns. He twists and cuts the aramid yarns to the proper length (Figure 21-8).

21.5.2.6 TRIM CENTRAL STRENGTH MEMBER

The installer cuts the central strength member at the distance indicated for the enclosure (Figure 21-8).

21.5.2.7 ATTACH TO ENCLOSURE

When the termination is complete, the installer places the cable in the enclosure, according to the instructions for the enclosure. He coils the service loop of buffer tube inside the enclosure. He coils the service loop of cable in a convenient location.

21.6 CONTINUITY TEST

After installing a cable, the installer checks the continuity of the cable. He can perform this check with a high intensity light, such as a 1 W LED flashlight, or a visual fault locator (VFL, 18.3.2). The installer performs this check by holding the light up to one end of each fiber. An installer at the other end of the cable checks for a 'sparkle' that indicates continuity.

21.7 ONE PAGE SUMMARY

21.7.1 PULLING

21.7.1.1 LOOSE TUBE

Choose a swivel
Set slitter blade depth
Make a test cut
Remove test cut
Remove jacket
Remove grease
Remove grease remover
Remove core materials
Attach swivel
Install cable
Cut off end
Check continuity

21.7.1.2 PREMISES CABLE

Attach swivel to pull rope
Slit jacket
Remove fibers
Remove central strength member
Attach cable to swivel
Pull cable
Cut off end
Check continuity

21.7.2 TERMINATION

21.7.2.1 LOOSE TUBE

Determine dimensions
Prepare work area
Pre-install cable
Access ripcord
Set slitter depth
Make test cut

© PEARSON TECHNOLOGIES INC.

Remove test cut
Remove jacket
Make ring cut
Remove grease
Remove grease remover
Remove buffer tube
Remove gel
Remove gel remover
Lubricate and separate fibers
Attach cable

21.7.2.2 PREMISES CABLE
Determine dimensions
Install cable into enclosure
Slit jacket
Separate fibers
Trim flexible strength members
Trim central strength member
Attach cable to enclosure

22 CONNECTOR INSTALLATION: EPOXY

Chapter Objectives: with the procedures in this chapter, you install epoxy SC connectors with low loss and high reliability. In addition, you achieve singlemode reflectance below -50 dB.

22.1 INTRODUCTION

These instructions enable the novice installer to achieve low power loss and high reliability.

These instructions enable installation of the SC connector onto four types of jacketed cables. Three of the cables, a single fiber, a zip cord duplex and a break out, have a 3mm jacket on each fiber.[61] The fourth type of cable is a premises cable with a 900μ tight buffer tube. Finally, these instructions enable installation onto loose tube cables with furcation kits, aka 'fan out' kits, on the fibers.

These instructions can apply to other connectors with changes to the dimensions (Figure 22-1 and Figure 22-4) and to the crimp nest diameter(s).

The method of connector installation is epoxy. The epoxy in this procedure requires heat curing for 10 minutes at 85° C. Other epoxies can be used with the appropriate cure times and temperatures.

22.2 MATERIALS AND SUPPLIES

The tools and supplies listed herein are required for installation of this connector by this method. We have chosen some of these tools and supplies on the basis of their superior performance. We have chosen others, on the basis of their price. Finally, we have chosen some of these tools and supplies based on their convenience.

Sources of equipment and supplies are: Fiber Optic Center Inc. (FOCI) 800-ISFIBER and Fiber Instrument Sales (FIS), 800-5000-FIS. The connector is a generic connector with a back shell that accepts a 900μ diameter buffer tube.

For installation of these connectors, the installer requires a connector installation tool kit, which includes the following

- Multimode or singlemode SC connectors
- Work mat (Clauss Fiber-Safe™)
- Tubing cutter (Ideal 45-162)
- Kevlar scissors (with ceramic blade)[62]
- Miller buffer tube and primary coating stripper (FO-103-S) or Clauss NoNik stripper (NN203)
- Syringe (Fiber Optic Center Inc., FOCI)
- Epoxy (FIS part number H05-100-R2 or equivalent; FOCI part number AB9112) μ85° C. connector curing oven (FIS part number F1-9772)
- Lens grade tissue (Kim-Wipes™ or equivalent)
- 98% isopropyl alcohol (Fiber Instrument Sales (FIS))
- Alternative to isopropyl alcohol: Alco pads or Opti-Prep Pads
- Connector cleaner (ElectroWash Px, from ITC Chemtronics)
- Wedge scriber (Corning Cable Systems Ruby Scribe, 3233304-01)
- Small plastic bottle (for bare fiber collection)
- Lens grade compressed gas (Stoner part number 94203 or equivalent)
- One SC polishing tool or one tool per film (FIS or FOCI)
- One hard rubber polishing pads, or one pad for each polishing film, except for the air polish film (FIS part number PP 575)

Multimode polishing requires these materials:

---

[61] At the time of this writing, jacket diameters include: 2.6mm, 2.0mm, and 1.6mm.

[62] Scissors with ceramic blades provide the best results, but have the highest cost.

- 12μ polishing film (FOCI part number AO12F913N100)
- 3μ polishing film (FOCI part number AO3T913N100)
- 1μ polishing film[9] (FOCI part number AO1T913N100
- Optional: 0.5 μ polishing film (FOCI part number AO05T913N100)
- Crimper for the connectors to be installed (0.137" and 0.190")
- 400 x connector inspection microscope with 2.5mm fixture (Westover Scientific)

Singlemode polishing requires these materials:

- 1μ diamond polishing film (Fiber Optic Center Inc., part number D1KT403N1
- 0.5μ diamond polishing film (Fiber Optic Center Inc., part number AO05F913N100
- 0.3μ alumina polishing film (Fiber Optic Center Inc., part number CA03F402N100
- Polishing extender solution (Fiber Optic Center Inc.)

22.3 PROCEDURE

22.3.1 SET UP OVEN

The installer plugs in the oven. If the oven has a temperature adjustment, he sets the temperature to that appropriate for the epoxy to be used. He allows the oven to heat up for at least 15 minutes or for the length of time indicated in the oven instruction sheet.

The films listed and instructions are for an epoxy that cures in 10 minutes at 85° C.

22.3.2 PREINSTALL CABLE

The installer installs the cable into the enclosure without permanently attaching the cable to the enclosure. He pulls the cable out of the enclosure to a work surface near the enclosure.

22.3.3 REMOVE JACKET

By the appropriate procedure (21), the installer removes the outer jacket of a premises cable or of a breakout cable. At this step, the installer does not remove the jacket of 3mm simplex or 3mm zip cord duplex.

22.3.4 INSTALL BOOTS AND SLEEVES

22.3.4.1 PREMISES CABLE

With the small end first, the installer installs the 0.9 mm boot on each buffer tube. He pushes the boots approximately 12" away from end of buffer tubes. He marks the length of buffer tubes to be removed (Figure 22-1).

Strip this much bare fiber

0.688"

Figure 22-1: Template for Length Bare Fiber

22.3.4.2 BREAKOUT AND SIMPLEX CABLE

With the small end first, the installer installs the boots on all fibers to be terminated. He pushes the boots approximately 12" away from end of cable.

The installer installs the crimp sleeves on all fibers to be terminated. The cable enters the small end of the crimp sleeve first. He pushes the crimp sleeve approximately 12" away from end of cable. He does not push the crimp sleeve under the boot.

22.3.4.3 ZIP CORD DUPLEX

The installer separates the two channels by pulling the channels apart. If necessary, the installer cuts the thin web that connects the channels. He installs a boot and crimp sleeve as in 22.3.4.2.

22.3.5 PREPARE EPOXY

The installer twists a needle onto the barrel of a syringe until the needle resists additional twisting. He removes the plunger from the barrel.

He checks the expiration date of the epoxy. If the date has not passed, he uses the epoxy.

He mixes the two-part epoxy by removing the separator and rubbing the package with a mixing roller. As an alternative method, the installer rubs the package over a dull edge, such as the edge of a table. While mixing, he moves all of the epoxy components from one end of package to the other a minimum of 15 times or until the color of the mixture is uniform.

The installer squeezes all the epoxy into one end of the package. He squeezes the epoxy away from one corner of package. He cuts off 1/8" of the corner. He squeezes as much epoxy as possible into the barrel of the syringe.

He inserts the plunger into the barrel 1/8"-1/4". He points the needle up, allowing all of the epoxy to run onto plunger. He presses the plunger into the barrel until all the air passes through the needle. He pulls the plunger back about 1/8", so that epoxy does not 'weep' from the needle. He places the needle on a non-absorbent material, such as the epoxy package. He wipes epoxy from the outside of the needle.

22.3.6 PREPARE END

The installer repeats this step for all fibers to be terminated at this location.

22.3.6.1 PREMISES CABLE

The installer can use one of three methods for removing the buffer tube and primary coating:

> NoNik stripper
> Miller stripper with one hole
> Miller stripper with two holes

If the installer uses a NoNik Stripper, he cleans the stripper by holding the stripper open. He snaps each cap once (Figure 22-2).

Occasionally, this method for the Clauss stripper does not work. In that case, the installer blows out the stripper with lens grade gas.

The installer grips the buffer tube firmly. The installer can grip by wrapping the cable around one finger 5 times (Figure 22-3) or by weaving the cable through his fingers.

The installer places the stripper on the buffer tube at the distance marked (Figure 22-1) or 1/2" from the end, whichever is shorter.

While holding the stripper at 90° to the fiber (

Figure 22-3), the installer pulls the stripper *in the direction of the arrow* on the stripper towards the fiber end slowly. He allows the buffer tube and primary coating to slide from the fiber. He does not force the buffer tube from the fiber.

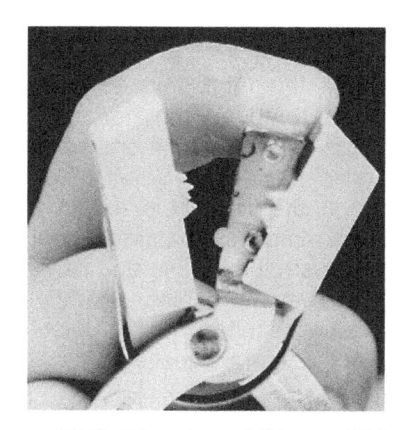

Figure 22-2: Cleaning of Clauss Stripper

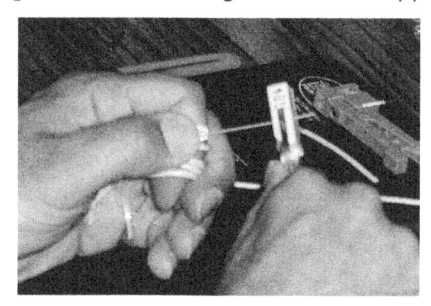

Figure 22-3: Fiber Straight In Stripper

This instruction indicates a maximum strip length of 1/2". US-made cables can be stripped to at least this distance. Some US-made cables allow a longer strip length without breakage. For such cables, you may strip to that increased distance.

If the installer uses a Miller Stripper with one hole, he cleans the stripper with lens grade gas. He grips the buffer tube firmly as described above.

The installer places the stripper on the buffer tube at the marked distance or at 1/2" from the buffer tube end, whichever is shorter. He holds the stripper at 45° to the fiber. He pulls the stripper towards the end of the fiber slowly, allowing the buffer tube to slide from the fiber. He does not force the buffer tube and primary coating from the fiber.

If necessary, the installer repeats this step until the length of bare fiber is as indicated in Figure 22-1. He inspects the fiber to ensure that there is no buffer tube or primary coating remaining on the fiber.

If the installer uses a Miller Stripper with two holes, he cleans the stripper with lens grade gas. He grips the buffer tube firmly as de-scribed above.

The installer places the large hole of the stripper at the total strip length (Figure 22-1). He removes the buffer tube by pulling the stripper to the end of the fiber slowly. He places the small hole of the stripper at the total strip length (Figure 22-1) and removes the primary coating by pulling the stripper to the end of the fiber slowly. With some cables, removal of the buffer tube results in simultaneous removal of the primary coating.

22.3.6.2 JACKETED FIBERS

22.3.6.2.1 REMOVE JACKET

The installer removes the inner jacket by placing the tubing cutter at the indicated distance from the cable end (Figure 22-4 and Figure 22-5). He rotates the cutter around the cable once or twice. He removes the cutter and slides the jacket from the cable.

Remove this much jacket

■■■■■■■■■■■ 1.25"

Leave this much buffer tube

■■■■■■ 0.625"

Leave this much aramid yarn

■■■ 0.313"

Figure 22-4: Template of Strip Lengths

22.3.6.2.2 STRIP BUFFER TUBE

The installer holds the aramid yarns back against the remaining jacket. He marks the buffer tube length to be left (Figure 22-4). He strips the buffer tube and primary coating as in 22.3.6.1.

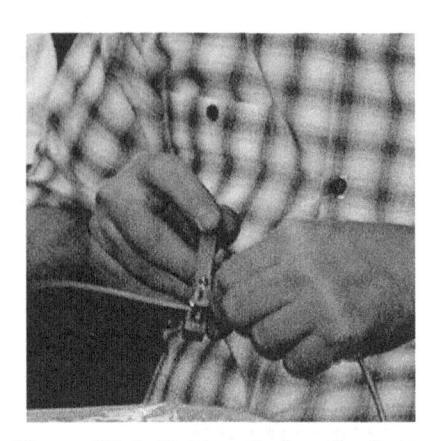

Figure 22-5: Tubing Cutter On Cable

22.3.6.2.3 TRIM STRENGTH MEMBERS

The installer separates the aramid yarn from the buffer tube. He twists the yarn. He cuts the yarn to the length in Figure 22-4 (Figure 22-6).

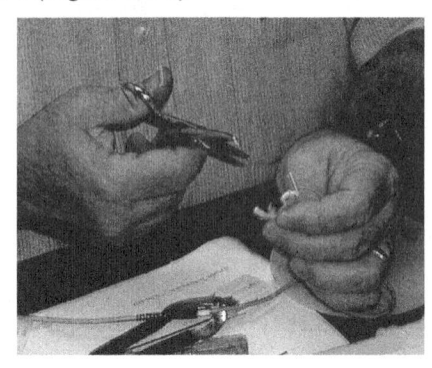

Figure 22-6: Cutting Aramid Yarn To Length

Some connectors allow trimming the strength member as the second step. For such connectors, the strength member length must be less than the buffer tube length.

22.3.6.3 CABLE CONTINUITY TEST

This test is optional, but recommended. This test works on both multimode and singlemode fibers to several thousand feet. The installer holds a high intensity light source close to the fibers at one end of the cable. A 1 W, single-LED flashlight works well. An installer at the opposite end views the fibers. Each fiber should glow or sparkle, indicating continuity.

22.3.6.4 FIBER STRENGTH TEST

The installer performs this step to troubleshoot fiber breakage. This step is optional. The installer performs this step for each fiber just prior to inserting it into a connector. The installer pushes each fiber against a lens grade tissue on the work surface. If the fiber bends without breaking, the installer has not damaged the cladding. If there is a single defect on the cladding, the fiber will break.

> SAFETY WARNING: the installer should not push a bare fiber against his finger. Doing so can result in the fiber penetrating the skin and breaking.

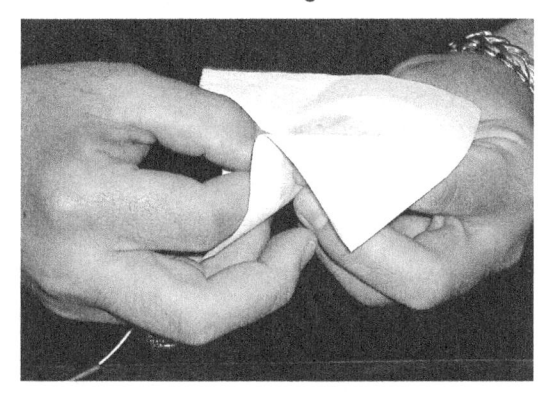

Figure 22-7: Fiber Cleaning

22.3.7 CLEAN FIBER

The installer performs this step for each fiber immediately prior to inserting it into a connector. The installer moistens a lens grade tissue with isopropyl alcohol. He folds the moist area around the fiber (Figure 22-7). He pulls the fiber through the fold twice. He inspects the fiber for particles on the cladding. If necessary, he cleans the fiber repeatedly until the cladding is free of all particles.

22.3.8 CONNECTOR TESTS

At this step, the installer can make two tests. Both these tests are troubleshooting steps for failure of fibers to fit into connectors or for fiber breakage. Both tests are optional. If necessary for troubleshooting or desirable for practice, the installer repeats these steps for each connector.

22.3.8.1 WHITE LIGHT TEST

While aiming the ferrule towards a light source, such as a window or ceiling light, the installer looks into the back shell of a connector. He should see a sparkle that indicates the presence of a fiber hole and the absence of contamination in the connector. Such contamination could block the path of the fiber.

22.3.8.2 DRY FIT TEST

While resting his hands together, the installer feeds the fiber into central tube in the back shell of the connector (Figure 22-8). As soon as the fiber enters the back shell, he twists the connector back and forth. As long as the fiber continues feeding into the back shell without bending, he continues twisting the connector and inserting the fiber. If the fiber bends, he withdraws the fiber approximately 1/8", twists the connector back and forth and feeds the fiber into the back shell.

Figure 22-8: Central Tube Of SC Connector

The installer must see the fiber protruding beyond the tip of the ferrule. This protrusion indicates that the fiber and the hole are compatible, that the fiber is sufficiently long and that there is nothing blocking the fiber path through the ferrule. The installer removes the fiber from the connector and places that connector near that fiber.

22.3.9 EPOXY INJECTION

Before each use, the installer wipes the epoxy from the outside of the needle. He places the needle into the connector back shell until the needle butts against the inside end of the ferrule. While maintaining pressure on the needle, he presses the plunger until the epoxy flows through the fiber hole in the end of the ferrule. As soon as the adhesive flows through the fiber hole, he removes the needle from the connector. He pulls the plunger 1/8" out of the syringe barrel to prevent 'weeping'. With experience, he can repeat this step for up to 12 connectors.

The installer can experiment with filling more than 12 connectors. However, all epoxies have a useful life, called the pot life. Should the epoxy harden excessively, the installer can experience two problems. First, the installer may not be able to feed the fiber through the ferrule. Second, the installer may break the fiber, resulting in loss of the connector.

This step is optional and recommended for installers with experience. With lens grade tissue, the installer wipes the tip of the ferrule to remove all of the epoxy from the tip.

Some installers will benefit from leaving the epoxy on the tip of the ferrule. This epoxy provides additional support for the fiber during polishing. However, this epoxy increases polishing time.

22.3.10 FIBER INSERTION

The installer inserts the fiber into the center tube in the back shell of a filled connector. As soon as the fiber enters the back shell, he twists the connector back and forth. As long as the fiber continues into the back shell without bending, he continues twisting the connector and inserting the fiber. If the fiber bends, the installer withdraws the fiber 1/8", twists the connector back and forth and feeds the fiber into the back shell.

When the buffer tube stops further motion of the fiber into the ferrule, he inspects the tip of the ferrule. He should see the fiber protruding beyond the tip of the ferrule. The amount of protrusion is not critical, as long as this length is not excessive. Excessive protrusion may result in the fiber breaking during insertion of the connector into the curing oven.

If he does not see the fiber protruding beyond the tip of the ferrule, he removes the fiber. The installer examines the fiber. If the fiber is shorter than it was after preparation the broken fiber is inside the connector. He discards the connector. If the fiber has not broken, the installer can reuse the same connector.

He repeats end preparation (22.3.6). If the time to prepare a new end is short, the installer can insert the new end into the same connector.

If the fiber is jacketed fiber design, the installer proceeds to 22.3.11. If the cable is a premises design, the installer proceeds to 22.3.12.2.

22.3.11 CRIMP SLEEVE

The installer repeats this step for each connector that has epoxy and has fiber protruding beyond the tip of the ferrule. He checks the length of the strength members. They should be flush with the end of the crimp sleeve or less than 1/16" longer than the crimp sleeve.

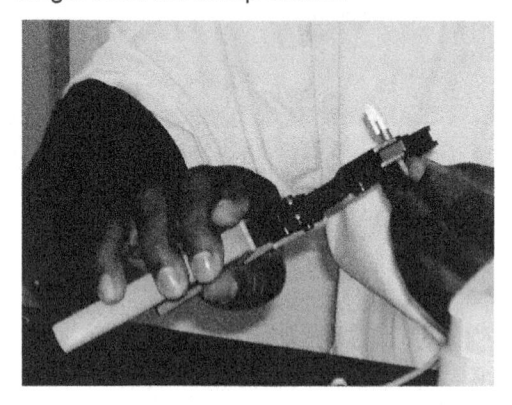

Figure 22-9: Crimping Crimp Sleeve

He slides the crimp sleeve up to and over the back shell of the connector. While sliding the sleeve over the back shell, he rotates the sleeve back and forth. He crimps the crimp sleeve with the crimp nest or nests specified by the connector manufacturer (Figure 22-9). He crimps the large diameter of the sleeve over the back shell of the connector and the small diameter of the crimp sleeve over the jacket.

22.3.12 CONNECTOR INSERTION

22.3.12.1 JACKETED FIBERS

While holding onto the crimp sleeve of the connector, the installer centers the connector over the oven port (Figure 22-10). Slowly, he lowers the connector into the port. If he feels any 'springiness' or resistance, the fiber is off center and touching the inside of the oven. If he detects such resistance, he lifts the connector 1/4", re-centers the connector in the oven port and reinserts slowly. If the springiness persists, the fiber protruding beyond the ferrule is excessively long.

Carefully, the installer trims the excess length without breaking the fiber below the end of the ferrule.

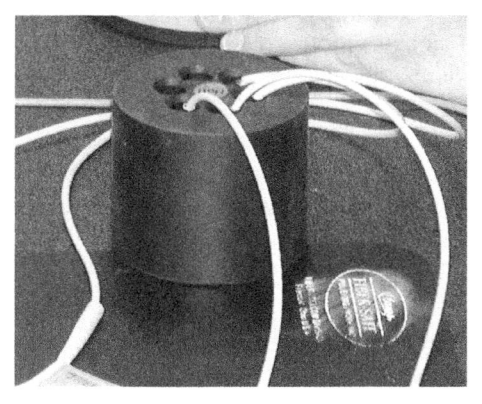

Figure 22-10: Connectors In Curing Oven

22.3.12.2 PREMISES CABLES

The installer slides the boot over the back shell. While squeezing the boot against the back shell of the connector, the installer inserts the connector into the oven without placing tension on the fiber. He records the cure start time for each connector or batch of connectors.

After the specified minimum cure time, he removes each connector from the oven. The cure time for the epoxy in this procedure is ten minutes. If no oven position is available for a new connector, the installer places the connector horizontally on the work surface in such a manner to prevent tension on the fiber.

The installer may tape the cables or buffer tubes to the work surface so that the cables do not slide out of the connectors. Should the cables slide, the fibers protruding beyond the end of the ferrule may break.

> Do not pause or delay installing fibers into filled connectors because the epoxy hardens continuously after being mixed.

> You may cure the epoxy for more than the minimum curing time as excessive cure time causes no problems. However, excessive temperature can cause cracking of the fiber.

Repeat this step until you have cured and allowed all connectors to cool. Do not force cool connectors.

22.3.13 REMOVE EXCESS FIBER

While watching the fiber at the end of the ferrule, the installer pulls gently on the jacket or buffer tube. The fiber should not move. If it moves, the epoxy has not cured. Once the epoxy has cured, the installer proceeds.

While resting his hands together, the installer places the wedge surface of a scriber onto the end of the ferrule (Figure 22-11). He moves the blade of the scriber up to the fiber so that the scriber gently touches the fiber at the top of the epoxy bead. The fiber should not move or bend.

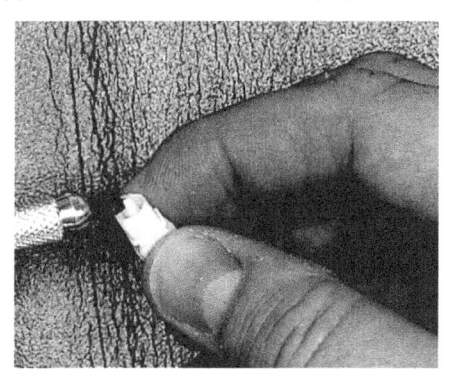

Figure 22-11: Scribing the Excess Fiber

The installer moves the scriber along the fiber 1/16" once. He should not break the fiber with the scriber. He should not 'saw' the fiber with the scriber.

The installer holds the connector with the fiber pointing up. He grips the connector with his thumb and forefinger lightly. He slides his thumb and forefinger up the connector towards and over the fiber. He grips and pulls the fiber away from the tip of the connector. The fiber should break easily. He places of the fiber in the fiber collection bottle.

If the fiber does not break, the installer scribed epoxy on the fiber and not the fiber. He repeats this step until the fiber breaks.

22.3.14 AIR POLISH

The installer holds the connector with the ferrule pointing up. He holds a 12μ polishing film at one edge with the dull (abrasive) side down. He places the opposite edge of the film above the connector. He curls the front and back edges of the film down. With a light pressure, he rubs the film against connector until the fiber is flush with the bead of epoxy (Figure 22-12). If he scribed the fiber just above the surface of the epoxy bead, this step will take less than 10 seconds. The installer does not remove the epoxy.

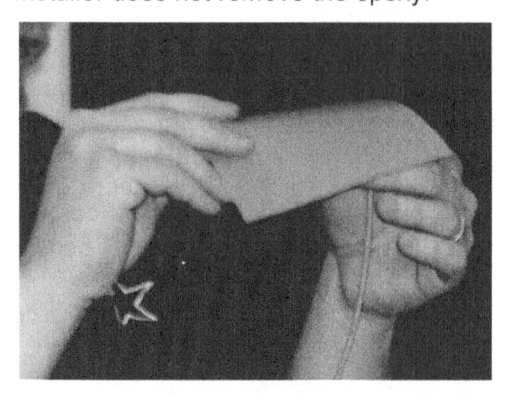

Figure 22-12: Air Polishing of Fiber

The installer checks the fiber by bringing a finger down onto the tip of the ferrule from the top. If he cannot feel the fiber, the fiber is flush, or nearly flush, with the epoxy.

Caution: do not rub a finger across the tip of the ferrule. If the fiber is not flush with the adhesive, the installer may snag and break the fiber. In addition, the broken fiber may pierce the installer's skin.

22.3.15 MULTIMODE POLISH

For single connectors, the installer follows the procedure in this section for that connector. For multiple connectors, the installer can perform each step on all connectors prior to performing the next step; i.e., 'batch processing'.

22.3.15.1 CLEAN CONNECTORS

The installer moistens a lens grade tissue with isopropyl alcohol. He wipes the ferrules of all connectors with this moistened tissue. This removes 12μ grit.

22.3.15.2 CLEAN EQUIPMENT

The installer can clean the polishing equipment with either lens grade gas or isopropyl alcohol. With lens-grade compressed gas, he cleans off the top surface of the polishing pads, both sides of the polishing films and the polishing tools. With isopropyl alcohol and lens grade tissues, the installer cleans his equipment in a similar manner. He places the polishing films, dull side up, on the pads.

22.3.15.3 FIRST POLISH

The purpose of the first polish is removal of the epoxy. Additional time after such removal will not improve the loss. In fact, additional time can result in increased loss.

The installer places the polishing tool, hereafter 'tool', on the 3μ film. He inserts the connector into the polishing tool. While holding both the connector and the tool (Figure 22-13), he moves the tool in a ½" high, figure-8 pattern slowly. If he feels scratchiness, the fiber is not flush with the epoxy.

Light pressure means that the installer holds the connector against the film without compressing the spring inside the SC connector. He maintains a light pressure and a slow, ½" high figure-8 pattern until the scratchiness ceases.

After the scratchiness ceases, the installer increases the size of the figure-8 pattern to cover the entire film. He does not increase the pressure. He continues polishing until the epoxy is completely removed.

Figure 22-13: Polishing

The installer may detect a change in the friction of the connector on the film. This change indicates the removal of the last of the adhesive. If he does not detect this change, he removes the connector from the tool periodically. He views the tip of the ferrule by reflecting light off the tip. When the epoxy is completely removed, the tip of the ferrule will be glass smooth and shiny. The tip will have the same appearance as that of the side of the ferrule.

22.3.15.4 CLEANING

With lens grade compressed gas or isopropyl alcohol and lens grade tissues, the installer cleans the connector and the tool.

22.3.15.5 SECOND POLISH

The purpose of the second polish is creation of a 'featureless,' low loss core (20.4). 'High pressure' for polishing means holding the connector in the polishing tool and pushing the inner housing down until the internal spring is fully compressed. The installer need not increase the pressure beyond the level at which the spring is fully compressed.

The installer places the polishing tool on the 1μ film and applies high pressure to the connector for ten, large figure-8 patterns. Large means using the full area of the film.

22.3.15.6 THIRD POLISH

This polish is optional. Some connector manufacturers recommend this third polish. This author has found no improvement in loss from this polish on multimode connectors. We have found only minor improvement in the microscopic appearance of the core.

The installer cleans the connector and polishing tool (22.3.15.4). The installer places the polishing tool on the 0.5μ film. He inserts the connector in the tool and applies high pressure to the connector for ten, large figure-8 patterns.

22.3.16 FINAL CLEANING

The installer slides the boot over the back shell. He cleans the connector with one of the following methods.

22.3.16.1 BEST METHOD

The installer moistens a lens grade, lint free tissue with isopropyl alcohol. He wipes the sides and tip of the ferrule with this tissue. He sprays a one-inch diameter area of connector cleaner, such as ElectroWash Px, onto a small pad of lens grade tissues. Three times, he wipes the tip of the ferrule from the wet area to the dry area.

Optional: the installer blows out the cap with lens grade gas. He installs the cap.

22.3.16.2 METHOD B

The installer moistens a tissue with isopropyl alcohol. He wipes the sides and the tip of the ferrule. Immediately, he wipes the tip of the ferrule with a dry tissue.

22.3.16.3 METHOD C

The installer wipes the sides and the tip of the ferrule with a pre-moistened lens grade tissue (Alco Pad or OptiPrep Pad). Immediately, he wipes the tip of the ferrule with a dry tissue. He installs the cap.

Optional: with lens grade gas, the installer blows out the connector cap. He installs the cap.

Optional: with lens grade gas, the installer blows out the connector cap. The installer installs the cap.

Do not use lens grade gas for final connector cleaning. This gas can leave water marks that increase loss and reflectance.

22.3.17 INSPECT CONNECTOR

The installer inspects and rates the connector according to the inspection procedure (20) If the connector is good, he installs the cap. If the connector is not good, he attaches a label to indicate a potential problem and installs the cap. If the problem is severe enough to indicate unacceptable loss, he replaces the connector.

22.3.18 WHITE LIGHT TEST

The installer performs a white light, continuity test as in 21.6 (Figure 22-14). The installer does not install connectors on the opposite end until all of the fibers have passed this continuity test. If one or more connectors fail this test, he troubleshoots these connectors to identify the problem. An OTDR test may be required.

The installer replaces the caps on all connectors that pass this test. He repeats this continuity test after he installs all connectors on the opposite end of the cable.

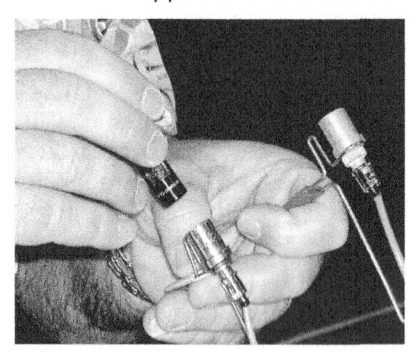

Figure 22-14: White Light Test

22.3.19 FINAL ASSEMBLY

The installer slides the boot over the back shell of the connector. The top of the inner housing has missing corners (Figure 22-15). The top of the outer housing has the key. He holds the top of the inner housing and the top of the outer housing up. He inserts the inner housing into the end of the outer housing that is opposite the end with the key (Figure 22-16). He wiggles the inner housing until it snaps through the outer housing (Figure 22-17).

Figure 22-15: Alignment Housings

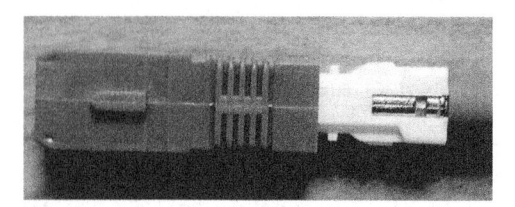

Figure 22-16: Insertion of Inner Housing

Figure 22-17: Inner Housing Fully Inserted

22.3.20 SINGLEMODE POLISHING

Singlemode polishing can be used for initial installation or for restoration of low reflectance. Because of high cost, field polishing of singlemode connectors by hand is done rarely. Pigtail splicing is preferred.

Our assumption is that the defects causing high reflectance are shallow. Although deep defects can be removed with 5µ diamond film, we recommend connector replacement as polishing cost can be excessive.

We provide this procedure, with the recommendation that it be used for initial installation when no other option is available. This procedure can be used for restoring a singlemode connector to low reflectance by replacing the 3µ film with a diamond film of 3µ.

22.3.20.1 CLEANING

As in 22.3.15.4, the installer cleans the polishing pads, the polishing tool and the following films:

> 3µ
> 1µ diamond
> 0.5µ diamond
> 0.3µ

22.3.20.2 FIRST POLISH

With the 3µ film the installer polishes the connector as in 22.3.15.3.

22.3.20.3 CLEANING

With isopropyl alcohol and lens grade tissues, the installer cleans the connector. He scrubs the tool with lens grade tissues moistened with isopropyl alcohol at least twice.

22.3.20.4 SECOND POLISH

The installer places the polishing tool on the 1μ diamond film. 'High pressure' for polishing means holding the connector in the polishing tool and pushing down the inner housing until the internal spring is fully compressed. The installer need not increase the pressure beyond the level at which the spring is fully compressed. The installer applies high pressure to the connector for 20 to 40, large figure-8 patterns. Large means using the full area of the polishing film.

The installer polishes for 20 strokes for new diamond film. He polishes 40 strokes after he has polished five connectors on the film. He can use the same film for at least 10 connectors.

22.3.20.5 CLEANING

The installer cleans the connector and polishing tool as in 22.3.15.4.

22.3.20.6 THIRD POLISH

With a high pressure, the installer polishes the connector for 20-40 large figure 8 strokes on the 0.5μ film. The installer polishes for 20 strokes for new diamond film. He polishes 40 strokes after he has polished five connectors on the film. He can use the same film for at least 10 connectors.

22.3.20.7 CLEANING

The installer cleans the connector and polishing tool as in 22.3.15.4.

> There must be no residue on the tool from previous polishing steps.

If the installer uses a single polishing tool for all films, he scrubs the tool with a toothbrush and isopropyl alcohol before each use on the 0.3μ film. Failure to do so will degrade the condition of the core to worse than that after the 0.5μ film.

An alternative polishing method is use of a separate tool for each film. This method reduces the time spent on cleaning, as the installer need clean only the connector.

22.3.20.8 FOURTH POLISH

The installer places the 0.3μ film on a polishing pad. The installer adds two drops of the polishing extender liquid to the film. With a high pressure, he polishes the connector for 20 large figure 8 strokes. He replaces this film after five connectors.

22.3.21 FINAL CLEANING

As in 22.3.16, the installer cleans the connector with isopropyl alcohol and lens grade tissue three times.

22.3.22 INSPECTION

The installer inspects the connector as in 22.3.17. In addition, he inspects the cladding for uniformity of appearance (Figure 20-6).

22.4 TROUBLESHOOTING

22.4.1 INSTALLATION

22.4.1.1 FIBER BREAKAGE DURING INSERTION

Potential cause: dirt or primary coating on fiber
Action: clean fiber

Potential cause: debris in connector
Action: perform white light check of connector

If the connector fails this test, the installer blows out connector with lens grade gas. If the connector still fails this test, he flushes the connector with isopropyl alcohol injected through the connector with a syringe needle. If the connector fails after this flush, he replaces the connector.

22.4.1.2 FIBER BREAKS OR DOES NOT FIT INTO CONNECTOR

Potential cause: primary coating not completely removed
Action: restrip fiber

Potential cause: epoxy has partially hardened
Action: replace connector

22.4.1.3 EPOXY FLOWS FROM BACK SHELL

Potential cause: excessive adhesive in back shell
Action: after epoxy flows from fiber hole in ferrule, immediately withdraw needle from connector

22.4.1.4 EPOXY ON OUTSIDE OF BACK SHELL

Potential cause: failure to wipe needle before each use
Action: wipe epoxy from outside of needle before each use

22.4.2 POLISHING

22.4.2.1 NON-ROUND CORE

Potential cause: fiber broken during first polishing due to incomplete air polishing
Action: replace connector and perform complete air polish

Potential cause: excessive pressure during first pad polishing
In this case, polishing sheared of the epoxy bead. This bead may be found on the first film.
Action: reduce polishing pressure

22.4.2.2 EXCESSIVE POLISHING TIME

Potential cause: large bead of epoxy due to failure to remove epoxy from tip of ferrule prior to insertion of fiber
Action: remove all epoxy from tip of ferrule after injecting epoxy

Potential cause: film worn out
Action: replace film

Potential cause: insufficient polishing pressure
Action: increase polishing pressure

22.4.2.3 NO APPARENT REDUCTION OF BEAD SIZE

Potential cause: dirt in ferrule hole of polishing tool
Action: clean hole in polishing tool with lens grade air or with pipe cleaner dipped in isopropyl alcohol.

22.4.2.4 FEW CORE AND CLADDING SCRATCHES

Potential cause: contamination of film by environment.
Action: replace film

During field installation, contamination of polishing films is common. Often, such contamination is unavoidable. The installer cleans the film only if no replacement film is available. In addition, he can move the polishing location away from air vents and any other source of airborne dust.

22.4.2.5 EPOXY SMEARS ON FIBER DURING POLISHING

Potential cause: uncured or incompletely cured epoxy due to insufficient time
Action: increase curing time. Monitor dwell time in oven with written log

Potential cause: low curing temperature
Action: check oven temperature

Potential cause: epoxy used past its expiration date
Action: discard epoxy

Potential cause: epoxy allowed to freeze
Action: discard epoxy[63]

22.4.2.6 CRACKED FIBER

Potential cause: excessive pressure during scribing
Action: scratch fiber lightly during scribing

Potential cause: excessive temperature during curing
Action: repair or replace oven

22.4.2.7 EPOXY FLOWS FROM BACK SHELL

Potential cause: excessive epoxy in back shell
Action: replace connector; after filling fiber hole in ferrule, immediately withdraw needle

---

[63] Some epoxy loses its ability to cure after it has been frozen. Store such epoxy so that it does not freeze.

22.4.2.8 EPOXY ON OUTSIDE OF BACK SHELL.

Potential cause: failure to wipe needle before each use.
Action: wipe epoxy from outside of needle before each use

22.4.2.9 EPOXY ON OUTSIDE OF INNER TUBE

Potential cause: failure to wipe needle before each use
Action: discard connector; wipe epoxy from outside of needle before each use

See 22.3.9

22.4.3 SINGLEMODE POLISHING

22.4.3.1 PITS REMAIN AFTER FIRST 20 STROKES ON NEW 1 µ FILM

Potential cause: insufficient polish pressure
Action: increase pressure as described in procedure

22.4.3.2 PITS REMAIN AFTER FIRST 20 STROKES ON 'OLD' 5 µ FILM.

Potential cause: film worn out
Action: replace film

22.4.3.3 HIGH REFLECTANCE WITH FEATURES NEAR CORE

Potential cause: features create roughness
Action: repolish starting with 3 µ diamond film

22.4.3.4 HIGH REFLECTANCE WITH NO FEATURES

Potential cause: roughness not visible at 400 x.
Action: repolish starting with 1 µ film

22.4.3.5 HIGH REFLECTANCE

Potential cause: dirt on connector outside of field of view
Action: re-clean and retest

22.4.3.6 PITS APPEAR AFTER POLISHING ON 0.3 µ FILM[64]

Potential cause: the polishing tool was contaminated with debris from the prior polishing films
Action: thoroughly clean the tool prior to polishing on 0.3 µ film

22.5 ONE PAGE SUMMARY

This summary is in three parts, one each for connector installation, multimode polishing and singlemode polishing.

22.5.1 INSTALLATION

Set up oven. Plug in. Allow to preheat.

Install cable through enclosure.

Pull cable through enclosure to work surface.

Remove jacket, Kevlar and central strength member to proper lengths.

Mix epoxy thoroughly and fill syringe.

Install boots on all buffer tubes.

Remove buffer tubes with stripper to proper lengths.

Optional Step: give all connectors a white light test.

Optional Step: give all connectors a dry fit.

For up to 12 connectors, inject epoxy through fiber hole and a small amount into back shell.

Optional Step: Wipe all epoxy off tip of ferrules.

Clean fiber with lens grade tissue and isopropyl alcohol.

While rotating fiber or connector, insert fiber into back shell until fiber bottoms out against ferrule. Do not allow fiber to bend.

For jacketed fiber, slide crimp sleeve over back shell and crimp sleeve. For premises cable, bring boot over back shell.

While keeping connector centered over oven port, insert connector into oven

---
[64] See Figure 21-6.

without breaking fiber. Allow to cure 10 minutes.

Remove connector from oven and allow to cool. Do not force cool in any way.

Scribe and remove excess fiber.

Air polish excess fiber flush with epoxy.

Proceed to multimode or singlemode polishing procedure.

22.5.2 MULTIMODE POLISH

Clean all connectors.

Clean pads, films and tool. Place 3 μ film on pad and tool on film.

Install connector into tool.

Polish in small figure 8 pattern with light pressure until scratchiness ceases.

When scratchiness ceases, polish in a large figure 8 pattern with slightly higher pressure until all the epoxy is gone. Polish the connector using the entire area of the film.

Clean the 1 μ film, the tool and the connector.

Install the connector into the tool. With moderate pressure, make 10 large figure 8 motions.

Clean the side and tip of the connector.

Proceed to connector inspection procedure (20).

Perform a white light continuity test after you install connectors on the first end of the cable.

Perform a white light continuity test after you install connectors on the second end of the cable.

Install outer housing.

Clean and install cap on ferrule.

22.5.3 SINGLEMODE POLISH

Scribe and air polish the fiber flush with the epoxy.

Clean the ferrules of all connectors.

Clean off the polishing pad, the polishing films and the polishing tool.

Polish on the 3 μ film to remove the epoxy.

With lens grade gas, clean the connector, pad, tool and the 1 μ diamond film.

With a heavy pressure, polish the connector on the 1 diamond μ film for 20-40 large figure 8 strokes.

With lens grade gas, clean the connector, pad, tool and the 0.5 μ diamond film.

With a heavy pressure, polish the connector on the 0.5 μ film for 20-40 large figure 8 strokes.

With lens grade gas, clean the connector, pad, tool and the 0.3 μ film thoroughly.

Place two or three drops of the polishing liquid on the film. With a heavy pressure, polish the connector on the 0.3 μ film for 20 large figure 8 strokes.

Clean the connector with alcohol and lens grade tissue three times.

Proceed to connector inspection (20).

Perform a white light continuity test after you install connectors on the first end of the cable.

Perform a white light continuity test after you install connectors on the second end.

Install outer housing.

Clean and install cap on ferrule.

Install boot on jacketed cable.

23 CONNECTOR INSTALLATION: HOT MELT ADHESIVE

Chapter Objectives: you ins(tall multimode, Hot Melt adhesive, ST-compatible connectors with low loss and high reliability on 900-μ tight tubes.

23.1 INTRODUCTION

This chapter applies to the installation of ST™-compatible and SC multimode, connectors onto fibers in 900-μ tight tubes. This tube can be in a premises cable or a loose tube cable with furcation tubes on each fiber.

Installation of connectors by this method on other cable types requires end preparation appropriate to the cable design.

The ST-™ compatible connector is shown. The SC connector installation method differs in the following ways:

- in the design of the holder
- the alignment of the flat surfaces of the boot to the flat surfaces of the inner housing
- the installation of the outer housing, which is the same as that in 22.3.19.

This method of connector installation is a hot melt adhesive. This adhesive is pre-loaded into the connector. The connector requires pre-heating prior to and cooling after installation.

This method reduces installation time. This method allows for salvage of damaged connectors through reheating. In addition, the preloaded adhesive creates a bead of adhesive on the tip of the ferrule with a predetermined size. These latter two characteristics enable the novice installer to achieve high process yield. Typical yield by first-time installers is 90-95 %. In field installations, reheating enables a 100% yield.

23.2 MATERIALS AND SUPPLIES

The tools and supplies listed herein are required for installation of this connector by this method. We have chosen some of these tools and supplies on the basis of their superior performance. We have chosen others on the basis of their price. Finally, we have chosen some of these tools and supplies based on their convenience.

For installation of these connectors, the installer requires a connector installation kit, which includes the following:

- 3M HOT MELT, ST-compatible multimode connectors, part number 6100 or SC connectors, multimode part number 6300W
- Work mat (Clauss Fiber-Safe™)
- Tubing cutter (Ideal 45-162)
- Kevlar scissors (with ceramic blade)
- Miller buffer tube and primary coating stripper (FO-103-S) or Clauss NoNik stripper (NN203)
- Hot Melt cooling stand or Hot Melt oven with built-in cooling ports
- Hot Melt holders for ST™-compatible (or SC) connectors
- Hot Melt oven
- 2μ polishing film for Hot Melt™ connectors (FIS, 3M part number 51144 85932, 254X Imperial, 6192A)
- Lens grade tissue (Kim-Wipes™ or equivalent)
- 99% isopropyl alcohol (Fiber Instrument Sales, FIS)
- Alternative to isopropyl alcohol: Alco pads or Opti-Prep Pads
- Connector cleaner (Electro Wash® Px, from ITC Chemtronics)
- Wedge scriber (Corning Cable Systems Ruby Scribe, 3233304-01)
- Small plastic bottle (for bare fiber collection)
- Lens grade compressed gas (Stoner part number 94203 or equivalent)
- ST-compatible polishing tool (FIS or FOCI)

© PEARSON TECHNOLOGIES INC.

- Two hard rubber polishing pads, (FIS part number PP 575)
- 12μ polishing film (FOCI part number AO12F913N100)
- For multimode polishing: 1μ polishing film (FOCI part numbers AO1T913N100)
- 400 x connector inspection microscope with 2.5mm fixture (Westover Scientific)

Scissors with ceramic blades provide the best results, but have the highest cost. Scissors with other types of blades will work. The polishing tool is also called a fixture and a puck. No pad is required for the 12μ air polish film. The films listed are for multimode polishing.

Sources of equipment and supplies: Fiber Optic Center Inc. (FOCI) 800-ISFIBER and Fiber Instrument Sales (FIS), 800-5000-FIS.

23.3 PROCEDURE

23.3.1 PREINSTALL CABLE

The installer installs the cable into the enclosure without permanently attaching the cable to the enclosure. He pulls the cable out of the enclosure to a work surface near the enclosure.

23.3.2 SET UP OVEN

The installer turns on the preheating oven and allows it to heat up for at least 15 minutes. He assembles the cooling stand (Figure 23-1).

23.3.3 LOAD HOLDERS

The installer removes the connector parts from the packages. He removes the connector caps. He aligns the key of a connector with the slot in a connector holder. He slides the connector into the holder. He rotates the retaining nut so that the holder retains the connector (Figure 23-2). The installer loads 4-6 connectors into holders. He places the holders into the cooling stand.

For SC connectors, he aligns the flat surfaces of the connector to the clips of a holder (Figure 23-3). He pushes the connector into the holder and rotates the connector 90° (Figure 23-4)

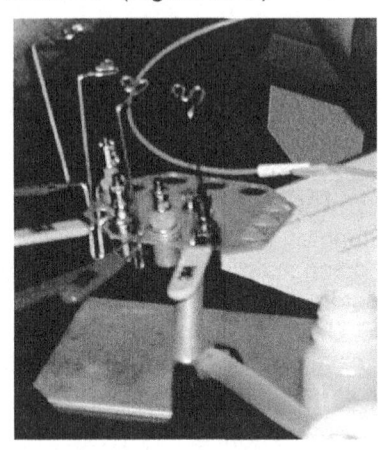

Figure 23-1: Hot Melt Cooling Stand

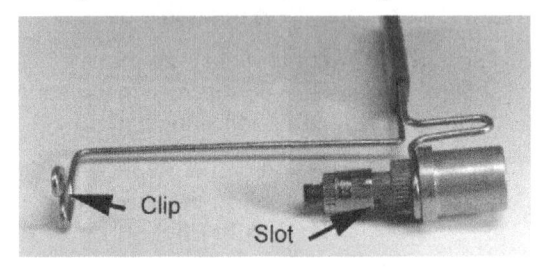

Figure 23-2: Loaded Connector Holder

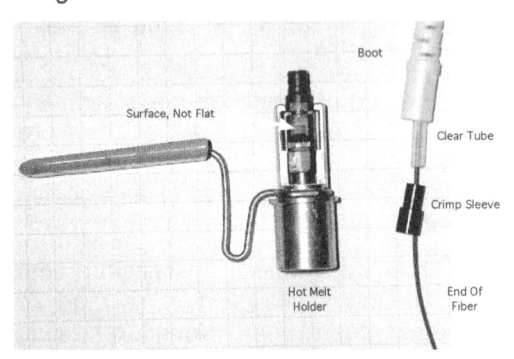

Figure 23-3: SC And Holder Prior to Rotation

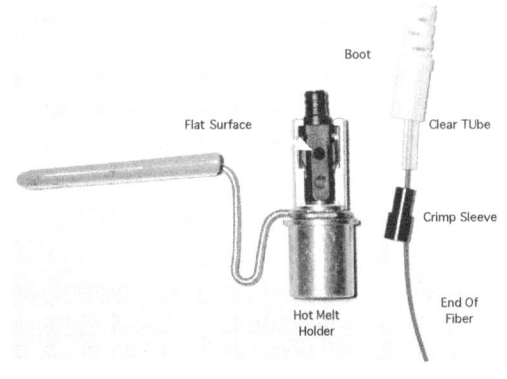

Figure 23-4: SC And Holder After Rotation

23.3.4 REMOVE OUTER JACKET

The installer removes the outer jacket of the premises cable (21.5.2.3).

23.3.5 INSTALL BOOTS

23.3.5.1 PREMISES CABLE

The installer installs the boot and a clear flexible tubing on each buffer tube. He pushes the boots and tubes approximately 12" away from end of buffer tubes. He marks the length of buffer tube to be removed (Figure 23-5).

▬▬▬▬▬▬▬▬▬

Strip this much bare fiber (1.06")

Figure 23-5: Template of Strip Length For Premises Cable (to scale)

For the SC connector, the installer the boot, a clear flexible tubing, and crimp sleeve on each buffer tube in the sequence indicated. The small ends of the boot and crimp sleeve slide on first.

23.3.6 PREPARE END

The installer repeats this step for all fibers to be terminated at this location.

23.3.6.1 NO-NICK STRIPPER

If the installer uses a No-Nick® Stripper, he cleans the stripper by holding the stripper open and snapping each cap once (Figure 23-6).

Occasionally, this method of cleaning the Clauss stripper does not work. In that case, the installer may blow out the stripper with lens grade gas.

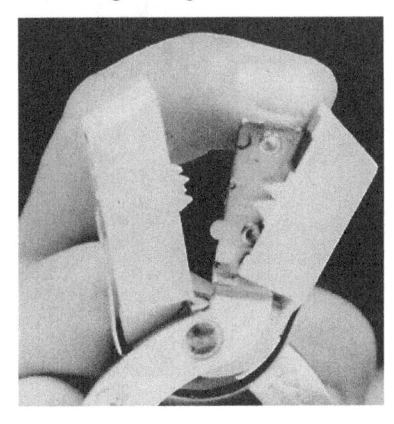

Figure 23-6: Cleaning Of Clauss Stripper

The installer grips the cable or buffer tube firmly. The installer can grip the cable by wrapping the cable around one finger 5 times (Figure 23-7) or by weaving the cable through his fingers. The installer places the stripper on the buffer tube at the marked distance (Figure 23-5) or 1/2" from the end of the buffer tube, whichever is shorter. While holding the stripper at 90° to the fiber (Figure 23-5), the installer pulls the stripper *in the direction of the arrow* on the stripper towards the fiber end slowly. He allows the buffer tube and primary coating to slide from the fiber. He does not force the buffer tube from the fiber.

This instruction indicates a maximum strip length of 1/2". US-made fiber optic cables can be stripped to at least this distance. Some cables allow a longer strip length without breakage. For such cables, you may strip to that increased distance.

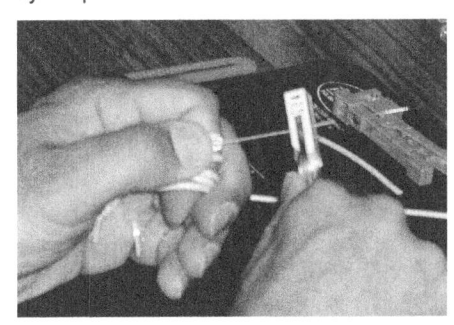

Figure 23-7: Fiber Straight In Stripper

23.3.6.2 MILLER STRIPPER

If the installer uses a Miller Stripper with two holes, he cleans the stripper with lens grade gas. He grips the buffer tube firmly as described above.

The installer places the large hole of the stripper at the total strip length (Figure 23-5). He removes the buffer tube by pulling the stripper to the end of the fiber slowly. He places the small hole of the stripper at the total strip length (Figure 23-5) and removes the primary coating by pulling the stripper to the end of the fiber slowly. With some cables, removal of the buffer tube results in simultaneous removal of the primary coating.

The installer can use a Miller Stripper with one hole. He cleans the stripper with lens grade gas. The installer grips the buffer tube firmly as described above.

The installer places the stripper on the buffer tube at the marked distance or at 1/2" from the buffer tube end, whichever is shorter.[13] He holds the stripper at 45° to the fiber. He pulls the stripper slowly towards the end of the fiber, allowing the buffer tube to slide from the fiber. He does not force the buffer tube and primary coating from the fiber.

If necessary, the installer repeats this step until the length of bare fiber is as indicated in Figure 23-5 (for premises cable) or the length of bare fiber is as indicated in Figure 22-4 (for jacketed cable). He inspects the fiber to ensure that there is no buffer tube or primary coating remaining on the fiber.

23.3.6.3 CABLE CONTINUITY TEST

This test is optional, but recommended. This test works on both multimode and singlemode fibers to several thousand feet.

The installer holds a high intensity light source close to the fibers in one end of the cable. A 1 W LED flashlight works well. An associate at the opposite end views the fibers. Each fiber should glow, indicating continuity.

23.3.6.4 FIBER STRENGTH TEST

This step is optional. The installer performs this step to troubleshoot fiber breakage. The installer performs this step for each fiber just prior to inserting it into a connector.

The installer pushes each fiber against a lens grade tissue on the work surface. If the fiber bends without breaking, the installer has not damaged the cladding. If there is a single defect on the cladding, the fiber will break.

> SAFETY WARNING: the installer should not push a bare fiber against his finger. Doing so can result in the fiber penetrating the skin and breaking.

23.3.7 CLEAN FIBER

The installer performs this step for each fiber immediately prior to inserting it into a connector. The installer moistens a lens grade tissue with isopropyl alcohol. He folds the moist area around the fiber (Figure 23-8). He pulls the fiber through the fold twice. He inspects the fiber for particles on the cladding. If necessary, he cleans the fiber repeatedly until the cladding is free of all particles.

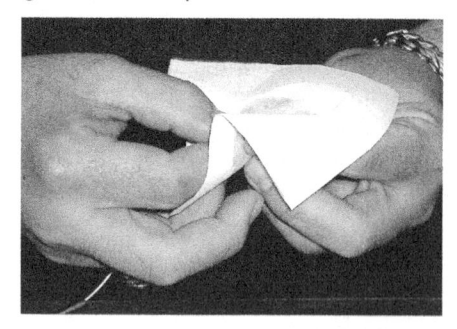

Figure 23-8: Cleaning Fiber With Moist Tissue

23.3.8 FIBER INSERTION

The installer places the loaded connector holders into the oven so that the wide flange of the holder rests against the oven-heating block (Figure 23-9). If the hot melt adhesive bubbles out of the back shell, he discards the connector: the connector has absorbed excessive moisture. After allowing the connector to heat for one minute, he removes one holder from the oven. He verifies that the boot and clear tubing are still on the jacket or buffer tube.

> Caution: the holders are hot enough to burn fingers severely!

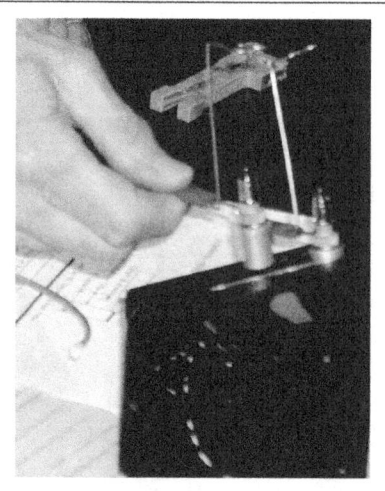

Figure 23-9: Connectors In Oven

The installer holds the connector holder so that he can insert the fiber into the back shell. He holds the cable near the end of the jacket or the end of the buffer tube. While twisting the cable or buffer tube back and forth slowly, he inserts the end of the fiber into the back shell. As long as the fiber continues feeding into the back shell without bending, he continues twisting and inserting the fiber.

For a jacketed fiber on an ST-™ compatible connector, the jacket enters the back shell. The aramid yarn will fold over the jacket and enter the back shell. A small drop of hot melt adhesive may come out of the back shell.

If the fiber bends, the installer withdraws the fiber 1/8", twists the fiber back and forth and inserts the fiber into the back shell. When the buffer tube stops further motion of the fiber into the ferrule (Figure 23-10), he presses the cable into a 'V' area of the holding clip (Figure 23-11). Note that the holder SC does not have a clip.

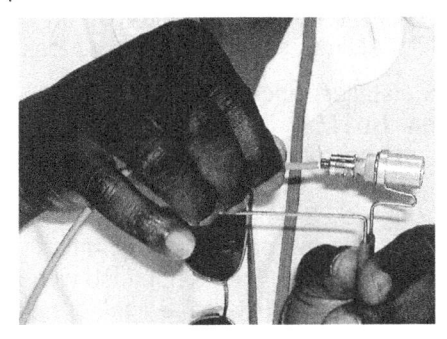

Figure 23-10: Cable Inserted

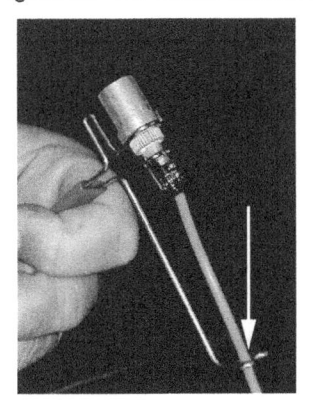

Figure 23-11: Cable In Clip

For a premises cable and the ST™-compatible connector, the installer slides the clear tubing into the back shell and the boot over the tubing and back shell. Without putting tension on the buffer tube, he places the holder in the cooling stand.

23.3.9 REMOVE EXCESS FIBER

The installer touches the holder carefully to ensure that it is cool enough to handle. Without breaking the fiber protruding beyond the bead of adhesive, the installer removes the connector from the holder. He should see the fiber protruding beyond the bead of adhesive.

For an SC connector, the installer pushes the connector into the holder, rotates the connector 90°, and slides the connector out of the holder.

The installer holds the wedge scriber with the flat surface up. While resting his hands together, the installer places the wedge surface of a scriber onto the ferrule end (Figure 23-12). He moves the blade of the scriber to the fiber so that the scriber gently touches the fiber at the top of the adhesive bead. The fiber should not move or bend.

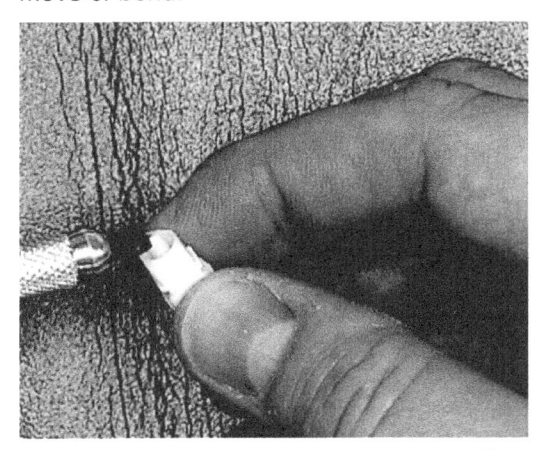

Figure 23-12: Scribing the Excess Fiber[65]

The installer moves the scriber along the fiber approximately 1/16. He should not 'saw' or break the fiber with the scriber.

The installer holds the connector with the fiber pointing up. He grips the connector with his thumb and forefinger lightly. He slides his thumb and forefinger up the connector towards and over the fiber. He grips and pulls the fiber away from the tip

---

[65] We have shown the SC connector.

of the connector. The fiber should break easily. He places of the fiber in the fiber collection bottle.

If the fiber does not break, the installer scribed the hot melt adhesive on the fiber and not the fiber. He repeats scribing and pulling the fiber until it breaks.

23.3.10 AIR POLISHING

The installer holds the connector with the ferrule pointing up. He holds a 12μ polishing film at one edge with the dull (abrasive) side down. He places the opposite edge of the film above the connector. He curls the front and back edges of the film down. With a light pressure, he rubs the film against connector until the fiber is flush with the bead of adhesive (Figure 23-13). If he scribed the fiber just above the surface of the adhesive bead, this step will take less than 10 seconds. The installer does not remove all the adhesive.

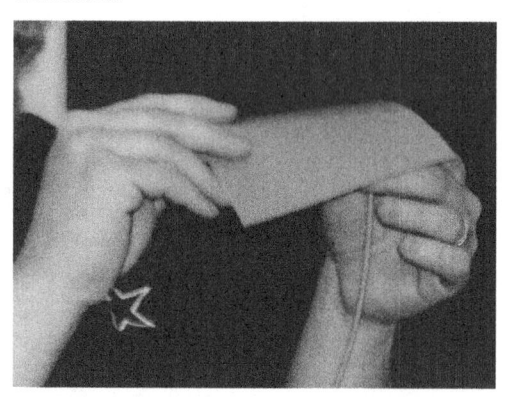

Figure 23-13: Air Polishing of Fiber

The installer checks the fiber by bringing a finger down onto the tip of the ferrule from the top. If he cannot feel the fiber, the fiber is flush, or nearly flush, with the adhesive.

> Caution: do not rub a finger *across* the tip of the ferrule. If the fiber is not flush with the adhesive, the installer may snag and break the fiber. In addition, the broken fiber may pierce the installer's skin.

23.3.11 MULTIMODE POLISHING

For a single connector, the installer follows the procedure in this section for that con-nector. For multiple connectors, the installer can perform each step on all connectors; i.e., a 'batch process'.

23.3.11.1 CLEAN CONNECTORS

The installer moistens a lens grade tissue with isopropyl alcohol. He wipes the ferrules of all connectors with this moistened tissue. This removes 12μ grit.

23.3.11.2 CLEAN EQUIPMENT

The installer can clean his polishing equipment with either lens grade gas or isopropyl alcohol. With lens-grade compressed gas, he cleans off the top surface of the polishing pads, both sides of the polishing films and the polishing tool. With isopropyl alcohol and lens grade tissues, the installer cleans his equipment in a similar manner. He places the polishing films, dull side up.

23.3.11.3 FIRST POLISH

The purpose of the first polish is removal of the adhesive. Additional time after such removal will not improve the loss. In fact, additional time results in increased loss.

The installer places the tool on the 2μ green, Hot Melt film. The installer inserts the connector into the polishing tool. While holding both the connector and the tool (Figure 23-14), he moves the tool in a ½" high, figure-8 pattern slowly. If he feels scratchiness, the fiber is not flush with the adhesive. He maintains a light pressure and a slow, ½" high figure 8 pattern until the scratchiness ceases.

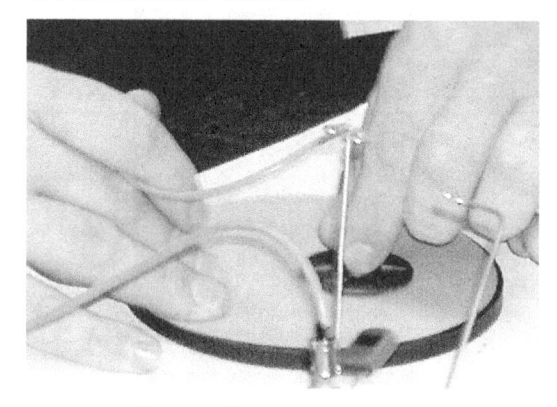

Figure 23-14: Polishing

> The installer does not push the connector towards the film. Doing so can cause the fiber to break.

After the scratchiness stops, the installer increases the size of the figure-8 pattern to cover the entire area of the film. He does not increase the pressure. He can increase the speed. He continues polishing until the adhesive is completely removed.

The installer may detect a change in the friction of the connector on the film. This change indicates the removal of the last of the adhesive. If he does not detect this change, he removes the connector from the tool periodically. He views the tip of the ferrule by reflecting light off the tip. When the adhesive is completely removed, the tip of the ferrule will be glass smooth and shiny. The tip will have the same appearance as that of the side of the ferrule.

23.3.11.4 CONNECTOR CLEANING

With lens grade compressed gas or isopropyl alcohol and lens grade tissues, the installer cleans the connector, and the polishing tool.

23.3.11.5 SECOND POLISH

This polish, improves microscopic appearance and simplifies inspection (20). This polish will not reduce multimode loss.

'High pressure' for polishing means holding the ST-™ compatible connector by its retaining nut and pushing the retaining nut down until the internal spring is fully compressed. The installer need not increase the pressure beyond this level. For an SC connector, the installer pushes the inner housing down until the internal spring is fully compressed.

The installer applies high pressure to the connector for ten, large figure 8 patterns. Large means using the full area of the polishing film.

23.4 SINGLEMODE POLISHING

See manufacturer instructions for this step. Singlemode polishing requires replacement of the 1μ, second polishing film with a 0.05μ film. Final polishing requires four to eight figure eights.

23.5 FINAL CLEANING

For the ST™-compatible connector, the installer slides the boot over the back shell. He cleans the connector with one of the following methods.

For the SC connector, installation of the tube, crimp sleeve, and boot occur in 23.9.

> Do not use lens grade gas for final cleaning of the connector. This gas tends to leave watermarks that can increase loss and reflectance.

23.5.1 BEST METHOD

The installer moistens a lens grade, lint free tissue with isopropyl alcohol. He wipes the sides and tip of the ferrule with this tissue. He sprays a one-inch diameter area of connector cleaner, such as ElectroWash Px, onto a small pad of lens grade tissues. Three times, he wipes the tip of the ferrule from the wet area to the dry area.

23.5.2 METHOD B

The installer moistens a tissue with isopropyl The installer moistens a tissue with isopropyl alcohol. He wipes the sides and the tip of the ferrule. Immediately, he wipes the tip of the ferrule with a dry tissue.

Optional: with lens grade gas, the installer blows out the connector cap and installs the cap.

23.5.3 METHOD C

The installer wipes the sides and the tip of the ferrule with a pre-moistened lens grade tissue (Alco Pad or OptiPrep Pad). Immediately, he wipes the tip of the ferrule with a dry tissue.

23.6 CAP INSTALLATION

Optional: the installer blows out the cap with lens grade gas.

The installer installs the cap.

23.7 INSPECT CONNECTOR

After polishing each connector, the installer inspects and rates the connector according to the connector inspection procedure (20.4) If the connector is good, he installs the cap. If the connector is not good, he attaches a label to indicate a potential problem and installs the cap. If the problem is severe enough to indicate unacceptable loss, he replaces the connector.

23.8 WHITE LIGHT TEST

When the installer has installed all connectors on the one end of a cable, he performs a white light, continuity test as in 21.6 (Figure 23-15). The installer does not install connectors on the opposite end until all of the fibers have passed this continuity test. If one or more connectors fail this test, he troubleshoots these connectors to identify the problem. An VFL or OTDR test may be required.

The installer replaces the caps on all connectors that pass this test. He repeats this continuity test after he installs all connectors on the opposite end of the cable.

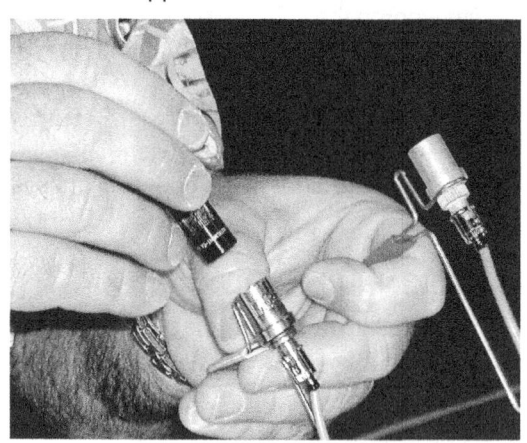

Figure 23-15: White Light Test

23.9 SC FINAL ASSEMBLY

The installer slides the crimp sleeve over the connector back shell. He slides the clear tube into the small diameter of the crimp sleeve. He crimps the large diameter of the crimp sleeve with the 0/190" crimp nest. He crimps the small diameter of the crimp sleeve with the 0.137" crimp nest. He aligns the flat surfaces of the boot with the flat surfaces of the inner housing. He slides the boot over the crimp sleeve. He installs the outer housing as in 22.3.19.

23.10 SALVAGE

23.10.1 PROCEDURE

When the installer installs the Hot Melt, SC connectors onto premises cables or the ST™ compatible on 3 mm jacketed cables, he can reheat the connector to repair damage. This salvage can be performed as long as the installer has not performed both of the following actions:

- Crimped the crimp sleeve
- Installed the SC outer housing over the inner housing

To salvage a damaged fiber end, the installer places the connector into the holder, reheats the connector, removes the cable, prepares the cable end, and inserts the new end. A new end is ready for polishing.

When the installer salvages a Hot Melt connector, the new end may have a very small adhesive bead, which provides minimal support to the fiber during polishing. Because of this reduction in bead size, this salvage procedure requires a change to the scribing and polishing techniques.

When scribing, the installer holds the scriber so that the blade touches the fiber slightly above the surface of the bead. After scribing, he air polishes without removing all the adhesive. He polishes with a modified technique: he uses a very light pressure, a ½" high 'figure 8' movement, and a very slow movement until the bead is completely removed. He polishes as in 23.3.11.5.

23.10.2 ALTERNATE PROCEDURE

The installer can eliminate the repreparation for salvage by modifying the technique in 23.3.8. Instead of fully inserting the fiber, he inserts the fiber fully, and then withdraws the cable by approximately 1/16". This modification leaves extra bare fiber buried in the adhesive.

After reheating, he pushes this extra fiber through the ferrule to create a new end for polishing.

23.11 TROUBLESHOOTING

23.11.1 INSTALLATION

23.11.1.1 FIBER BREAKAGE

Multimode fiber breaks upon insertion into connector
Potential cause: dirt on fiber
Action: clean fiber

Fiber breaks or does not fit into connector

Potential cause: primary coating not completely removed
Action: restrip the fiber

Potential cause: adhesive is too cool for insertion.
Action: reheat connector

23.11.1.2 ADHESIVE FLOWS FROM BACK SHELL

Potential cause: moisture in adhesive
Action: discard connector

23.11.2 POLISHING

23.11.2.1 NON-ROUND CORE

Potential cause: broken fiber due to excessive polishing pressure
Action: check the 2 μ polishing film. If abrasive has been torn from backing, fiber protruding above adhesive snagged on film and broke. Increase air polish time

Potential cause: excessive pressure during polishing sheared off bead of adhesive Check 2 μ film for small bead of adhesive
Action: reduce polishing pressure

Potential cause: all adhesive removed by excessive air polishing
Action: check the connector before final polishing. If the tip of the ferrule is mirror smooth without any dull or colored film or bead of adhesive, no adhesive remains
Action: salvage connector and reduce air polish time

23.11.2.2 EXCESSIVE POLISHING TIME ON EITHER FILM

Potential cause: film worn out
Action: replace film

23.11.2.3 NO APPARENT REDUCTION OF BEAD SIZE

Potential cause: dirt in ferrule hole of polishing tool prevents tip of ferrule from contacting film
Action: clean hole in polishing tool with lens grade air or with pipe cleaner dipped in isopropyl alcohol

23.11.2.4 A FEW CORE AND CLADDING SCRATCHES

Potential cause: contamination of film by environment
Action: replace film. Clean film only if no replacement film is available.
Action: move polishing location away from air vents or any other source of airborne dust

23.11.2.5 A FEW SCRATCHES ON CORE

Scratches extend across core and cladding

Potential cause: contamination of film
Action: replace film. Clean film only if no replacement film is available.

Potential cause: incomplete polishing
Action: complete polishing

23.11.2.6 CRACKED FIBER

Potential cause: excessive pressure during scribing
Action: scratch fiber lightly during scribing

23.11.2.7 HIGH LOSS

With no cause apparent from microscopic inspection

Potential cause: over polished on either film.
Action: polish of 2 μ film only until adhesive is gone. Polish on the 1 μ film for 10 strokes

23.12 ONE PAGE SUMMARY

This summary is in two parts, one each for connector installation and multimode polishing

23.12.1 INSTALLATION

Install cable through enclosure

Pull cable through enclosure to work surface

Set up oven and cooling stand

Load connectors into holders

Remove jacket, Kevlar and central strength member to proper lengths

Perform cable continuity test

For ST-™ compatible and SC connectors and 900-µ tight tube, install boots and clear plastic tube on all buffer tubes. For jacketed cables, install boots and crimp sleeves

Strip buffer tubes to proper lengths

Preheat connector

Clean the fiber with lens grade tissue and isopropyl alcohol

With rotation, insert fiber into back shell until fiber bottoms out against ferrule

Do not allow fiber to bend

For jacketed cable and SC connectors, slide the crimp sleeve or boot over back shell

For jacketed fibers, press the jacket into the clip of the connector holder

For jacketed fiber. crimp the crimp sleeve

Scribe and remove the excess fiber

Air polish the excess fiber flush with the adhesive

Proceed to multimode or singlemode polishing procedure

23.12.2 MULTIMODE POLISHING

Clean all connectors

Clean the connector ferrule, pad, film and tool.

Place 2µ film on the pad and the tool on the film

Install the connector into the tool

Slowly polish in a ½" high figure 8 pattern with light pressure until the scratchiness ceases

Polish in large figure 8 pattern with light pressure until all the adhesive is gone, using the entire area of the film

Clean the 1µ film, the tool and the connector

Install the connector into the tool With high pressure, make 10 large figure 8 motions

Clean the side and tip of the connector

Clean and install the cap

Inspect the connector

Perform a white light test of the cable

After installing connectors on the second end, perform a white light test of the cable

24 CONNECTOR INSTALLATION: CLEAVE AND CRIMP

Chapter Objectives: you learn to install SC, LC, and ST-™ compatible connectors by a 'cleave and crimp' method and achieve low loss and low labor cost.

24.1 INTRODUCTION

This Corning Inc. product, the UNICAM™, requires no adhesive and no polishing. This installation method has the lowest installation time of all methods. While this method is fast, its' apparent simplicity hides subtleties. These subtleties determine the insertion loss, the process yield, and the total installed cost. Attention to these subtleties results in consistent and acceptable loss and yield.

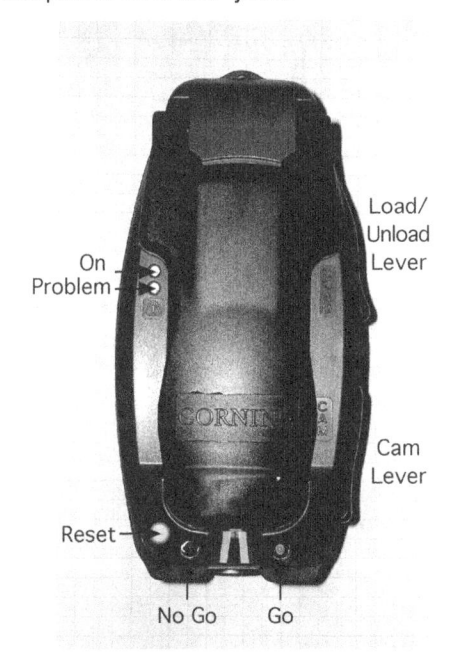

Figure 24-1: Unicam™ VFL-Crimper Tool

At the time of this writing, Corning has developed at least four generations of installation tools. The fourth generation consists of a VFL-crimper tool (Figure 24-1), an improved cleaver (Figure 24-2), and the latest generation cleaver (Figure 24-3). The VFL tool injects light into the connector. When a fiber in inserted into the connector, this tool detects any light that escapes from the connector. With colored LEDs, the tool indicates that the escaping light power level is acceptable or excessive. This author has worked with this tool for the last three years. The results have been impressive.

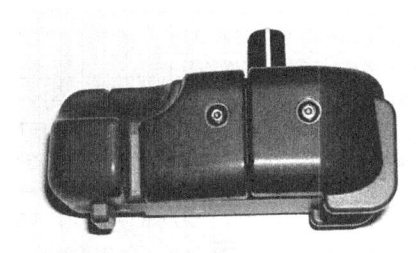

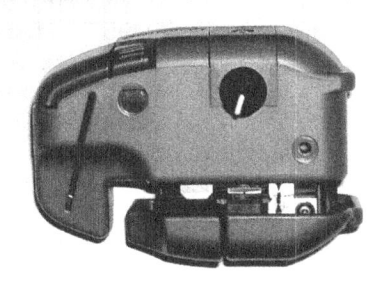

Figure 24-2: Prior Generation Unicam™ Cleaver

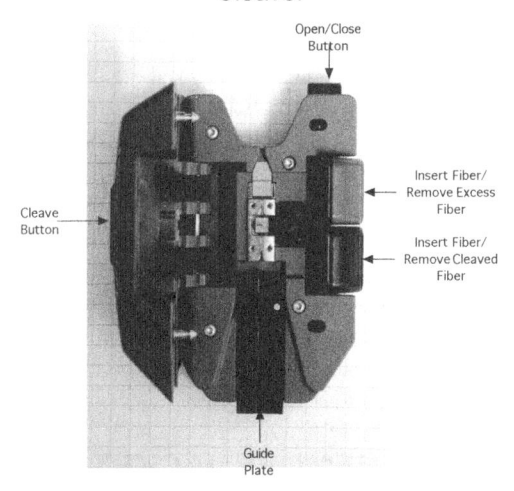

Figure 24-3: Latest Unicam Cleaver, Open

In this chapter, we provide the instructions for use of the latest generation tools,[66] including the latest generation cleaver (Figure 24-3). The instructions are written

---
[66] 'Latest' means at the time of this writing.

for training. However, with adjustments to the cable end preparation dimensions, they can be followed for field installation.

24.2 TOOLS AND SUPPLIES REQUIRED

- Jacket stripper, Ideal tool (45-162)
- Tight tube and primary coating stripper (Clauss No-Nick® 203 µ or Miller stripper)
- Kevlar® cutters
- Unicam cleaver (Figure 24-3)
- Premises cable, at least 6' long
- Unicam connectors with the same core diameter as in the cable above: 95-000-50 (ST-™ compatible, 62.5 µ, composite ferrule); 95-000-40 (SC, 62.5 µ, composite ferrule); 95-050-51 (ST-™ compatible, 50 µ, zirconia ferrule); 95-050-41 (SC, 50 µ, zirconia ferrule); 95-000-51 (ST-™ compatible, 62.5 µ, zirconia ferrule); 95-000-41 (SC, 62.5 µ, zirconia ferrule)
- 99% isopropyl alcohol or Electro-Wash Px
- Lens grade tissues
- Connector installation tool (Figure 24-1)
- Fiber collection bottle

24.3 PROCEDURE

For field installation, the installer repeats this procedure on all connectors on one end of the cable. He repeats this procedure on all connectors on the opposite end.

For training, the installer installs a connector on one buffer tube. He repeats this procedure on the opposite end of the same color buffer tube.

24.3.1 REMOVE JACKET

For field installation, the installer prepares the cable end to the dimension specified for the enclosure. For training, the installer prepares the cable ends to expose a minimum buffer tube length of 18".

24.3.2 LOAD TOOL

The installer opens the installation tool by pivoting the top cover (Figure 24-4). He pushes the reset button (Figure 24-1) to set the crimp jaw of the tool to the starting position (Figure 24-4). The crimp jaw will not be open, as in Figure 24-5.

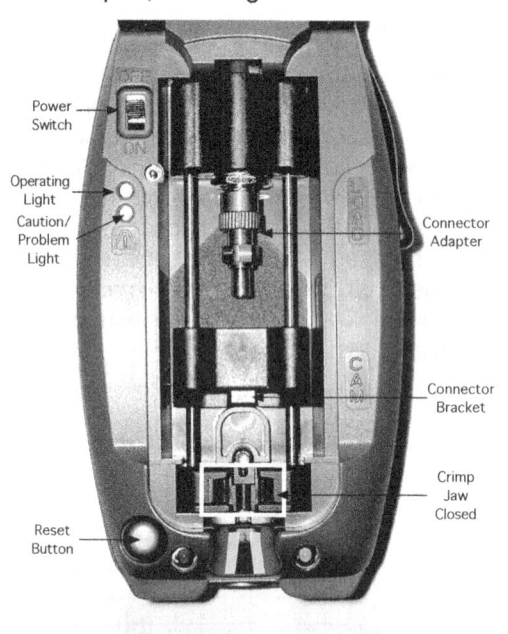

Figure 24-4: Open UniCam Installation Tool, Crimp Jaw Closed

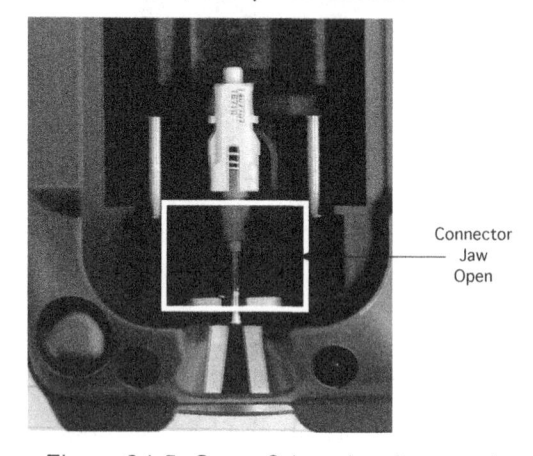

Figure 24-5: Open Crimp Jaw Incorrect

He removes the caps from the ferrule and the feed-in tube on the back end of the connector (Figure 24-8).

If necessary, he adjusts the position of the connector cam: it should be at 90° to the date code surface (Figure 24-7)

Professional Fiber Optic Installation 24-3

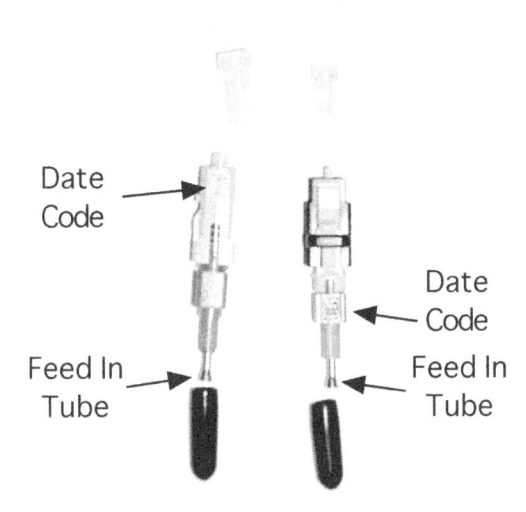

Figure 24-6: Caps Removed, SC And LC

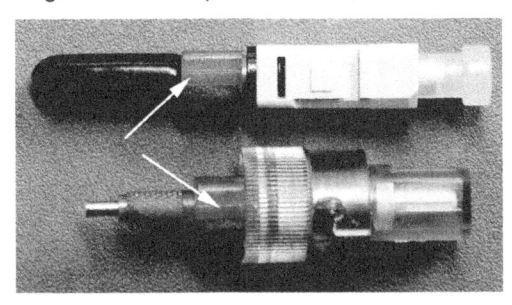

Figure 24-7: Cams In Correct Position, SC And ST™-compatible

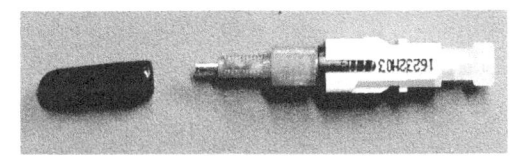

Figure 24-8: Date Code On Top Surface

The installer positions the connector so that the date label for the SC, (Figure 24-8), 'UP' label, for ST-™ compatible, or the retaining clip for the LC, is up, or facing out of the tool.

While holding the connector in the up position and pushing the load/unload lever (Figure 24-1) all the way in, the installer inserts the connector under the connector bracket (Figure 24-9).

The connector should be below the connector bracket, as shown in Figure 24-9. The connector should not be above the connector bracket, as shown in **Error! Reference source not found.**.

The installer releases the load/unload lever slowly. The connector feed in tube will slide through the crimp jaw as shown in Figure 24-9. The adapter may not stay on the ferrule (Figure 24-9).

If the adapter does not stay on the ferrule the installer pushes a adapter fully onto the ferrule. (Figure 24-11). The ferrule should not be visible, as is shown in Figure 24-13.

The feed in tube should be completely through the crimper jaw, as in Figure 24-9. It should not be as shown in Figure 24-13.

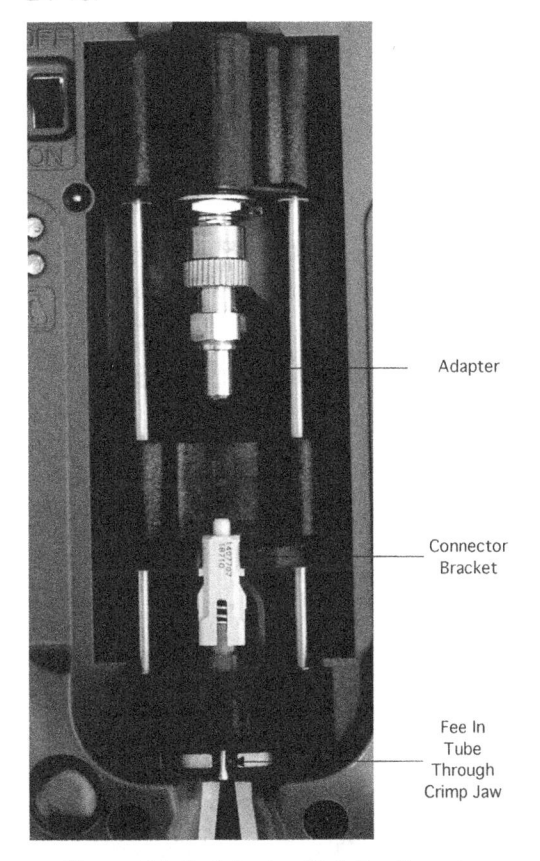

Figure 24-9: Adapter Not On Ferrule

The installer turns on the tool and closes the cover. The green light should be lit (Figure 24-4). The yellow problem/caution light should not flash (Figure 24-4).

If the yellow light flashes, the adapter may not be on the connector completely (Figure 24-12). The installer pushes the adapter down completely onto the connector (Figure 24-11).

© PEARSON TECHNOLOGIES INC.

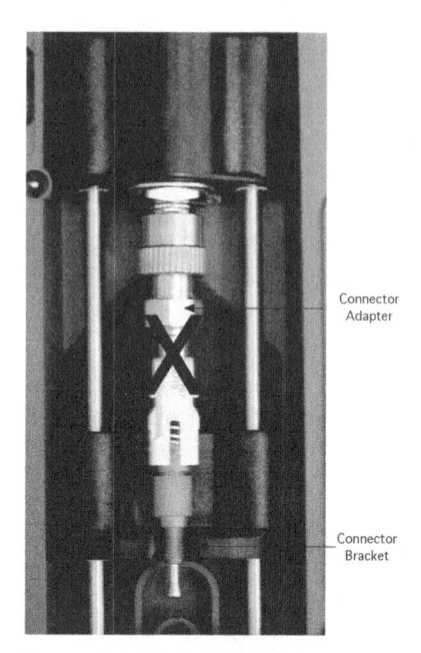

Figure 24-10: Incorrect Connector Location

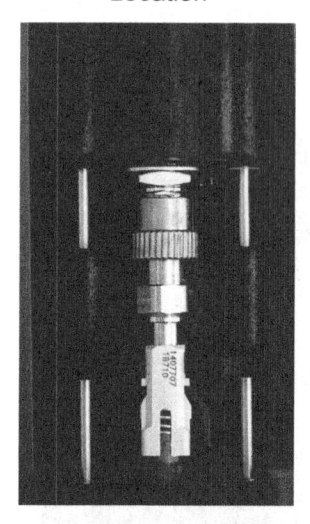

Figure 24-11: Adapter Fully On Ferrule

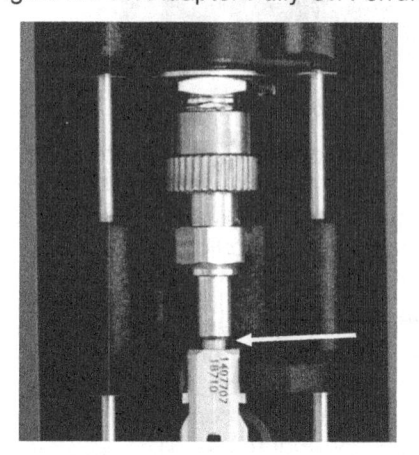

Figure 24-12: Adapter Not Full On Ferrule

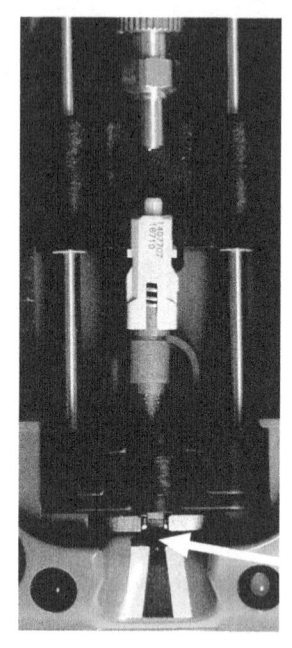

Figure 24-13: Improper Lead In Tube Position

If the yellow light continues flashing, the installer replaces the batteries.

24.3.3 PREPARE FIBER END

The installer installs a boot on the buffer tube, small end first. He marks the buffer tube at 40mm from the end. With a NoNik striper, the installer strips the buffer tube and primary coating in increments 1/2" (12.7 mm) or less. He strips the buffer tube and primary coating to expose 40 mm of bare fiber.

With a Miller stripper, the installer strips as described in 23.3.6.2.

Note that the fiber can be longer than 40mm as excess fiber length does not cause a problem. However, if the fiber is shorter than 40mm, the installer may need to open the cleaver to remove the excess fiber. Such removal is a nuisance.

24.3.4 PREPARE CLEAVER

The installer pushes the open lever to open the cleaver (Figure 24-3). Before the first and after every 10-20 cleaves, he blows out the grooves of the cleaver with lens grade gas. He closes the cleaver and inserts the 900µ guide plate (Figure 24-3).

24.3.5 CLEAN FIBER

The installer moistens a lens grade tissue with isopropyl alcohol. He folds the moist area of the tissue around the fiber. He pulls the fiber through the fold twice. He inspects the fiber to verify lack of contamination of the fiber surface (Figure 24-14).

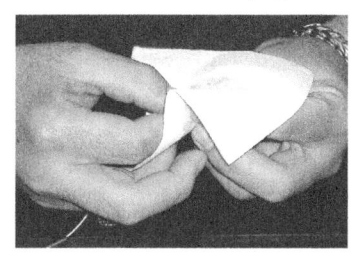

Figure 24-14: Cleaning Fiber With Moist Tissue

Caution
Do not place the fiber down.

24.3.6 CLEAVE FIBER

The installer presses and holds in the red and black buttons on the side of the cleaver (Figure 24-15). He slides the fiber and buffer tube along the center of the guide plate into the cleaver until the fiber stops. He releases the two buttons. He presses and releases the large blue button on the top of the cleaver. The fiber is now cleaved.

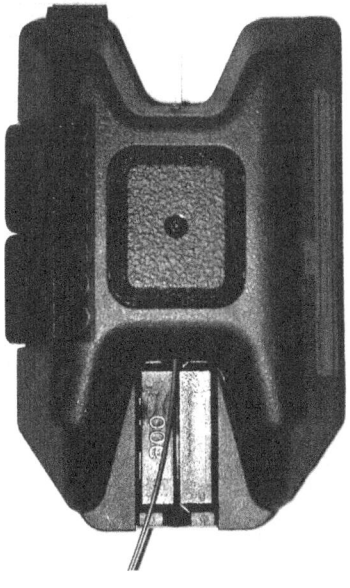

Figure 24-15: Cleaver With Fiber installed

Caution: Avoid Contaminating End
Do not place the cleaved fiber onto any surface. Do not touch the sides or the end of cleaved fiber against anything. Do not clean the cleaved fiber. Do not delay the next step.

24.3.7 INSTALL FIBER

The installer pushes in the black button of the cleaver and removes the cleaved fiber. Without touching the end of the fiber to anything, he inserts the fiber into the feed in tube on the back shell of the connector. He puts enough pressure on the fiber so that it bends slightly (Figure 24-16).

While watching the red and green status lights on the bottom of the installation tool, the installer pushes in the cam lever very slowly. If the green, GO, light glows, he pushes in the cam lever completely until he hears a click. He proceeds to 24.3.8.

Slight Bend

Figure 24-16: Fiber Slightly Bent

If the yellow, NO GO, light on the bottom of the installation tool glows, he releases the cam level, removes the fiber, and re-inserts the fiber. He repeats this step. If the yellow light continues to flash, he removes the fiber, pushes in the cam lever completely, and pushes the reset button. He repeats this step. If the yellow light continues to flash, he begins troubleshooting (24.6).

24.3.8 CONNECTOR REMOVAL

The installer pushes in the load/unload lever completely, pulls the connector down from the adapter and removes the fiber from the installation tool. He pushes the reset button (Figure 24-4).

He pushes the boot onto the back shell of the connector (Figure 24-17).

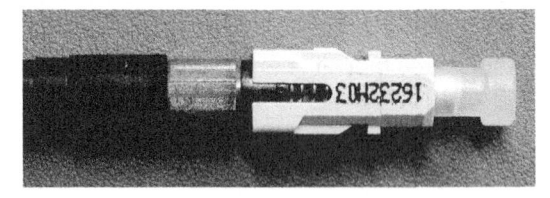

Figure 24-17: SC Boot Installed

24.3.9 VFL EVALUATION

The installer can use a VFL to provide a preliminary indication of the loss of a crimp and cleave connector. In addition, this test can indicate which end is high loss. This author's experience is that:

- No glow indicates low loss
- A faint glow does not necessarily indicate an unacceptable high loss connector

A bright glow (a Christmas tree light!) indicates unacceptable high loss (Bottom, Figure 24-18).

Caution: do not install the connector directly into the VFL, as it will launch light into the cladding. This cladding light will cause almost all connectors to glow.

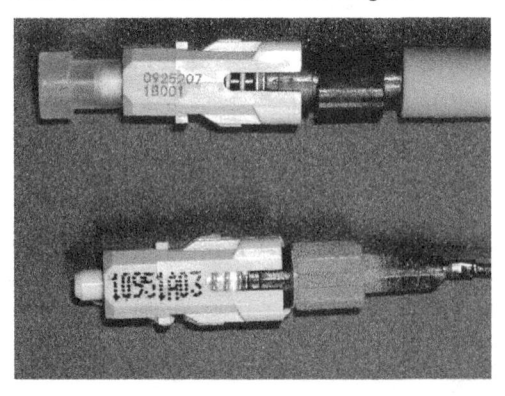

Figure 24-18: VFL Test

For an SC connector, this test should be done prior to installation of the outer housing (24.3.10). The installer places a VFL on one end of a fiber. The installer views the connector on the opposite end. If he sees a red glow, light is escaping from the core and the loss of the connector can be high. The installer can place a tag on this connector to indicate the potential for high loss.

The VFL is placed on the opposite end because it launches light into the cladding. Such light can be lost without causing high loss in the connector. This cladding light can escape, incorrectly indicating high loss. Even if low loss, cleave and crimp connectors will almost always glow at the end on which the VFL in installed.

An alternative method is insertion of a short patch cord between the VFL and the connector under test. This cord may allow the cladding light to be lost prior to the connector. If the outer housing is on an SC connector, the installer can use this method.

24.3.10 FINAL ASSEMBLY

For an LC connector, the installer removes the installation housing (Figure 24-19). He slides the trigger so that it is over the release lever.

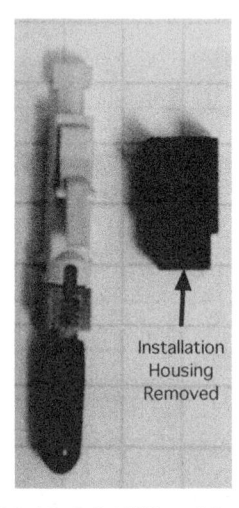

Figure 24-19: LC Without Installation Housing

For the SC connector, the installer positions the date code on the inner housing and the key on the outer housing in the up position (Figure 24-20). He inserts the connector into the outer housing from the end opposite the key until the inner housing snaps through the opposite end (Figure 24-21). He places a cap onto the ferrule end of the connector.

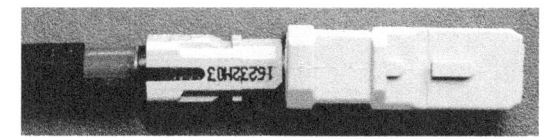

Figure 24-20: Proper Alignment Of Date Code and Key

Figure 24-21: Inner Housing Fully Inserted

24.4 ST-™ COMPATIBLE PROCEDURE

Installation of the ST-™ compatible connector is almost the same as that of the SC connector. The only difference is in the connector and its orientation in the installation tool. The installer installs the ST-™ compatible connector in the tool with the cap on the ferrule. The date code and the word 'UP' on the cap will be up.

After installation of the ST™-compatible connector, the installer removes the black loading sleeve from the connector. He removes this sleeve by pushing the loading sleeve towards the connector and rotating it until it stops. He slides the loading sleeve away from the connector body.

24.5 TEST LOSS

When he has installed connectors on both ends of a buffer tube or on both ends of all buffer tubes, the installer tests the insertion loss (14.3 or 14.4).

24.6 TROUBLESHOOTING

24.6.1 BLINKING YELLOW LIGHT

Symptom: several connectors in a row exhibit a blinking yellow light when fiber is fed into feed in tube.

Potential cause: index matching gel not uniformly distributed across core

Action: remove fiber from connector; push cam lever in until it clicks; push reset/red button; insert same fiber into feed in tube

24.6.2 HIGH LOSS

Potential cause: contamination of fiber from placing of surface or touching fiber after cleaving
Action: do not place fiber on any surface or touch fiber after cleaving

Potential cause: dirty, worn or damaged scribing blade
Action: clean blade; if problem persists, rotate or replace blade

Potential cause: dirt on end of fiber
Action: install in an environment that does not have floating dust or particles in air

Potential cause: bad cleave due to dirt on fiber
Action: clean fiber prior to cleaving

Potential cause: bad cleave due to dirt on cleaver
Action: clean cleaver pads, scribing blade, and fiber groove

24.7 ONE PAGE SUMMARY

Remove caps from both ends of connector
Push reset button
Open installation tool cover
Push in and hold load/unload lever fully
Insert ferrule into adapter
Release load/unload lever slowly, guiding feed in tube through crimp jaw
Verify feed in tube is through crimp jaw
Turn on installation tool; verify that green light glows

Prepare cable end to expose buffer tube
Install boot on buffer tube
Mark buffer tube at 40mm
Strip the buffer tube to the 40 mm mark
Open and clean cleaver; clean before first and after every 10-20 cleaves
Insert 900µ guide plate into cleaver
Clean the fiber
Push in and hold the red and black buttons on cleaver
Insert the fiber into the cleaver
Verify fiber appears at opposite end of cleaver
Release both buttons
Push blue cleave button on top of cleaver
Hold fiber and push in black button; remove fiber from cleaver

Insert the fiber into lead in tube until fiber bends slightly
Slowly push the cam lever in until you see a green or red light at bottom of tool
If you see a green light, push the cam lever in completely until you hear a click
If you see a red light, release the cam lever; prepare a new fiber end
Push in load lever completely
Pull connector down and out of tool
Push the reset button
Slide boot over lead in tube
Optional: VFL test
Install connector into outer housing (SC)
If necessary, install cap on connector
When connectors are installed on both ends, test cable.

25 MID-SPAN SPLICING

Chapter Objectives: you learn to perform all the activities of mid-span splicing: fusion and mechanical splicing on multimode and singlemode fibers, installation of splices in trays, and trays and cable into an enclosure.

25.1 INTRODUCTION

In this chapter, we present the procedure for single fiber, mid span splicing. There are procedures for fusion (25.8.1) and mechanical splicing (25.8.2) on either singlemode or multimode fiber.

This procedure is written for a specific enclosure, as it must be. For an enclosure different from that herein, the dimensions for cable end preparation, fiber routing, buffer tube routing, and final assembly specified in the instructions for that enclosure should be used.

To perform mid span splicing, the installer needs two cables. To enable testing of the splice, the cables must be longer than the dead zone of the OTDR. For practical purposes, the recommended length is at least three times the dead zone. For training, this author recommends cables with at least 12 fibers in one or two buffer tubes.

Mid span splicing requires ten steps:

- Cable end preparation
- Enclosure preparation
- Cable attachment
- Buffer tube attachment
- Splicing
- Fiber coiling
- Buffer tube coiling
- Tray attachment
- Enclosure finishing
- Testing

25.1.1 TOOLS AND SUPPLIES REQUIRED

- Jacket slitter tool (Clauss RCS-20)
- Tight tube and primary coating stripper (Clauss No-Nick® 203 µ or Miller stripper)
- Kevlar® cutters
- High performance cleaver, Fujikura CT07, CT04, CT20 or CT30
- Two loose tube cables
- Gel/grease remover
- Paper towels or clean rags
- Fusion splicer with heat shrink oven
- Splice covers, either heat shrink (FIS P/N F1-1002 or F1-100240) or adhesive (FIS P/N F1-FS40, F1-FS60)
- 3M Fibrlok™ II splices (P/N 2529)
- Fibrlok™ assembly tool (P/N 2501)
- 98% isopropyl alcohol
- Lens grade tissues
- Foam fiber optic swabs (FIS P/N F1-0005)
- Lens grade gas (Stoner 94203 or equivalent)
- OTDR
- Launch cable
- Outdoor fiber optic splice enclosure (FIS P/N 1120-F, 2000-F)
- Two splice trays
- Splice holders compatible with splice trays
- Cable ties for buffer tubes and for pigtails
- Visible Feature Locator
- Optional: index matching gel
- Optional: reusable mechanical splice, Elastomeric splice, (FIS)

See 21.2 for sources of items not identified in this section.

The end of the launch cable requires a connector of the same type as that on the end of the cable. Alternatively, the cable can be terminated with a bare fiber adapter.

The splice trays must be compatible with the enclosure. The splice holders must be compatible with the trays and the splice cover or mechanical splice.

25.2 CABLE END PREPARATION

25.2.1 DIMENSIONS

From the enclosure instructions, the installer identifies the following dimensions (Figure 25-1):

- Length of jacket to be removed
- Length of strength member to remain
- Length of central strength member to remain
- Length of buffer tube to remain

25.2.2 END PREPARATION

The installer prepares the cable end to those dimensions (Figure 25-1), according to the procedure in Chapter 21. This process may include replacement of the buffer tube with a tubing material that is compatible with the bend radius of the enclosure.

Before cutting and removing the first buffer tube, the installer verifies the buffer tube length. To do so, the installer marks the buffer tube at the distance indicated in the enclosure instructions. The installer places the cable into the enclosure. He routes one buffer tube from the cable entrance to the opposite end of the enclosure, back to the entrance end and into the location of the splice tray. If the buffer tube length is correct, the mark on the buffer tube will be inside the tray by approximately 1" (Figure 25-2). If the mark on the buffer tube is inside the tray by more than 1", the installer reduces the buffer tube length by the appropriate amount.

If the mark is outside the tray, the installer adjusts the loop of buffer tube until the mark is inside the tray. He checks the buffer tube to ensure that there are no sharp bends. If there are no sharp bends, he uses the buffer tube length indicated in the enclosure instructions. If the buffer tube has a sharp bend, the installer increases the buffer tube length by the amount necessary to eliminate all sharp bends.

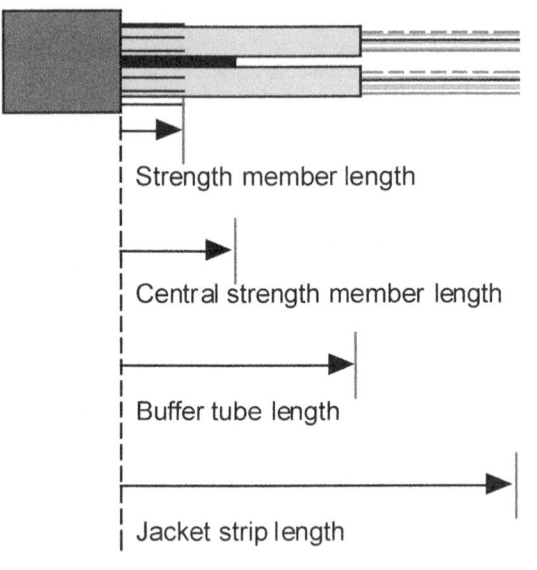

Figure 25-1: Cable End Preparation Dimensions

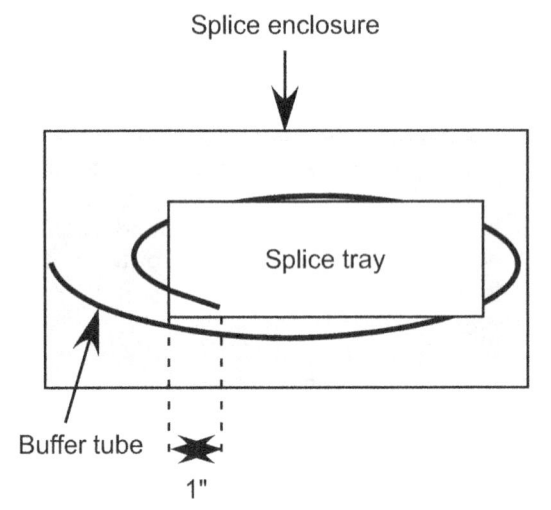

Figure 25-2: Proper Buffer Tube Length

25.3 ENCLOSURE PREPARATION

The installer prepares the enclosure according to the instructions for that enclosure. These instructions will address at least the following aspects:

- Moisture seals around cables
- Moisture seals for enclosure
- Installation of splice holders
- Grounding and bonding hardware

There are at least three types of cable moisture seals. The first type consists of a plug with an inner diameter equal to the outer diameter of the cable Figure 25-3).

The second type consists of parts assembled to create a seal (Figure 25-4). The third type consists of a solid rubber plate, through which the installer drills holes for cables (Figure 25-5).

Figure 25-3: 7 and 12 mm Plugs

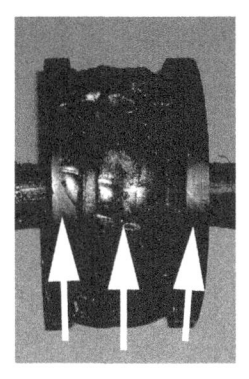

Figure 25-4: Moisture Seal Components

The moisture seal(s) for the enclosure may include plugs for unused cable entrances (Figure 25-6) and gasket material for use between the enclosure halves.

If splice holders are permanently attached to the trays, the installer installs the holders. Finally, the enclosure may include hardware for grounding or bonding (Figure 25-7). If the cables contain conductive materials, the installer bonds the cables with such hardware.

Figure 25-5: End Cap Requiring Drilling

Figure 25-6: Enclosure Plugs

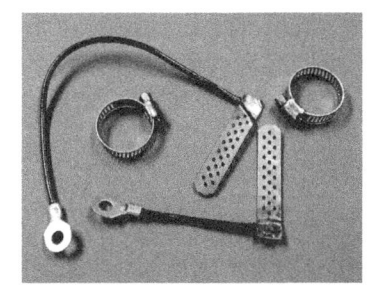

Figure 25-7: Grounding And Bonding Hardware

25.4 CABLE ATTACHMENT

The installer attaches the cable strength members of both cables to the same end of the enclosure according to the instructions for that enclosure. This procedure is for the preferred configuration of butt splicing (13.5.1). Figure 25-8 and Figure 25-9 contains an examples of attachment hardware.

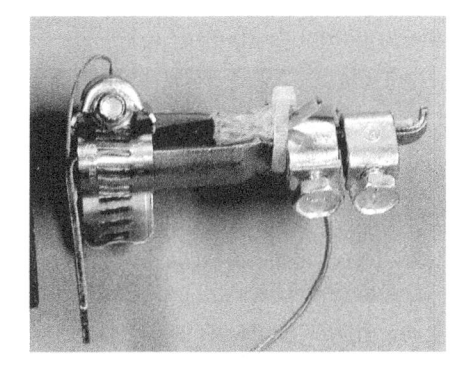

Figure 25-8: Strength Member Attachment Hardware

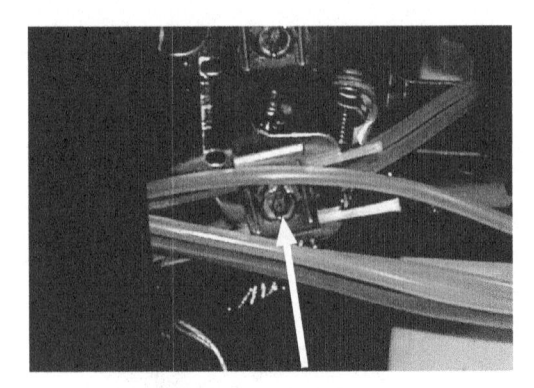

Figure 25-9: Attachment Mechanism

25.5 BUFFER TUBE ATTACHMENT

Unless there is an overriding reason, use the same color buffer tube from both cables.[67] The installer selects the same color buffer tube from each cable. He straightens both buffer tubes to eliminate any twisting. While straight, he routes one buffer tube from each cable to the same end of one splice tray.

Hereafter, we will use the term buffer tube with the understanding that a furcation tube or spiral wrap may replace the buffer tube. The installer feeds one buffer tube from each cable into opposite sides of the same end of the tray. By the method specified in the enclosure or splice tray instructions, he attaches both buffer tubes to the tray (Figure 25-10).

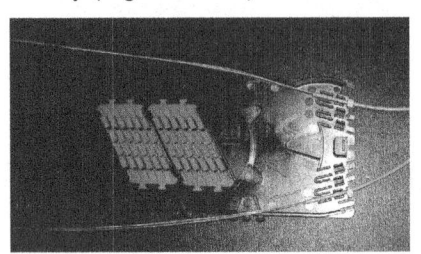

Figure 25-10: Buffer Tubes Attached To Tray

Some trays require pinching of a tab over the buffer tube. Other trays require a cable tie around the buffer tube and the tray (Figure 25-10). With a separate tray for each pair of buffer tubes, the installer repeats this step for all buffer tubes.[68]

25.6 FIBER LENGTH VERIFICATION

The installer routes the fibers from the tray to the splicer. If the fibers reach the splicer, the fibers are sufficiently long. If they do not, he adjusts the location of the splicer so that the fibers reach the splicer without any tension or sharp bends.

The lengths of fibers outside the two buffer tubes may be different. The difference depends on the orientation of the splice holder. Splice holders have three possible orientations to the long axis of the tray:

➢ Parallel
➢ Perpendicular
➢ 45°

With an approximately 45* orientation (Figure 25-10), the two groups of fibers will have slightly different lengths.

Each fiber within a buffer tube may have a slightly different length. Review instructions for the splice tray.

For example, trays with a capacity of 42 splices with a perpendicular orientation can have splice holders for six splices. Each group of six fibers will be about 3" shorter than the previous group.

The installer reviews the instructions for the splice tray to determine the fiber lengths. The fibers must end at a specific location. If the length is not correct, there can be a bend radius violation that increases the splice loss.

The installer coils the fibers from one buffer tube inside the tray so that the size of the coil is as large as possible. If the fibers end beyond the location stated in the instructions, the installer trims them to proper length. The installer repeats this step for the fibers from the second buffer tube, but the fibers will coil in the direction

---

[67] In many systems, connections of buffer tubes of different colors and fibers of different colors are common.

[68] Some trays have capacities of more than 12 splices. For such trays, the installer attaches multiple buffer tubes to both sides of the same end of the tray.

opposite to that of the fibers from the first buffer tube (Figure 25-11).

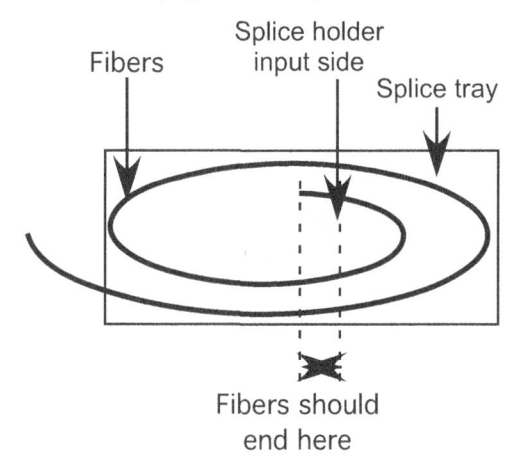

Figure 25-11: Proper Fiber Length

If the fibers are long, the installer trims them so that they are at the splice holder. If the fiber ends do not reach the input side of the splice holder, the installer tightens the coil of fibers until the ends reach the input side of the splice holder. He checks the bend radius of the fibers (Figure 25-12). If the fibers do not exhibit any tight bend radius in the tray, he uses this fiber length.

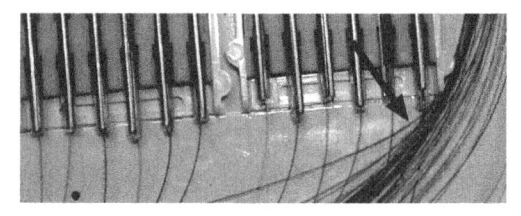

Figure 25-12: Potential Bend Radius Violation

If the fibers appear to take a tight bend radius, the installer lengthens fibers from both buffer tubes by removing additional jacket and buffer tube. If the installer lengthens the fibers, he rechecks the buffer tube length (25.2.2).

The installer removes the fibers from the tray. He untwists all fibers from each other back to the buffer tube. Without twisting or crossing the fibers, he routes the fibers from each buffer tube to opposite sides of the splicer or splicing tool.

25.7 OTDR SET UP

If the intent is to test each splice prior to placing the splice in the tray, the installer sets up the OTDR. If the intent is to test the splices after all splices have been made and placed in the tray, he may perform this set up at a later time. However, performing this set up at the beginning of a splicing session enables easy identification of the location of the splice. Remember that 0 dB splices are not unusual. Finding the location of a 0 dB fusion splice requires bending the fiber in the splice tray. Such bending requires care.

The installer sets up the OTDR at the end of one of the cables. He attaches the launch cable to the OTDR. He enters the index of refraction, the maximum length of the cable to be tested, the time of the test, the wavelength of the test, and the pulse width necessary to obtain a loss measurement in an acceptable time (15.8.1).

The installer connects the OTDR to the first fiber to be spliced. He can make this connection with a launch cable with the same connector type as on the end of the cable. Alternatively, he can make this connection to the cable with a reusable mechanical splice or a bare fiber adapter.

The installer makes a trace of the first fiber to be spliced in order to determine the distance to the cable end. He records that distance.

25.8 SPLICING

25.8.1 FUSION SPLICING

25.8.1.1 SPLICER SET UP

The installer plugs in and turns on the splicer. He opens all covers. He inspects the grooves. With a dry, lint free, fiber optic swab, he cleans the grooves by swabbing from the inside end of the grooves to the outside end (Figure 25-13).

Caution
If the installer swabs in the opposite direction, he may contaminate the mirrors and optics with dust. If he cleans the optics with a swab moistened with isopropyl alcohol, the alcohol may contaminate the mirrors and optics with watermarks. If he uses lens grade gas, the cold gas may create watermarks on the optics. Such

contamination results in error messages from the splicer software.

Figure 25-13: Splicer Grooves

25.8.1.2 SPLICE MENU

The installer examines the active menu in the splicer (Figure 25-14). If the active menu is for the fiber to be spliced, he proceeds. If the desired menu is not, he activates the appropriate menu according to the splicer instructions.

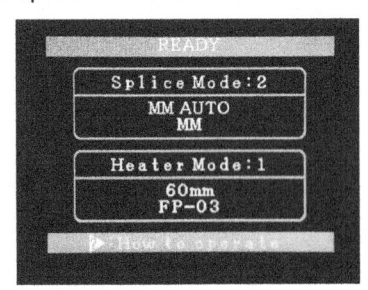

Figure 25-14: Active Fiber Menu

25.8.1.3 MAXIMUM CLEAVE ANGLE

The installer examines the maximum cleave angle (Figure 25-15). If the value is the installer wants, he proceeds. If not, he enters the appro-priate value according to the splicer instructions. While some splicing tech-nicians use a maximum of 2°, this author uses 1.5°. High performance cleavers do not consistently exceed this value unless the scribing blade needs rotation or replacement.

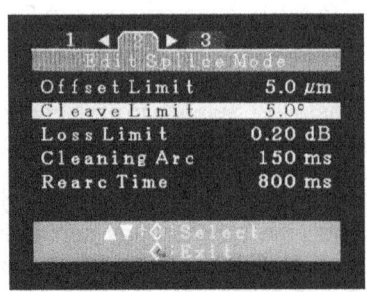

Figure 25-15: Cleave Angle Setting

25.8.1.4 SPLICE COVER LENGTH

The installer examines the splice cover length setting (Figure 25-16). If the value is the one the installer wants, he proceeds. If not, he enters the appropriate value according to the instructions for the splicer.

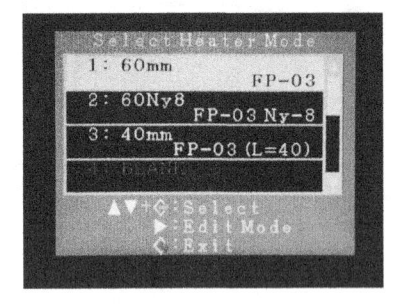

Figure 25-16: Splice Cover Length Setting

25.8.1.5 OPTICS CHECK

The installer checks the screen image. The image should be free from spots. If spots or dust exist, he cleans the mirrors according to the splicer instructions.

25.8.1.6 CLEAVER PREPARATION

The installer opens, cleans and sets up the cleaver according to the instructions for that cleaver.

25.8.1.7 COVER INSTALLATION

If the installer uses an adhesive splice cover, he ignores this step. The installer places a heat shrink splice cover on one fiber.

25.8.1.8 FIBER STRIPPING

The cleaver determines the strip length. The installer follows the instructions for the cleaver. Fujikura cleavers require a strip length of 1.5-1.75". In a single strip, the installer strips the appropriate length of primary coating from the fiber (22.3.6.2.2).

The installer moistens a lens grade tissue with isopropyl alcohol. He makes a fold in the tissue. He pulls the fiber through the fold at least twice (Figure 22-7). He examines the fiber to ensure that there are no particles on the cladding. If necessary, he re-cleans the fiber.

25.8.1.9 FIBER CLEAVING

From the splicer instructions, the installer determines the cleave length. The installer uses the appropriate cleave length. For some Alcoa Fujikura splicers, he cleaves the fibers at 12 mm.

For the 900μ fiber, the cleave length may be approximately 16 mm. See instructions for the splicing machine.

> Caution
> After cleaving, do not clean the fiber.

25.8.1.10 PLACE FIBER IN SPLICER

Following the instructions for the splicer, the installer sets the splicer to 'start'. From its side of the splicer, he places the first cleaved fiber in the holders so that the fiber is on its side of the electrodes (Figure 25-17).

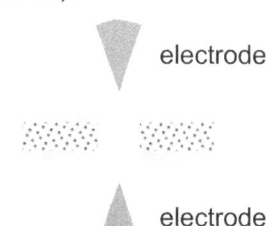

Figure 25-17: Proper Fiber Location For Splicing

25.8.1.11 SECOND FIBER

The installer repeats Steps 25.8.1.8-25.8.1.10 with the fiber from the second buffer tube.

25.8.1.12 FUSING

To begin the splicing process, the installer pushes the 'set', 'fuse', or 'start' button. Some splicers allow the option of reviewing and accepting the cleave angles and align-ment prior to fusing. If this option is active, the installer pushes the 'fuse' button several times.

25.8.1.13 ESTIMATED LOSS REVIEW

Most splicers have a splice loss estimation feature. The splicer displays the estimated splice loss. If this value is acceptable, the installer proceeds to 25.8.1.14..

Some fusion splicers allow refusing of the fiber. If the estimated loss is high, refusing may result in reduced loss. This author's experience is that current-generation fiber from the same fiber manufacturer rarely needs refusing. However, fiber from different manufacturers or fibers produced in the 1980s and early 1990s may need refusing to obtain acceptable loss. If the final value is unacceptable, the installer breaks the splice and repeats Steps 25.8.1.8-25.8.1.12

25.8.1.14 COVER INSTALLATION

For heat shrinkable splice covers, the installer follows the instructions in 25.8.1.14.1. For adhesive covers, he follows the instructions in 25.8.1.14.2.

25.8.1.14.1 HEAT SHRINK COVER

The installer slides the splice cover up to the fiber holder. He opens the fiber holder on the side opposite the splice cover. He opens the holder on the side with the splice cover. While holding the fiber at the holder opposite to the splice cover, he centers the bare fiber in the cover.

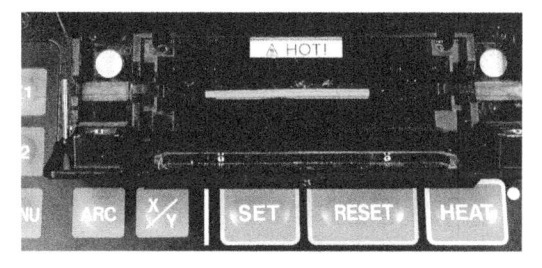

Figure 25-18: Splice Cover In Heating Oven

The installer places the cover in the heating oven (Figure 25-18). At the same time, he gently tugs on both fibers to ensure that the fibers are straight in the cover. He closes the oven cover. He turns on the oven. He allows the oven to run its cycle. While waiting for the end of this cycle, the installer prepares the next fibers.

After the heating is complete, the cover may stick to the oven surface. To release the cover, the installer uses a toothpick or tweezers to move the splice cover towards one side of the oven. This action will free the cover from the oven surface. Once the cover is free, the installer removes the splice cover from the oven by holding the fibers. He checks the cover to see that the bare fiber is inside the cover. He places the splice in the splice tray. After each splice, the installer proceeds to 25.9 or 25.10, depending on the experience of the installer. The installer repeats 25.8.1.7-25.8.1.14 until he has spliced all fibers in one tray.

> **Caution**
>
> Do not grip the splice cover tightly. Do not grip the splice cover on the area of the bare glass. Doing so may break the fiber inside the cover.

25.8.1.14.2 ADHESIVE SPLICE COVER

The installer removes the protective tape from the adhesive surface (Figure 25-19, Top). With both hands, the installer centers the bare fiber above the center of one side of the cover. He lowers the fiber to the surface slowly (Figure 25-19, Middle). He folds one side of the cover over the other (Figure 25-19, Bottom).

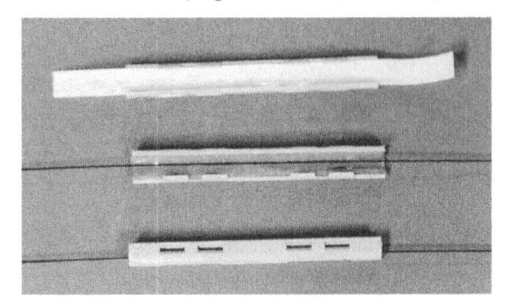

Figure 25-19: Adhesive Splice Cover

After each splice, the installer proceeds to 25.9 or 25.10, depending on the experience of the installer. The installer repeats 25.8.1.7-25.8.1.14 until he has spliced all fibers in one tray.

25.8.2 MECHANICAL SPLICE

These instructions are for the Fibrlok™ II splice from 3M (P/N 2529). If the installer uses a different mechanical splice, he modifies the procedure in this section.

25.8.2.1 SPLICE TOOL PREPARATION

The installer cleans the 'feed-in' grooves of the tool with a fiber optic swab moistened with isopropyl alcohol (Figure 25-20). He lifts the closing arm (Figure 25-21). With the closing button of the mechanical splice up (arrow in Figure 25-21), he inserts the splice into the tool (Figure 25-21). He sets the position of the foam clamps for 250μ fiber (Figure 25-22) or for 900μ tight tube (Figure 25-23).

25.8.2.2 CLEAVER PREPARATION

The installer opens, cleans and sets up the cleaver according to the instructions for that cleaver.

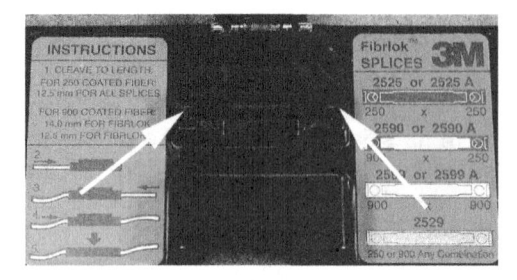

Figure 25-20: Splice Tool Feed-In Grooves

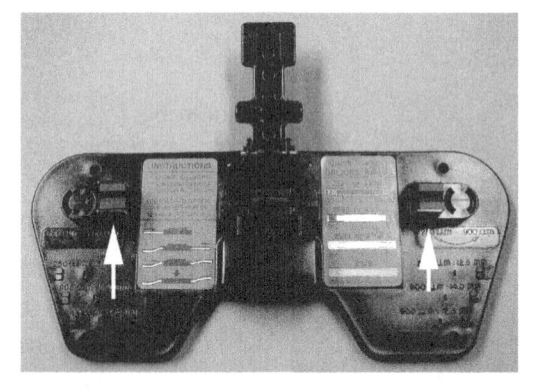

Figure 25-21: Position Of Holders For 250μ Primary Coating

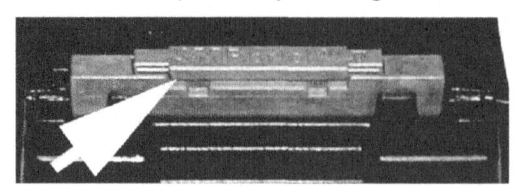

Figure 25-22: Open Splice In Tool

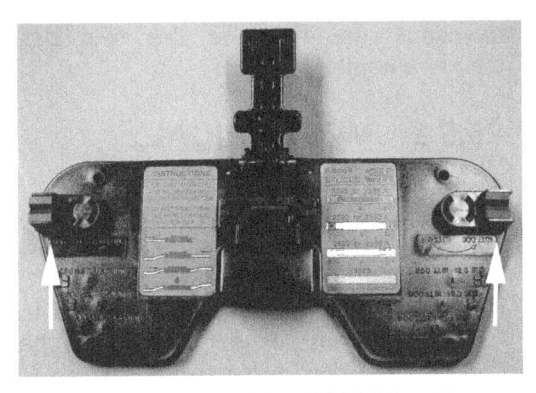

Figure 25-23: Position Of Holders For 900µ Tight Tube

25.8.2.3 FIBER STRIPPING

The cleaver determines the strip length. The installer follows the instructions for the cleaver. Some strippers require a strip length of 1.5-1.75". The installer strips the appropriate length of primary coating from the fiber.

The installer moistens a lens grade tissue with isopropyl alcohol. He makes a fold in the tissue (Figure 23-8). He pulls the fiber through the fold at least twice. He examines the fiber to ensure that there are no particles on the cladding. If necessary, he re-cleans the fiber.

25.8.2.4 CLEAVE FIBER

For the Fibrlok™ II splice, the installer uses a cleave length of 12.5 mm. For a different mechanical splice, the installer cleaves to the length indicated in the instructions for that splice.

> **Caution**
>
> After cleaving, the installer does not clean the fiber, does not place fiber on any surface and does not delay inserting the fiber into the splice.

25.8.2.5 FIBER INSERTION

The installer inserts the first fiber into one side of the mechanical splice until the fiber stops. The installer places the primary coating or buffer tube in the foam clamp. From outside the foam clamp, the installer gently pulls on the fiber so that the fiber is straight in the area between the holder and the splice (Figure 25-24). The installer grips the fiber near the splice and gently pushes the fiber into the splice.

If the fiber moves, he checks the fiber between the splice and the foam clamp. It should be straight.

Figure 25-24: First Fiber In Tool

The installer cleaves and inserts the second fiber into the opposite end of the splice until the fiber stops. The installer places the primary coating or buffer tube in the foam clamp. From outside the foam clamp, the installer gently pulls on the fiber so that the fiber is straight in the area between the holder and the splice (Figure 25-24). The first fiber will be bowed.

The installer grips the second fiber near the splice and gently pushes the fiber into the splice. If the fiber moves, he checks this fiber between the splice and the foam clamp. It should be straight.

While holding the first fiber between the holder and the splice, the installer pushes the first fiber into the splice until the amount of bow of both fibers is equal (Figure 25-25).

The two bows need not be in the same direction. That is, one can be up and the other down. The requirement is that the two bows have the same deflection from straight.

Figure 25-25: Splice With Even Bows

The installer lowers the tool arm onto the splice button (Figure 25-26). He pushes the arm so that the button (Figure 25-22) moves into the splice and is level with the rest of the splice (Figure 25-27).

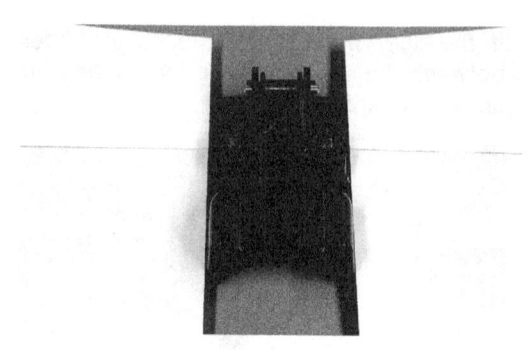

Figure 25-26: Arm In Closing Position

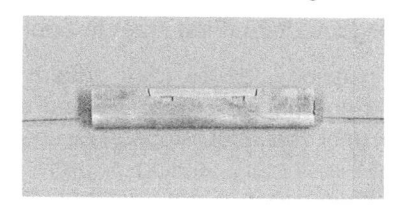

Figure 25-27: Closed Fibrlok II™ Splice

After each splice, the installer proceeds to 25.9. The installer repeats this step until he has spliced all fibers in the tray.

25.9 TEST LOSS
The installer can perform splice loss testing at various times:

- After splicing
- After installation of the splice cover
- After installation of the splice in the splice tray
- After closing the enclosure

The most meaningful time for testing is the last, after closing the enclosure. This time of testing is appropriate for experienced installers, who are able to perform the entire procedure without making errors.

For training, the trainee can perform testing at any or all of the times indicated. Such testing can indicate the step at which an error has occurred.

25.9.1 FUSION SPLICES
The installer tests the loss with an OTDR. If the loss is acceptable, the installer proceeds to 25.10. If the loss is not acceptable, the installer cuts the fibers and makes a new splice. For final testing, the installer tests the splice in both directions (15.7.4.3). If the splice loss is positive, the installer tests in the opposite direction and averages the two loss measurements.

25.9.2 MECHANICAL SPLICES

25.9.2.1 THE FIBRLOK™ II SPLICE
The installer tests the loss with an OTDR. If the loss is acceptable, he proceeds to 25.10. If the loss is not acceptable, he cuts the fibers and replaces the splice.

Note that the Fibrlok II splice can be opened and reused. Such reuse requires a 3M tool for this purpose. There are two risks of reuse. The first risk is contamination of the index matching gel. The second is breakage of the fiber inside the splice. In both situations, the loss will be high.

25.9.2.2 RE-ENTERABLE SPLICES
If the installer uses a splice that allows opening of both ends, he can tune the splice to achieve acceptable loss. To tune, the installer opens one side of the splice, rotates one fiber, closes the splice, and tests the splice. He repeats this process until he obtains an acceptable low loss measurement.

If the loss is not acceptable, the installer repeats this tuning on the second fiber until he achieves a low loss measurement. If this measurement is not acceptable, he replaces the splice.

If the installer has a VFL and a mechanical splice that lets him see the fiber ends, he can tune the splice without multiple OTDR tests. To tune the splice, he connects a VFL to the end of one of the fibers. The installer observes the glow at the mechanical splice. As described above, he tunes one or both of the fibers until there is no glow.

25.10 FIBER COILING
The installer holds the splice so that the fibers are parallel to the sides of the tray (Figure 25-28). He lays the fibers in the tray from the input end to the second end.

He twists the fibers at the second end. He lays the fibers in the tray from the second end to the input end. He twists the fibers in the direction opposite to that of the first twist (Figure 25-29).

Figure 25-28: Fibers Parallel To Sides

Figure 25-29: Twisted Fibers At Second End

The installer repeats this process until almost all the fiber is in the tray (Figure 25-30).

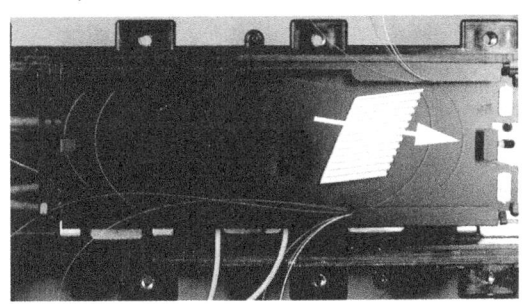

Figure 25-30: Fibers Reversed Twisted At Input End

The first position is the position on either end of the splice holder (Figure 25-31 and Figure 25-32). The installer places the splice cover, or mechanical splice, in the first position. He places additional splices in the tray in sequence from that position.

The installer repeats this step for all fibers in the tray. He places the cover on the splice tray (Figure 25-33).

An alternative method requires placing the splice holder outside of the tray. When the holder is filled with splices, the installer coils all fibers and places the splice holder into the tray.

Figure 25-31: First Splice Holder Position

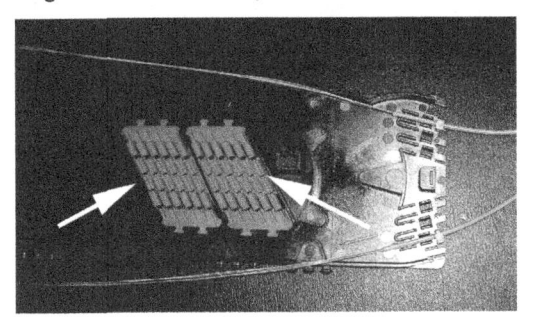

Figure 25-32: Starting Positions Of Splice Holder

Figure 25-33: Tray With Cover

25.11 BUFFER TUBE COILING

After the installer has installed all splices in a tray, he moves the tray so that the buffer tubes run parallel to the sides of the enclosure (Figure 25-34).

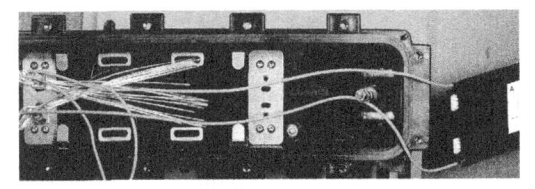

Figure 25-34: Starting Position

He places the buffer tubes into the enclosure, under any structure that supports the trays. He twists the tray upside down (Figure 25-35). He places the tray on its support structure (Figure 25-36).

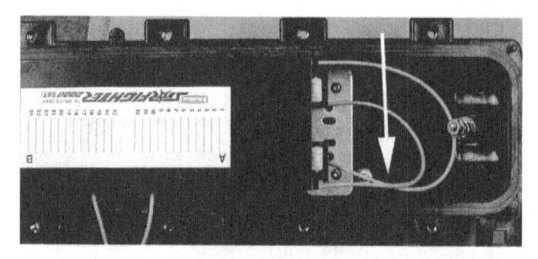

Figure 25-35: Twisted Buffer Tubes

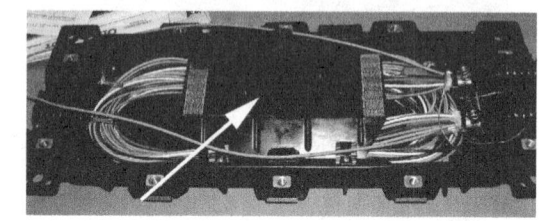

Figure 25-36: Tray Support Structure

If the length of buffer tube requires more than one routing of the buffer tubes from one end to the other, the installer twists the buffer tubes at one end of the enclosure, places the buffer tubes parallel to the sides of the enclosure back towards the input end of the enclosure (Figure 25-36). He twists the tubes by flipping the tray, and places the tray in the enclosure.

25.12 TRAY ATTACHMENT

After the installer has installed all trays onto the support structure, he attaches the trays to that structure in the manner indicated by the enclosure instructions.

Some trays attach to the enclosure with a bolt, which feeds through a bolt hole in the tray (Figure 25-37). Some trays attach with Velcro™ bands (Figure 25-38). Finally, Some trays attach with rubber holders (Figure 25-39).

25.13 ENCLOSURE FINISHING

In accordance with the instructions for the enclosure, the installer performs the following steps. He verifies that the buffer tubes are restricted to the inside of the enclosure. He cleans all moisture seal surfaces. He installs moisture seals and gaskets. He assembles the enclosure housing parts. If the enclosure instructions require use of a specific torque, he tightens the bolts with a torque wrench. By pressurizing the enclosure, he tests moisture seals for leaks with a soap solution. If so instructed, he releases the pressure.

The installer installs the enclosure in its location. If the enclosure has weep holes, he installs the enclosure so that the weep holes are on the bottom of the enclosure.

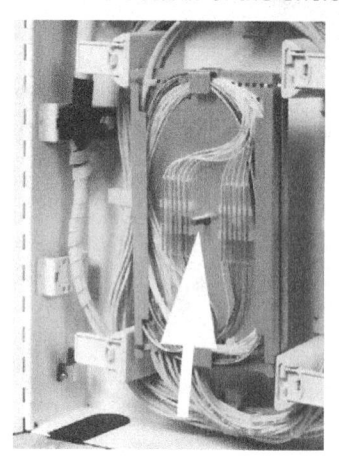

Figure 25-37: Bolt Hole For Tray Attachment

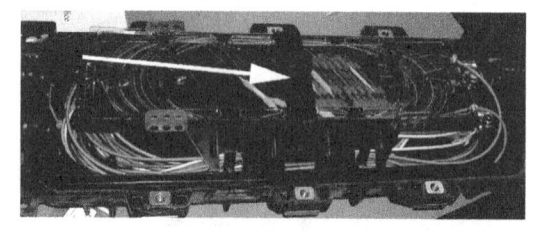

Figure 25-38: Attachment With Velcro Bands

Figure 25-39: Tray Attachment With Rubber Holders

25.14 TROUBLESHOOTING

25.14.1 BAD CLEAVES

Potential cause: dirt on scribing blade
Action: clean blade with swab and isopropyl alcohol.

Potential cause: dirt in alignment grooves
Action: clean grooves with lens grade gas

Potential cause: fiber from previous cleave is on pads on underside of the outer cover
Action: inspect and clean pads with isopropyl alcohol and lens grade tissue

Potential cause: loose alignment pad
Action: tighten screws on alignment pad

Potential cause: worn scribing blade
Action: rotate blade to next position or replace blade

Potential cause: missing or misplaced alignment shim
Action: return cleaver to factory for service

Potential cause: worn bearings in scribing arm
Action: return cleaver to factory for service

25.14.2 FUSION SPLICING

25.14.2.1 FAILURE TO ALIGN FIBERS

If the installer has problems with fiber alignment, he cleans the splicer grooves until he eliminates this problem.

Potential cause: dirt in fiber holders
Action: clean holders with fiber optic swab or fine paintbrush; brush from center towards outside of holder

Potential cause: cleave length below minimum value
Action: check cleave length; re-cleave

Potential cause: primary coating in final alignment groove
Action: reposition fiber so that cladding is in final groove (Figure 25-17)

25.14.2.2 HIGH LOSS SPLICE

Potential cause: high cleave angle
Action: reduce cleave angle acceptance value on splicer

Potential cause: incorrect menu active
Action: reset menu for fibers being spliced

Potential cause: dirty or worn electrodes

Action: clean or replace electrodes, according to splicer instructions.

Note: an indication of dirty or worn electrodes is the sound of the arc. If the sound is uniform and low, the electrodes are good. If the noise is loud and uneven, as though the electrodes are 'spitting', the electrodes need cleaning or replacement

25.14.2.3 NON UNIFORM SPLICE DIAMETER

Potential cause: overrun set incorrectly
Action: reset overrun in menu

Potential cause: incorrect menu active
Action: reset menu for fibers being spliced

25.14.2.4 WEAK SPLICE

Potential cause: arc current and/or time set low
Action: increase overrun value or change menu to that for a different fiber

Potential cause: incorrect menu active
Action: reset menu for fibers being spliced

25.14.2.5 ROUND BALL ON FIBER ENDS

Potential cause: arc current and/or time set high
Action: reduce value

Potential cause: incorrect menu active
Action: reset menu for fibers being spliced

25.14.3 MECHANICAL SPLICING

25.14.3.1 HIGH LOSS SPLICE

Potential cause: fiber not in center of splice
Action: ensure equal amount of bow on both fibers

Potential cause: bad cleave
Action: see 25.14.1

Potential cause: dirt on fiber end
Action: review procedure to ensure no contamination of end between cleaving and insertion; see Caution in 25.8.2.4

Potential cause: incorrect cleave length
Action: correct cleave length

25.15 ONE PAGE SUMMARY

25.15.1 CABLE END PREPARATION

Determine the end preparation dimensions
Prepare the cable end
If necessary, replace the buffer tubes or place routing tubing on fibers

25.15.2 ENCLOSURE PREPARATION

Install moisture seals
Install bonding
Attach the cables to enclosure
Verify buffer tube length
Attach the buffer tubes to tray
Verify the fiber length

25.15.3 SET UP OTDR

Set OTDR parameters
Run trace to determine splice location

25.15.4 FUSION SPLICING

Set up the splicer
Activate the appropriate splice menu
Set the maximum cleave angle
Set the splice cover length
Check the optics
Prepare the cleaver
Install the splice cover
Strip and clean the fiber
Cleave the fiber
Place the fiber in the splicer
Strip, clean and cleave the second fiber
Place second fiber in splicer
Fuse the fibers
Review the estimated loss
Center bare fiber in cover
Shrink the splice cover

25.15.5 MECHANICAL SPLICE

Prepare the splice tool
Prepare the cleaver
Strip and clean the fiber
Cleave the fiber
Place the first fiber in the mechanical splice
Strip, clean and cleave the second fiber
Place the second in the mechanical splice
Make the bows equal
Close splice

25.15.6 TEST LOSS

25.15.7 COIL FIBER

25.15.8 COIL BUFFER TUBE

25.15.9 ATTACH TRAY

25.15.10 FINISH ENCLOSURE

25.15.11 TEST LOSS

26 APPENDICES

26.1 INDICES OF REFRACTION

26.1.1 MULTIMODE

	Wavelength=	850	1300
Corning	Clearcurve	1.480	1.479
OFS	LaserWave/OM4/OM3	1.483	1.479
Prysmiam/Draka		1.482	1.477

26.1.2 SINGLEMODE

	Wavelength=	1310	1550
Corning	28e+	1.4676	1.4682
	LEAF		1.4680
OFS	Allwave	1.4670	1.4680
	TruWave Reach	1.4710	1.4700
Prysmiam/Draka	G.652	1.4670	1.4680
	Teralight/G.655	1.4682	1.4683

26.2 BACKSCATTER COEFFICIENTS

26.2.1 MULTIMODE

	Wavelength=	850	1300
Corning	Clearcurve	-68.0	-76.0
OFS	LaserWave/OM4/OM3	-68.4	-75.8

26.2.2 SINGLEMODE

	Wavelength=	1310	1550	1625
Corning	28e+	-77.0	-82.0	
	LEAF		-81.0	-82.0
Prysmiam/Draka	G.652	-79.4	-81.7	-82.5
	Teralight/G.655	-77.4	-80.4	-81.3

26.3 RI INACCURACIES
Error in distance to event, in meters

	IR error	% error	100 m	500 m	1000 m	5000 m	10000 m
1.482	RI error	% error	100m	500m	1000m	50000m	10000m
1.483	0.001	0.0674309	0.07	0.34	0.67	3.37	6.74
1.484	0.002	0.1347709	0.13	0.67	1.35	6.74	13.48
1.485	0.003	0.2020202	0.20	1.01	2.02	10.10	20.20
1.486	0.004	0.2691790	0.27	1.35	2.69	13.46	26.92
1.487	0.005	0.3362475	0.34	1.68	3.36	16.81	33.62
1.488	0.006	0.4032258	0.40	2.02	4.03	20.16	40.32
1.489	0.007	0.4701142	0.47	2.35	4.70	23.51	47.01

26.4 DISTANCE INACCURACIES
Difference between fiber length and cable length, in meters

fiber excess length	buffer tube excess length	100 m	500 m	1000 m	5000 m	10000 m
0.01%	1.91%	1.92	9.60	19.19	95.97	191.94
0.02%	1.91%	1.93	9.65	19.29	96.47	192.94
0.03%	1.91%	1.94	9.70	19.39	96.97	193.94
0.04%	1.91%	1.95	9.75	19.49	97.47	194.94
0.05%	1.91%	1.96	9.80	19.59	97.97	195.94
0.06%	1.91%	1.97	9.85	19.69	98.47	196.94
0.07%	1.91%	1.98	9.90	19.79	98.97	197.94
0.08%	1.91%	1.99	9.95	19.89	99.47	198.94
0.09%	1.91%	2.00	10.00	19.99	99.97	199.94
0.10%	1.91%	2.01	10.05	20.09	100.47	200.94

26.5 GLOSSARY

905 SMA: see SMA 905

adapter: a device for mating two connectors. Also known as mating adapter, barrel, bulkhead and feed through

APD: avalanche photodiode. This device converts an optical signal to an electrical signal.

armor: a layer of material, usually stainless steel, which is placed around a cable core to prevent damage from gnawing rodents and from crush loads.

attenuation: the loss of optical power or intensity as light travels through a fiber. It is expressed in units of decibels. When used to describe fibers or cables, it is expressed as a rate in decibels/kilometer. In this book, attenuation refers to reduction in signal strength in the fiber or cable.

back reflection: an outdated term that was used to mean return loss or reflectance. See reflectance and return loss.

back shell: the portion of a connector in back of the retaining nut or latching mechanism.

bandwidth: a measure of the transmission capacity of an analog transmission system.

bandwidth-distance product: the product of the length of a fiber and the analog bandwidth that the fiber can transmit over that length. It is expressed in units of MHz-km for multimode fibers. It is not the relevant parameter for laser optimized fibers. It is not used for singlemode fibers.

bend radius, long term: see bend radius, minimum unloaded.

bend radius, minimum loaded: the smallest radius to which a cable can be bent during installation at the maximum recommended installation load without any damage to either the fiber or the cable materials. Typically, this radius is 20 times the diameter.

bend radius, minimum unloaded: the smallest radius to which a cable can be bent without any damage to either the fiber or the cable materials while the cable is unloaded. Typically, this radius is 10 times the diameter for the life of the cable.

bend radius, short term: see bend radius, minimum loaded.

Biconic: a type of connector.

binding tape: a cable component. This tape holds buffer tubes together during jacket extrusion.

binding yarn: a cable component. This yarn holds buffer tubes together during jacket extrusion.

bit error rate: a measure of the accuracy of a digital fiber optic system. The BER is the rate of errors produced by the optoelectronics. Abbreviated as BER.

bit rate: the data transmission rate of a digital transmission system.

boot: a plastic device that slides over the cable and the back shell. It serves to limit the radius of curvature of the cable as the cable exits the back shell.

break out: a type of cable composed of sub cables, each of which contains a single fiber.

buffer coating: obsolete term. See primary coating.

buffer tube: a layer of plastic that surrounds a fiber or a group of fibers.

bulkhead: see adapter.

butt coupling: a method of transmitting light from one fiber to another by precise mechanical alignment of the two fiber ends without the use of lenses.

BWDP: bandwidth-distance product.

cable: the structure that protects an optical fiber or fibers during installation and use.

cable core: the structure of fibers, buffer tubes, fillers and strength members that reside inside the inner most jacket of a cable.

cable end boxes: the enclosures placed on the end of a cable to protect the buffer tubes and fibers.

cap: a plastic structure that protects the end of a connector ferrule from dust and damage when the connector is not in use.

CBT: see central buffer tube

central buffer tube: a cable design in which all fibers reside in a single, centrally located buffer tube.

central strength member: a strength member that resides in the center of a cable.

chromatic dispersion: the spreading of pulses of light due to rays of different wavelengths traveling at different speeds through the core. See spectral width.

cladding diameter: the outer diameter of the cladding. It is measured in micrometers.

cladding non-circularity: the degree to which the cladding and the core deviate from perfect circularity.

cladding: the region of an optical fiber that confines the light to the core and provides additional strength to the fiber.

cleaver: device to create a flat and perpendicular surface on end of a fiber.

cleaving: the process of creating a fiber end that is flat and perpendicular to the axis of the fiber.

coating: see primary coating

concentricity: the degree to which the core deviates from being in the exact center of the cladding.

cone of acceptance: the cone defined by the critical angle or the numerical aperture. This is the cone within which all of the light enters and exits a fiber.

core diameter: the diameter of the region in which most of the light energy travels. It is measured in micrometers.

core offset: see concentricity.

core: the region of an optical fiber in which most of the light energy travels.

coupler: a device that allows two separate optical signals to be joined for transmission on a single fiber.

crimp ring: the device that is deformed around the back shell of a connector. The crimp ring traps the strength member of the cable, providing acceptable cable-connector strength.

critical angle: the maximum angle to the axis of a fiber at which rays of light will enter a fiber and experience total internal reflection at the core-cladding boundary.

crush load, maximum recommended: the recommended maximum load that can be applied to a fiber optic cable without any permanent change in the attenuation of the cable. This can be specified as a long-term or a short-term crush load.

cut-off wavelength: the wavelength, below which a singlemode fiber will not transmit a single mode. Below this wavelength, the singlemode fiber will transmit multimode manner.

D4: a connector type

design: used in this text to refer to a cable

dielectric: having no components that conduct electricity in the cable.

differential modal attenuation (DMA): the mechanism by which rays in multimode fibers experience differing attenuation rates due to their mode, or location in the core.

differential modal delay (DMD): a measurement of dispersion in a multimode fiber; DMD is the measurement used for bandwidth measurement of laser-optimized for fiber.

dispersion: the spreading of pulses in fibers.

ESCON™: a connector type.

expanded beam coupling: a method of transmitting light from one fiber to another with lenses.

FC: a connector type.

FC/PC: a connector type.

FDDI: fiber data distributed data interface.

feed through: see adapter.

ferrule: that portion of the connector which a aligns a fiber. A ferrule is present in all connector types except the Volition™.

fiber: the structure that guides light in a fiber optic system.

Fiber Jack: see Opti-Jack™.

filled and blocked cable : a type of cable in which all empty space is filled with compounds to prevent moisture ingress. Filling refers to a gel material in buffer tubes. Blocked refers to grease outside of buffer tubes.

fillers: cable materials that fill otherwise empty space in a cable.

Fresnel reflection: the reflection that occurs when light travels between two media in which the speed of light differs.

FRP strength member: a fiberglass reinforced plastic or epoxy rod that is used as a dielectric strength member in cables. The term 'plastic' may refer to either a polymer plastic or an epoxy.

fusion splice: a splice made by melting two fibers together.

gel-filling compound: a compound placed inside a loose buffer tube to prevent water from contacting the fiber(s) in that buffer tube.

graded index: a type of multimode fiber, in which the chemical composition of the core is not uniform.

HCS™: a hard clad silica fiber.

heat shrink tubing: tubing placed on the back shell of a connector.

HiPPI: high speed, parallel processor interface.

index of refraction (IR, h): the ratio of the speed of light in the material to the speed of light in a vacuum. This is a dimensionless number.

inner duct: a corrugated plastic pipe in which fiber optic cables are placed.

inner jacket: any layer of jacketing plastic that is not the outer-most layer.

installation strength, maximum recommended: the maximum load that can be applied along the axis of a cable without any damage to the fibers .

installation temperature range: the temperature range within which a cable can be installed without damage.

jacket: a layer of plastic in a cable. It can be an outer jacket or an inner jacket.

jumper: a short length of a single fiber cable with connectors on both ends.

Kevlar®: an aramid yard produced by DuPont Chemical. It is used to provide strength in fiber optic cables.

keying: a connector mechanism by which ferrules are prevented from rotating.

laser diode: a semiconductor that converts electrical signals to optical signals.

laser optimized (LO) a multimode fiber that is designed to enable long transmission distance at 1 Gbps-100 Gbps.

latching mechanism: device that retains a connector to a receptacle or an adapter.

LC: a SFF connector with a 1.25 mm ferrule.

LX.5: a SFF connector with a 1.25 mm ferrule and a built-in dust cover.

LED: light emitting diode. A semiconductor that converts electrical signals to optical signals.

lensed coupling: see expanded beam coupling.

loose buffer tube: a buffer tube with space between the outer diameter of the primary coating and the inner diameter of the buffer tube.

loose tube: a cable design in which the fiber floats loosely inside an oversized tube.

loss: the end to end reduction in optical power as light travels through a fiber, connectors or splices. In this book, the 'loss' refers to reduction of optical power at a splice or a connector pair.

material dispersion: the spreading of pulses of light due to rays of light traveling through different regions of the core.

maximum recommended installation load: see installation strength, maximum recommended.

mechanical splice: a mechanism that aligns two fiber ends precisely for efficient transfer of light from one fiber to another.

MFPT: a multiple fiber per (buffer) tube cable design. This design usually has 6 or 12 fibers per loose buffer tube.

mini-BNC: a connector type.

minimum recommended long term bend radius: see bend radius, minimum unloaded.

minimum recommended short term bend radius: see bend radius, minimum loaded.

MPO: the generic term for a 4-24 fiber, connector with a single ferrule. The usual form of use is MPO/MPT.

MPT: the proprietary term for a 4-12 fiber, connector with a single ferrule. The usual form of use is MPO/MPT.

modal dispersion: the spreading of pulses of light due to rays traveling different paths through a multimode fiber.

modal pulse spreading: see modal dispersion.

mode: a paths in which light can travel in a fiber core. Mode translates roughly to 'path.'

mode field diameter: the diameter within which the light energy field travels in a singlemode fiber.

monomode: see singlemode. This term is used outside of North America.

MT-RJ: a duplex SFF connector with a single ferrule.

MU: a SFF connector with a 1.25 mm ferrule and a size of approximately half that of the SC connector.

multimode: a method of propagation of light in which all of the rays of light do not travel in a path parallel to the axis of the fiber.

NA: see numerical aperture.

numerical aperture (NA): the sine of the critical angle. The NA is a measure of the solid angle within which rays of light will enter and exit the fiber.

offset, core: see core offset.

optical amplifier: a device that increases the signal strength without an optical to electrical to optical conversion.

optical coupling: see expanded beam coupling.

optical power budget: the maximum loss of optical power which transmitter-receiver pair can withstand while still functioning at the specified level of accuracy.

Opti-Jack™: a duplex SFF connector.

optical return loss: see return loss.

optical rotary joint: a rotating joint that allows transmission of light from a stationary fiber to a rotating fiber.

optical switch: a switch that can direct light to more than one output path.

optical time domain reflectometer (OTDR): a test device that creates a map of the loss of signal strength of an optical path.

optical waveguide: another term for an optical fiber.

optoelectronic device: any device that converts a signal from electrical to optical domain or vice versa.

optoelectronics: see optoelectronic device.

OTDR: optical time domain reflectometer or optical time domain reflectometry.

ovality: a measure of the degree to which a fiber deviates from perfect circularity. See cladding non-circularity.

passive component: a device that manipulates light without requiring an optical to electrical signal conversion.

patch panel: a sheet of material that contains adapter(s).

PCS: a plastic clad silica fiber.

PD: photodiode. This device converts an optical signal to an electrical signal.

pigtail: a length of fiber or cable that is permanently attached to a connector or an optoelectronic device.

ping pong: a type of transmission that results from using a single LED as both a transmitter and receiver.

plug: another term for connector.

POF: plastic optical fiber. A fiber with a plastic core and a plastic cladding.

polishing fixture: device to hold connector perpendicular to fiber end;

polishing puck: see polishing fixture

polishing tool: see polishing fixture

primary coating: a layer of plastic placed around the cladding by the fiber manufacturer.

primary coating diameter: the diameter of the layer of plastic that is placed around the fiber by the fiber manufacturer.

pull-proof: a performance characteristic of connector types. A connector type is pull-proof when tension on the cable attached to the connector does not produce an increase in the loss of the connector.

pulse dispersion: see dispersion.

pulse spreading: see dispersion.

receptacle: the device within which an active device is mounted. The receptacle is designed to mate with a specific connector type.

reflectance: a measure of the ratio of reflected power to incident power for a single device, such as a connector or mechanical splice. Reflectance is measured in units of dB.

reinforced jacket: two layers of plastic that are separated by strength members.

repeatability: the maximum change in loss between successive measurements of the loss of a connector pair.

retaining nut: device that attaches a connector to receptacle or to an adapter

return loss: the ratio of incident power to power reflected or back scattered for an entire.

ribbon: a structure on which multiple fibers are precisely aligned.

SAP: see super absorbent polymer

super absorbent polymer: a material that absorbs moisture by converting it to a gel; used in fiber optic cables to provide moisture resistance; incorporated into cables as tapes, yarns, threads and powders.

SC: a connector type.

SFF: see 'small form factor'.

shrink tubing: plastic that covers back shell of connector

simplex: a single fiber cable.

singlemode: a method of propagation of light in which all of the energy of light travels the same path length

slotted core: a cable design with a core containing helical slots.

SMA 905: a connector type

SMA 906: a connector type

small form factor: a type of connector with the characteristic of small size that enables doubling of density in patch panels, enclosures, and switches.

spectral width: the measure of the width of the output power-wavelength curve at a power level equal to half the peak power.

splice enclosure: a structure that encloses and protects splice trays and cable ends.

splice tray: a structure that encloses and protects fibers.

splice: a device for permanent alignment of two fiber ends.

splitter: a device that creates multiple optical signals from a single optical signal.

spot size: the size of the area of an LED or laser diode from which light is produced.

ST®: the first of a series of connector types designed by ATT. Other types from ATT are the ST-II and the ST-II+.

star core: see slotted core.

ST-compatible: a connector with a type which is compatible with the ST® connector.

step index: a type of multimode fiber, in which the chemical composition of the core is uniform.

storage temperature range: the temperature range within which a cable can be stored without damage.

strength members: those elements of a cable design that provide strength.

type: the sum of characteristics that differentiates one connector type from another type.

TECS™: a technically enhanced clad fiber similar to hard clad silica fiber.

temperature operating range: the range of temperature within which the cable can be operated during its lifetime without degradation of its properties.

tight tube: a design in which the tube does contact the entire circumference of the fiber. A tight tube can contain only 1 fiber.

total internal reflection: the mechanism by which optical fibers confine light to the core.

type: used in this text to refer to a fiber or connector

use load, maximum recommended: the maximum longitudinal load that can be applied to a cable during its entire lifetime without damage.

VCSEL: vertical cavity, surface emitting laser. A type of relatively low cost multimode light source that transmits at 1-10 Gbps.

Volition™: a duplex SFF connector with no ferrules.

water blocking compound: a compound placed in the interstices between buffer tubes or between jackets in a cable.

wavelength division multiplexer: a passive device that combines optical signals with different wavelengths on different fibers light onto a single fiber. The wavelength division de-multiplexer performs the reverse function.

wavelength: a measure of the color of the light. It is a length stated in nanometers (nm) or in micrometers (μ).

wiggle proof: a connector performance characteristic. A connector type is wiggle proof if lateral pressure on the back shell does not produce an increase in the loss.

window: the wavelength range within which a fiber is designed to provide specified performance.

Zip cord®: a two-fiber cable with a figure 8 cross-section that allows each of the two fibers to be separated in the same manner as a lamp cord.

Acronyms

ADSS	all dielectric self support
APC	angled physical contact connector
ATM	Asynchronous Transfer Mode
BER	bit error rate
CATV	Cable TV
CWDM	coarse wavelength division multiplexing
D4	a connector
DAC	dual attachment concentrator
DAS	dual attachment station
DS	dispersion shifted
DS-NZD	dispersion shifted, non-zero dispersion
DWDM	dense wavelength division Multiplexing
EDFA	erbium doped fiber amplifier
ESCON	Enterprise System Connection
FC	a connector type
FDDI	fiber data distributed interface
FOLS	Fiber Optic LAN Section of the TIA
FTTD	fiber to the desk
FTTH	fiber to the home
FTTP	fiber to the premises
FTTX	any of the above three
GI	graded index
HDPE	high density polyethylene
IP	Internet Protocol
LAN	local area network
LD	laser diode
LEAF™	large effective area fiber (Corning Inc.)
LED	light emitting diode
LSA	least squares analysis
LX.5	a connector type
MFPT	multiple fiber per tube
MIC	media interface connector
MM	multimode
MT-RJ	a duplex connector type
MU	a SFF connector type
NA	numerical aperture
NDS	non dispersion shifted
NEC	National Electrical Code
NIST	National Institute of Science and Technology
OFC	optical fiber cable, conductive, horizontal rated
OFCP	optical fiber cable, conductive, plenum rated
OFCR	optical fiber cable, conductive, riser rated
OFN	optical fiber cable, non conductive, horizontal rated
OFNP	optical fiber cable, non conductive, plenum rated

OFNR optical fiber cable, non conductive, riser rated
OPBA optical power budget available
OPBR optical power budget requirement
OSNR optical signal to noise ratio
PC   physical contact
PMD   polarization mode dispersion
PON   passive optical network
SAN   storage area network
SAP   super absorbent polymer
SAC   single attachment concentrator
SAS   single attachment station
SC   a connector type
SDH   Synchronous Digital Hierarchy
SI   step index
SM   singlemode
SONET synchronous optical network
STP   shield twisted pair
UPC   ultra physical contact
UTP   unshielded twisted pair
VCSEL vertical cavity, surface emitting laser
WDM   wavelength division multiplexing
ZDW   zero dispersion wavelength

26.6 CHAPTER 13 ANSWERS

1. The supervisor rents the splicer with the estimation feature. If the splicer has no estimation feature, the OTDR operator cost is $1200 for 24 man hours. Each splice will need to be tested prior to installing the splice in the splice tray. The OTDR technician sits idle while the splicing technician splices. With the estimation feature, the OTDR technician can test the completed splices in one day. With the estimation feature, the cost of the OTDR technician will be $400, an $800 reduction in labor cost, a $300 savings.

26.7 CHAPTER 19 ANSWERS

1. Insertion loss acceptance value= 3.775 dB

   OTDR acceptance values
   Attenuation rate: 1.1 dB/km
   Connector loss: 0.525 dB/pair
   Splice loss: 0.125 dB/splice

Loss in	dB/km		#km		dB
Cable	1.5	*	2.0	=	3.00
	dB/pair		#pairs		
Connector	.75	*	3	=	2.25
	dB/splice		#splices		
Splice	0.15	*	0	=	0.0
			Total	=	5.25

   Table 26-1: Calculation, Maximum Loss

Loss in	dB/km		#km		dB
Cable	0.7	*	2.0	=	1.4
	dB/pair		#pairs		
Connector	0.3	*	3	=	0.9
	dB/splice		#splices		
conductive	0.1	*	0	=	0.0
			Total	=	2.3

   Table 26-2: Calculation, Typical Loss

18. Insertion loss acceptance value= 10.975 dB

    OTDR acceptance values
    Attenuation rate: 0.425 dB/km
    Connector loss: 0.525 dB/pair
    Splice loss: 0.125 dB/splice

Loss in	dB/km		#km		dB
Cable	0.5	*	20.0	=	10.0
	dB/pair		#pairs		
Connector	.75	*	4	=	3.0
	dB/splice		#splices		
Splice	0.15	*	3	=	0.45
			Total	=	13.45

    Table 26-3: Calculation, Maximum Loss

Loss in	dB/km		#km		dB
Cable	0.35	*	20.0	=	7.0
	dB/pair		#pairs		
Connector	0.3	*	4	=	1.2
	dB/splice		#splices		
Splice	0.1	*	3	=	0.3
			Total	=	8.5

    Table 26-4: Calculation, Typical Loss

19. Insertion loss acceptance value= 2.5077 dB

OTDR acceptance values
Attenuation rate: 3.25 dB/km
Connector loss: 0.525 dB/pair
Splice loss: 0.125 dB/splice

Loss in	dB/km		#km		dB
Cable	3.5	*	.257	=	.900
	dB/pair		#pairs		
Connector	.75	*	4	=	3.0
	dB/splice		#splices		
Splice	0.15	*	1	=	0.15
			Total	=	3.536

Table 26-5: Calculation, Maximum Loss

Loss in	dB/km		#km		dB
Cable	0.7	*	.257	=	.180
	dB/pair		#pairs		
Connector	0.3	*	4	=	1.2
	dB/splice		#splices		
Splice	0.1	*	1	=	0.1
			Total	=	1.48

Table 26-6: Calculation, Typical Loss

20. Insertion loss acceptance value= 2.038 dB

OTDR acceptance values
Attenuation rate 3.25 dB/km
Connector loss: 0.525 dB/pair
Splice loss: 0.125 dB/splice

Loss in	dB/km		#km		dB
Cable	3.5	*	.304	=	1.067
	dB/pair		#pairs		
Connector	.75	*	2	=	1.5
	dB/splice		#splices		
Splice	0.15	*	0	=	0.0
			Total	=	2.56

Table 26-7: Calculation, Maximum Loss

Loss in	dB/km		#km		dB
Cable	3.0	*	.304	=	.9128
	dB/pair		#pairs		
Connector	0.3	*	2	=	0.6
	dB/splice		#splices		
Splice	0.1	*	0	=	0.0
			Total	=	1.51

Table 26-8: Calculation, Typical Loss

21. Insertion loss acceptance value= 1.3849 dB

OTDR acceptance values
Attenuation rate: 1.1 dB/km
Connector loss: 0.525 dB/pair
Splice loss: 0.125 dB/splice

Loss in	dB/km		#km		dB
Cable	1.5	*	.304	=	.457
	dB/pair		#pairs		
Connector	.75	*	2	=	1.5
	dB/splice		#splices		
Splice	0.15	*	0	=	0.0
			Total	=	1.957

Table 26-9: Calculation, Maximum Loss

Loss in	dB/km		#km		dB
Cable	0.7	*	.304	=	.2128
	dB/pair		#pairs		
Connector	0.3	*	2	=	0.6
	dB/splice		#splices		
Splice	0.1	*	0	=	0.0
			Total	=	.8128

Table 26-10: Calculation, Typical Loss

22. Insertion loss acceptance value= 12.06875 dB

OTDR acceptance values
Attenuation rate: 0.425 dB/km
Connector loss: 0.525 dB/pair
Splice loss: 0.125 dB/splice

Loss in	dB/km		#km		dB
Cable	0.5	*	20.750	=	10.375
	dB/pair		#pairs		
Connector	.75	*	5	=	3.75
	dB/splice		#splices		
Splice	0.15	*	5	=	0.75
			Total	=	14.875

Table 26-11: Calculation, Maximum Loss

Loss in	dB/km		#km		dB
Cable	0.35	*	20.750	=	7.2625
	dB/pair		#pairs		
Connector	0.3	*	5	=	1.5
	dB/splice		#splices		
Splice	0.1	*	5	=	0.5
			Total	=	9.2625

Table 26-12: Calculation, Typical Loss

23. Insertion loss acceptance value= 3.06025 dB

OTDR acceptance values
Attenuation rate: 3.25 dB/km
Connector loss: 0.525 dB/pair
Splice loss: 0.125 dB/splice

Loss in	dB/km		#km		dB
Cable	3.5	*	0.257		.900
	dB/pair		#pairs		
Connector	.75	*	4	=	3.0
	dB/splice		#splices		
Splice	0.15	*	1	=	0.15
			Total	=	4.05

Table 26-13: Calculation, Maximum Loss

Loss in	dB/km		#km		dB
Cable	3.0	*	0.257	=	.77
	dB/pair		#pairs		
Connector	0.3	*	4	=	1.2
	dB/splice		#splices		
Splice	0.1	*	0	=	0.10
			Total	=	2.07

Table 26-14: Calculation, Typical Loss

24. Insertion loss acceptance value= 2.038 dB

   OTDR acceptance values
   Attenuation rate: 1.1 dB/km
   Connector loss: 0.525 dB/pair
   Splice loss: 0.125 dB/splice

Loss in	dB/km		#km		dB
Cable	3.5	*	0.304	=	1.06
	dB/pair		#pairs		
Connector	.75	*	2	=	1.5
	dB/splice		#splices		
Splice	0.15	*	0	=	0.0
			Total	=	2.56

Table 26-15: Calculation, Maximum Loss

Loss in	dB/km		#km		dB
Cable	3.0	*	0.304	=	.912
	dB/pair		#pairs		
Connector	0.3	*	2	=	0.6
	dB/splice		#splices		
Splice	0.1	*	0	=	0.0
			Total	=	1.512

Table 26-16: Calculation, Typical Loss

26.8 ADVANTAGES AND DISADVANTAGES OF 1, 2, AND 3 LEAD INSERTION LOSS TEST METHODS

26.8.1.1　ONE LEAD METHOD

The one lead method has two advantages. The first advantage is that the method results in an insertion loss value that is close to the loss that a transmitter-receiver pair will experience. With such a value, this method provides the best indication that the link will work. The other two methods underestimate this loss.

The second advantage is of reduced uncertainty. The launch power of this method is well known, as all the power exiting the source test lead is measured.

The two lead launch power has increased uncertainty relative to that of the one lead method. This increased uncertainty is due to the somewhat arbitrary alignment of the mated connectors during the setting of the input power level. It is possible that the mated reference connectors will align with the connectors on the ends of the cable under test more precisely than they align with each other. In this case, the power delivered to the power meter through the cable under test can be higher than the power delivered to the power meter through the reference leads. In other words, the loss will be positive. A positive loss is a useless number, as the best possible loss is 0.0 dB. This author has observed positive losses between 0.0 and 0.2 dB on short cables.

26.8.1.2　TWO LEAD METHOD

The two lead method has the advantage of allowing use of an imperfect, or lightly damaged, power meter test lead. Any excess loss due to imperfections in either reference lead will be compensated by the input power level measurement. In contrast, a lightly damaged meter test lead will bias all one lead measurements higher than reality. In this case, the installer may reject low loss links.

26.8.1.3　THREE LEAD METHOD

The three lead method enables testing of cable segment loss without the connector loss. This type of test enables adding the segment losses to estimate the total link loss.

26.8.1.4　1 VS. 2 LEAD METHODS

Neither the one nor two lead methods bias the interpretation of the insertion loss measurement. As long as the test is performed properly, both methods will result in the same interpretation. That is, both methods will indicate that the loss is acceptably low or unacceptably high.

26.9 ALL CONNECTOR TYPES

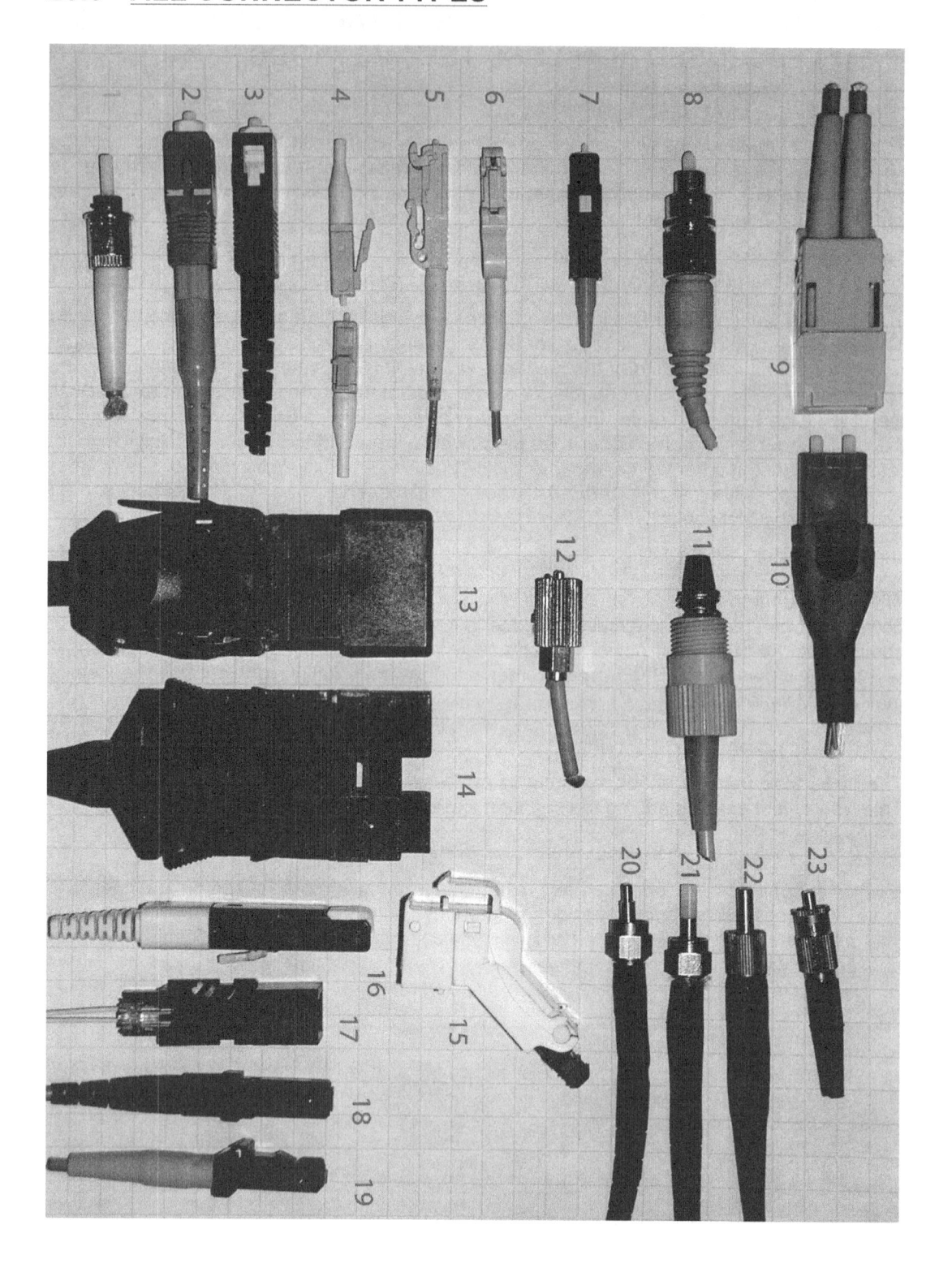

26.10 STUDY GUIDE FOR CFOT EXAMINATION, VERSION 12

If the student is able to answer the review questions indicated below, he will pass the FOA CFOT examination. That is not to indicate that the questions indicated are on the examination. Rather, the subjects of the questions are included in the examination.

Chapter=	1	2	3	4	5	6	7	8	9	10	11	12	13	14	15	16	17	18
Question #																		
1			√	√	√		√	√										√
2	√	√	√		√		√	√	√		√							
3		√	√	√	√	√					√							
4		√	√		√	√	√	√			√				√			
5			√		√		√	√							√			
6	√				√										√			
7		√		√											√			
8														√	√	√		
9		√		√							√				√			
10		√		√	√									√	√			
11			√	√											√			
12				√										√	√			
13				√														
14					√													
15				√	√													
16				√	√													
17				√	√													
18			√	√														
19			√															
20			√	√														
21		√	√	√	√													
22		√	√		√													
23		√	√	√														
24		√	√	√	√													
25			√	√	√													
26				√	√													
27				√	√													
28			√	√														
29			√															
30			√															

Chapter=	1	2	3	4	5	6	7	8	9	10	11	12	13	14	15	16	17	18
Question #																		
31			√		√													
32																		
33				√	√													
34				√														
35				√														
36				√														
37					√													
38					√													
39			√		√													
40			√		√													
41			√		√													
42																		
43																		
44																		
45																		
46																		
47																		
48																		
49			√															
50			√															
51			√															
52																		
53																		
54																		
55																		
56																		
57																		

26.11 CFOT PREPARATION PUZZLES

To receive answers, send email to: fiberguru@ptnowire.com. Put 'puzzle answers-2016' in subject box. We will send a pdf with answers.

Light And Fiber
PROFESSIONAL FIBER OPTIC INSTALLATION, V.9
Chapters 1-3, 15

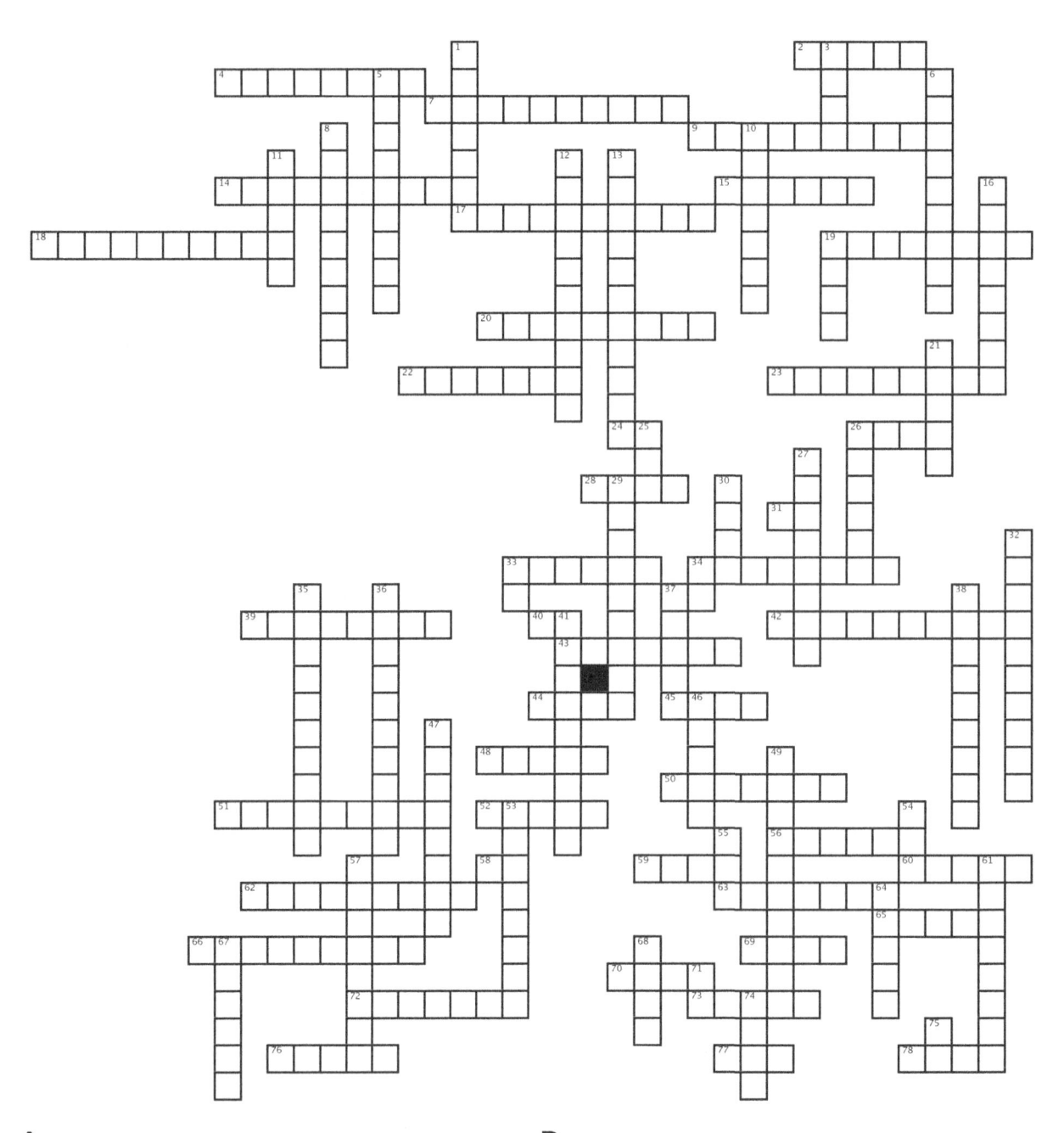

Across

2. minimizing ____ ____ is the most important concern of installer (first word)
4. in (14 Across) fibers, some of the optical power travels in the _____

Down

1. A (14 Across) core is _____ than a (8 Down) core
3. acronym for test equipment that enables viewing the loss of power along a fiber

Across

7. 1310 is a _____
9. the technical name for the mechanism describing change in width of optical pulse as pulse travels through fiber
14. telephone systems use ____ fibers
15. unit of fiber (27 Down)
17. one of two mechanisms that reduces (9 Across) in (8 Down) fiber
18. 1550 is a wavelength used on _____ fibers
19. region of fiber that confines light to center of fiber
20. as (7 Across) increases, (13 Down) _____
22. material of first optical fiber developed
23. Fiber type in which light can take many paths
24. abbreviation for units of measure for (7 Across)
26. 50 is a _____ diameter
28. acronym for term that indicates capacity of (8 Down) fibers developed
31. During testing, the installer must match the _____ of the fiber under test to that in the test leads
33. first word of old term for the layer that protects fiber
34. in (8 Down) fibers, optical power _____ at (26 Across)- (4 Across) boundary
37. acronym for the second type of (8 Down) fiber developed
39. 125 μ is a _____ _____ (first word)
40. acronym for type of fiber optimized for use with VCSELs
42. first word of term that is a measure of the speed of light in fiber
43. first word in layer that protects fiber
44. the number of causes of (9 Across)
45. First word for first type of fiber with a single composition in core
48. what the installer does to the outer layer of the fiber
50. One of two types of reflections
51. 850 nm is a wavelength used on _____ fibers
52. the type of light source used on (8 Down) fiber for transmission at and above 1 Gbps

Down

5. unit in measure for wavelength
6. first word of acronym for term that indicates capacity of (8 Down) fibers developed
8. many data systems use ____ fibers.
10. the second type of (14 Across) fiber developed was dispersion _____
11. second word of term indicating speed of light in fiber
12. excessive (9 Across) results in signal _____
13. term that describes the loss of power in fiber
16. Not (14 Across), but another name for fiber in which light travels in a single path
19. acronym indicating 3-16 wavelengths traveling in fiber
21. The capacity of (8 Down) fibers is _____ than that of (14 Across) fibers.
25. acronym for region of fiber with small core in which most of optical power travels
26. (33 Down) fibers _____ be used as test leads
27. 125 μ is a _____ _____ (second word)
29. technical name for optical fiber
30. (46 Down) reflection occurs at ____ ____ boundary (first word)
32. third word for the acronym for the wavelength of maximum capacity
33. acronym for fiber with reduced sensitivity to power loss when bent
34. abbreviation for term that indicates speed of light in fiber
35. determines both (9 Across) and (13 Down)
36. second word for the acronym for the wavelength of maximum capacity
37. material of most fibers
38. fibers are designated by at least two _____
41. second word indicating type of fiber optimized for use with VCSEL
46. Second type of reflection
47. type of (9 Across) that results from non crystalline or amorphous structure of fiber material
49. fiber type in which light takes a single path

Across

56. first word for second type of fiber with multiple compositions in center
58. acronym for the technical term that is created by difference in composition in (68 Down) and (4 Across) of fiber
59. first word for the acronym for the wavelength of maximum capacity
60. largest type of (9 Across)
62. first word for technical term that is created by the difference between the compositions of (4 Across) and (68 Down) of fiber
63. Technical name for (50 Across) reflection
65. second word of fiber term that is created by difference in composition in the (4 Across) and (68 Down) of fiber
66. second largest type of (9 Across) that results from a characteristic of the transmitter
69. region of fiber in which most of light energy travels
70. minimizing ____ ____ is most important concern of installer (second word)
72. second word in layer that protects fiber
73. second word for first type of fiber with single composition in core
76. second word of the region of fiber with small core in which most of the optical power travels
77. acronym for the wavelength of maximum capacity
78. first word of acronym for region of fiber with small core in which most of the optical power travels

Down

53. (46 Down) reflection occurs at ____ ____ boundary (second word)
54. acronym indicating two wavelengths traveling in fiber
55. acronym for first fiber developed
57. first word of the characteristic of fiber that is created by the difference in composition in (4 Across) and (68 Down) of fiber
61. second word for technical term that is created by difference in composition in center and second layer of fiber
64. first word indicating type of fiber optimized for use with VCSEL
67. OM3 and OM4 have _____ bandwidth or capacity than OM1 and OM2
68. During testing, the installer must match the _____ of the fiber under test to that in the test leads
71. acronym for first type of fiber developed
74. acronym indicating up to 200 wavelengths traveling in fiber
75. OM3 and OM4 are ____ fibers.

Cables
PROFESSIONAL FIBER OPTIC INSTALLATION, V.9
Chapters 4, 10, 11

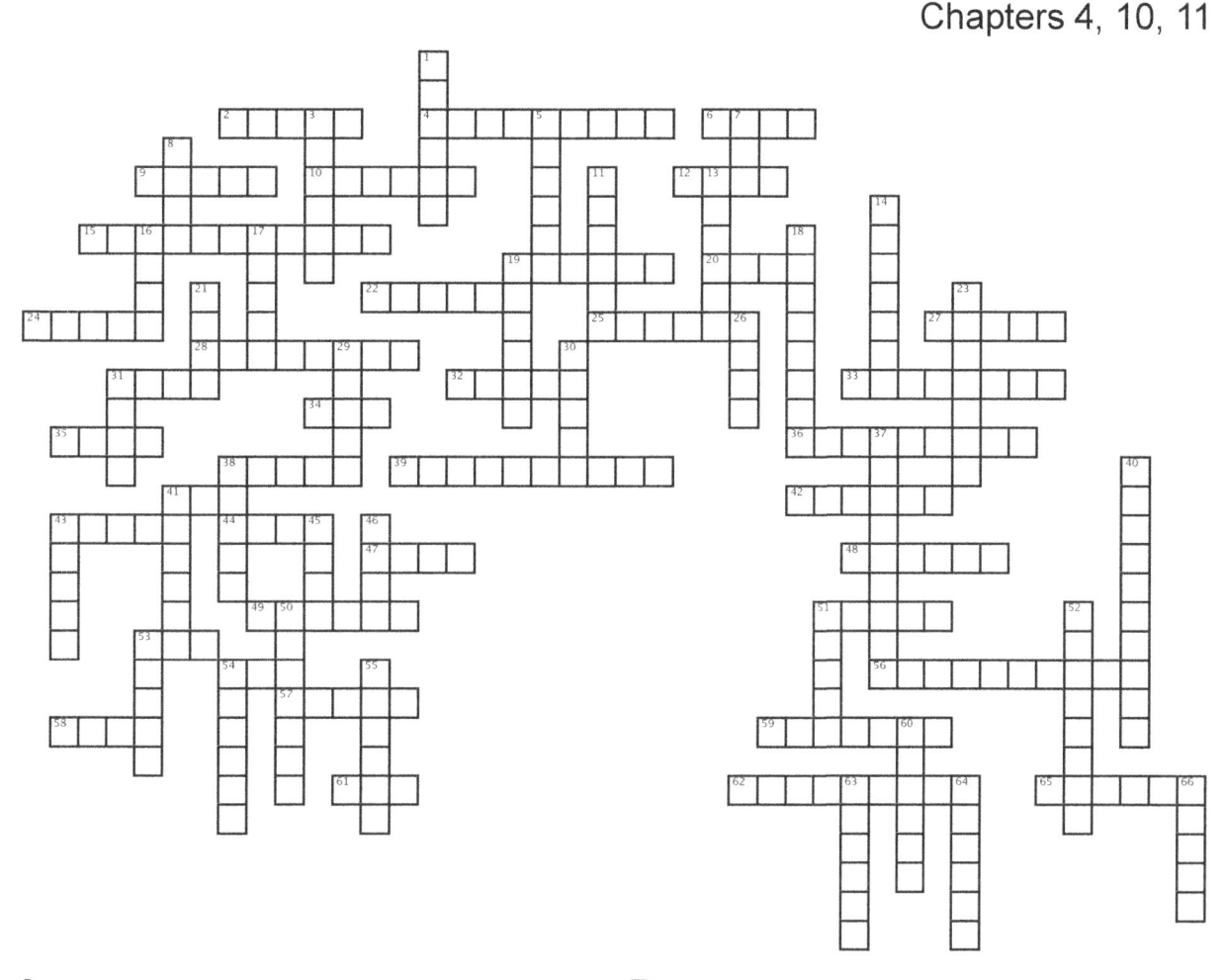

Across

2. a _____ tube contains only one fiber
4. a type of cable containing both conductors and fibers
6. To avoid fiber breakage, the installer must limit the _____ applied to the cable
9. fifth color in the color code sequence
10. The first word of the first layer placed on the fiber by the cable manufacturer
12. an (acronym) cable is (40 Down) and can be installed anywhere inside a building and be in compliance with the electrical code
15. Subjecting a cable to an excessively high or low temperature results in increased _____
19. ninth color in the color code sequence

Down

1. the outermost layer of a cable
3. a cable with both multimode and singlemode fibers
5. second color in the color code sequence
7. an (acronym) cable is (40 Down) and can be installed in a single floor inside a building and be in compliance with the electrical code
8. first color in the color code sequence
11. tenth color in the color code sequence
13. When being terminated, a multiple fiber per tube cable requires a _____ kit on the fibers
14. The third word of (54 Across) is _____

Across

20. an (acronym) cable is (40 Down) and can be installed in a single floor or bewteen floors inside a building and be in compliance with the electrical code
22. First word of a method to reduce load when pulling cable into a conduit
24. a _____ tube can contain more than one fiber
25. During installation, the minimum bend radius of a cable is ___ times the cable diameter
27. Second word of a method to reduce load when pulling cable into a conduit
28. the first word of the structural element that prevents excessive stretching of the fibers
31. The one hundred and thirty first fiber in loose tube cable has a _____ color
32. The ____ term bend radius is larger than the (43 Down) term bend radius.
33. The name of the most commonly used indoor cable type
34. seventh color in the color code sequence
35. The time to prepare a cable with (54 Across) is _____ than that with (41 Across) and (41 Down)
36. The cable (28 Across) (42 Across) is attached to the _____ at the cable ends
38. fourth color in the color code sequence
39. A common (28 Across) (42 Across) is often white and is
41. old outdoor cables achieved moisture resistance with _____ inside buffer tubes
42. the second word of the structural element that prevents excessive stretching of the fibers
43. The first word of (54 Across) is _____
44. twelfth color in the color code sequence
47. The designation of an indoor cable that has conductive elements and can be placed anywhere
48. Cable structural material used as a manufacturing aid or to keep cable round
49. color of jacket on indoor multimode cables

Down

16. The second word of the first layer placed on the fiber by the cable manufacturer
17. The structural element that provides crush and rodent resistance
18. Exceeding the installation load can result in fiber _____
19. color of jacket on indoor singlemode cables
21. eleventh color in the color code sequence
23. The cable bend radius is a _____ value
26. In a (59 Across) (24 Across) (16 Down) cable, the fiber bundles are held together by a color coded _____
29. third color in the color code sequence
30. A conductive (28 Across) (42 Across) is
31. The one hundred and thirty first fiber in loose tube cable has a buffer tube with a _____ color
37. A material used to reduce load when pulling cable into a conduit
38. eighth color in the color code sequence
40. A cable which has no conductive materials is _____
41. old outdoor cables achieved moisture resistance with _____ outside buffer tubes
43. During installation, the (54 Down) between the pull rope and the cable has a ____ pin to avoid fiber breakage.
45. color of jacket on indoor LO cables
46. When stored, a cable cannot be bent smaller than its ____ term radius.
50. structual element that eases removal of jacket
51. A group of fibers that is installed into a preinstalled tube with air pressure is known as _____ fiber
52. When entering a buidling, cable with (17 Down) must be
53. The number of colors used to indicate fiber type of indoor cables
54. During installation, the installer places a ____ between the pull rope and the cable to avoid twisting.
55. When being spliced mid-span, cable with (17 Down) must be
60. A common (28 Across) (42 Across) is _____ yarn

Across

51. The most common color of an outdoor cable
53. After installation, the minimum bend radius of a cable is ___ times the cable diameter
54. current generation cables achieve moisture resistance with this (acronym)
56. maximum number of fibers in a (63 Down) is
57. fibers in a loose tube are identified by their ___
58. A common jacket material in outdoor cables
59. A cable with all fibers in a single (24 Across) (16 Down) is a _____ loose tube cable
61. acronym for the National Electric Code
62. The second word of (54 Across) is ___
65. Indoor singlemode cables have the color ___

Down

63. Twelve fibers aligned on a tape form a ___
64. The number of fibers commonly found in a (24 Across) buffer tube
66. sixth color in the color code sequence

Connectors & Splices
PROFESSIONAL FIBER OPTIC INSTALLATION, V.9
Chapters 5, 6, 12, 13, 20-25

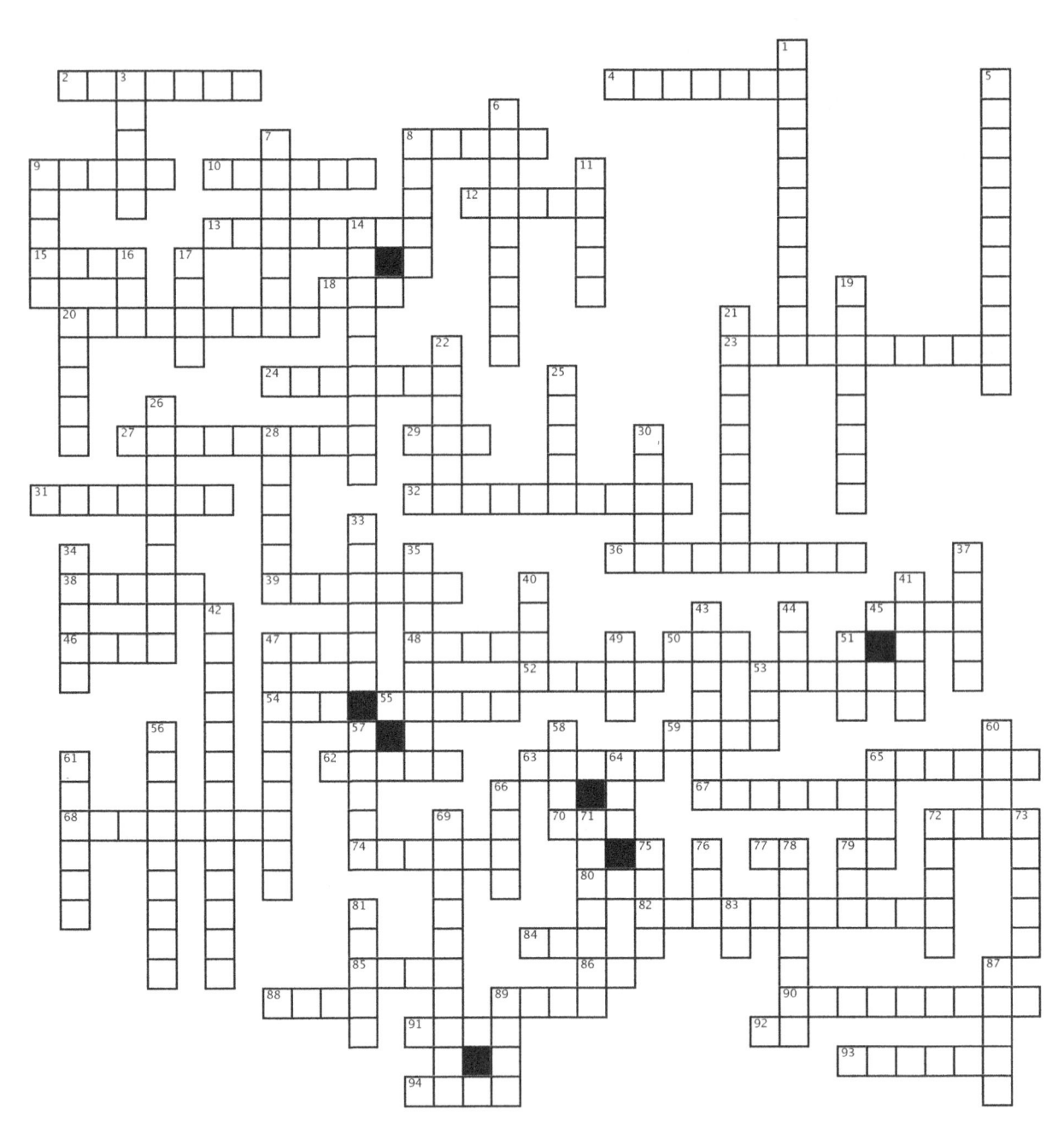

Across

2. Another term for (65 Across) splicing is glass _____
4. The second type of (65 Across) splicing is _____ alignment

Down

1. The (74 Across) on a ferrule tip has two benefits. One is reduced _____
3. (51 Across) stands for (first word)

Across

8. If the (45 Across) of a properly installed connector is not (48 Across), part of the (45 Across) is _____ the surface of the ferrule
9. The ferrule surface of a properly installed connector is
10. A name of a popular connector that requires no polishing is _____
12. A connector that requires no (6 Down) has a pre-installed _____
13. All (32 Across) splices contain an _____ _____ _____ (second word)
15. LO connectors have the color _____
18. A heat shrink (87 Down) has a _____ _____ (second word)
20. When installing connectors on a patch cord, the strength members are attached to the _____ of the connector
23. The (44 Down) is created by two
24. Creating a (9 Down) fiber requires _____ _____ (second word)
27. Connectors are used for _____ connections
29. connectors have consistent power loss from insertion to insertion because of the _____
31. When properly installed, connectors do not exceed their _____ loss
32. The second type of splicing is _____
36. One type of (65 Across) splicing is _____ _____ (second word)
38. All (32 Across) splices contain an _____ _____ _____ (first word)
39. Most fiber optic connectors have one fiber and are called _____
45. One type of (65 Across) splicing is _____ alignment
46. One connector installation method uses a ___ ___ adhesive (second word)
47. In both connector installation and splicing, the fiber must be _____ and _____ (first word)
48. a properly installed connector (45 Across) is _____
50. One connector installation method uses a ___ ___ adhesive (first word)
52. A heat shrink (87 Down) has a _____ _____ (first word)
53. One type of (65 Across) splicing is _____ alignment
54. When not in use, a connector should have its _____ installed

Down

5. Another of the characteristics of a properly installed connector (72 Across) is
6. The (94 Across) supports the (81 Down) during _____
7. A short length of buffer tube with a connector on one end is a _____
8. If not the colors (15 Across) or (20 Down), the multimode connector color is _____
9. The cladding of a properly installed connector is _____
11. When installing connectors on a patch cord, the strength members are attached to the (20 Across) of the connector by a ___ ___ (first word)
14. Creating a (9 Down) fiber requires _____ _____ (first word)
16. The acronym for a (25 Down) connector is
17. The (54 Across) keeps _____ off the tip.
19. Most (65 Across) splicers display an _____ of the splice loss
20. If not LO, connectors for this same core diameter have the color _____
21. (51 Down) stands for (third word)
22. When installing connectors on a patch cord, the strength members are attached to the (20 Across) of the connector by a ___ ___ (second word)
25. the connector color that has no reflectance is _____
26. Splices are used for _____ connections
28. The (89 Across) controls the fiber ___ ___ (second word)
30. the first word of (76 Down) is
33. connectors with two fibers are _____
34. (6 Down) is done with multiple
35. The _____ of the connectors aligns the fibers
37. In both connector installation and splicing, the fiber must be _____ and _____ (second word)
40. most (6 Down) is done on _____
41. (93 Across) splicing aligns singlemode fiber _____
42. Loss loss splices require fiber ends that are _____ and _____ (second word)
43. Current connectors make _____

Across

55. A properly installed connector has a _____ (72 Across)
59. Connectors exhibit loss when in a _____
62. A properly installed connector core is _____ with the ferrule
63. Early connector (6 Down) was performed on a flat _____
65. The first type of splicing is _____
67. When properly installed, connectors exhibit _____ loss
68. (32 Across) splicing aligns the fiber _____
70. There are ___ types of (65 Across) splicing
72. The loss of a connector is determined mostly by the condition of the _____
74. Today's connectors have a _____ on the fip of the ferrule
77. The unit of measure of splice (66 Down) is _____
79. a common connector type
80. Whenever possible, the installer should inspect connectors with a microscope in ___ ways
82. Connectors with a (65 Down) ferrule tip have high _____
84. There are _____ types of splicing.
85. If not (25 Down), (69 Down) connectors have the color _____
86. The splice is placed in a splice holder with the (18 Across) ___
88. According to the standards, there are _____ connector colors
89. The installer installs a _____ on the (20 Across) of all connectors
90. (51 Down) stands for (second word)
91. Active splicing aligns the (45 Across) fiber _____
92. a common, push on, pull off, connector type
93. One type of (65 Across) splicing is _____
94. (73 Down) and (50 Across) (46 Across) installation methods leave a _____ on the ferrule tip

Down

44. (65 Across) splicing is performed by an electrical _____
47. The first of the (64 Down) ways to inspect a connector is with _____
49. All (32 Across) splices contain an _____ _____ _____ (third word)
51. The second type of (65 Across) splicing is _____ (acronym)
53. If (40 Down) are not used in (6 Down) the step is called _____ (6 Down)
56. (32 Across) splicing aligns singlemode fiber _____
57. Another of the characteristics of properly installed core is _____
58. Low loss splices require fiber ends that are _____ and _____ (first word)
60. The (45 Across) of a properly installed connector has ___ characteristics
61. the third word of (76 Down) is _____
64. There are _____ types of (65 Across) splicing.
65. The first connectors had a _____ ferrule tip
66. The (28 Down) on a ferrule tip has two benefits. The second is reduced _____
69. (16 Down) connectors are for ___ fibers
71. The second of the (70 Across) ways to inspect a connector is _____ (47 Down)
72. connectors serve to align the _____ of the fibers
73. The earliest connectors used _____ to retain the fiber
75. the second word of (76 Down) is _____
76. current small connector types are (acronym)
78. an early fiber optic telephone connector type
79. an early type of fiber optic data connector
81. The (94 Across) supports the _____
83. a popular (76 Down) connector type
87. Fusion splices are protected in a _____
89. The (89 Across) controls the fiber _____ _____ (first word)

Optoelectronics And Testing
PROFESSIONAL FIBER OPTIC INSTALLATION, V.9
Chapters 8, 14, 15

Across
2. EF stands for (second word)
4. The cable between an OTDR and the cable under test is known as a _____ cable

Down
1. (40 Across) stands for (third word)
3. The _____ _____ calibrates the OTDR to create accurate attenuation rate measurements (first word)

Across

5. The ____ ____ calibarates the OTDR to create accurate attenuation rate measurements (second word)
7. Without a second launch cable, the OTDR ____ cannot measure the loss of the far end connector from the near end
8. Usually, insertion loss measurements in opposite directions are ____
12. On an OTDR trace, fiber _____ creates (3 Down)
14. During insertion loss testing, the installer matches three characteristics of the test leads to those of the cable under test. One of these characteristics is _____
15. The OTDR trace shows three types of features. A second is ____
16. A properly installed cable segment has a ____ line
17. Properly made fusion splices and bend radius violations can create ____
18. If the opposite of (32 Across) insertion loss value is observed, the installer should expect ____ on the cable
20. To measure the loss of the near end connector with an OTDR, the launch cable needs to be _____ than the (44 Across) zone
21. The (4 Across) cable _____ the OTDR port
22. As the wavelength increases, the fiber (19 Down) to (18 Across) ____ (second word)
23. (36 Down) stands for (third word)
31. A test source is ____
32. Insertion loss values at a long wavelength are ____ than those at a short wavelength
33. (40 Across) stands for (fourth word)

Down

6. On an OTDR trace, the slope of the (3 Down) line is the ____ ____ (first word)
9. The mandrel used to establish HOML conditions is sized to the diameters of the ____ and ____ (second word)
10. The OTDR trace shows three types of features. One is ____
11. A HOML launch conditions may, or may not, require a ____
13. A fiber end can create ____
14. During insertion loss testing, a _____ power meter is used
19. As the wavelength increases, the fiber ____ to (18 Across) (22 Across)
24. If the maximum distance setting on the OTDR is less than the length of the cable under test, the installer will ____ see the far end of the cable
25. A splice with a positive loss on an OTDR is known as a ____
26. Singlemode transmitters have ____ ____ (second word)
27. In the insertion loss test, the meter simulates the ____
28. The mandrel used to establish HOML conditions is sized to the diameters of the ____ and ____ (first word)
29. The technical name for a (73 Across) is a ____ (13 Down)
30. Wavelength is measured in units of ____
34. The loop on a singlemode test lead removes _____ in the _____ (second word)
35. A (16 Across) (10 Down) indicates ____ loss
36. An installer uses a ____ (acronym) to find locations of excessive power loss in a splice case or near the end of a cable
38. During an insertion loss test, the input power level is measured with ____ lead(s)
40. (36 Down) stands for (first word)

Across

37. During insertion loss testing, the installer matches three characteristics of the test leads to those of the cable under test. One of these characteristics is _____
39. In the insertion loss test, the source simulates the ____
40. EF launch conditions come close to simulating the launch conditions of a ____ transmitter
42. Singlemode transmitters have ____ ____ (first word)
44. A (13 Down) and (17 Across) create ____ zones
46. During insertion loss testing, the installer matches three characteristics of the test leads to those of the cable under test. One of these characteristics is _____
47. The OTDR trace shows three types of features. A third is ____
48. During an insertion loss test, the output power level is measured with ____ lead(s)
50. EF launch conditions are used to test fiber with a _____ micron core diameter at 850 nm
52. when performing a singlemode insertion loss test, the installer puts a ____ in the test lead connected to the source.
55. (40 Across) stands for (first word)
56. HOML stands for (third word)
57. HOML testing is performed on _____ fiber
59. The (52 Across) in the singlemode test lead removes _____ in the _____ (first word)
60. A properly installed connector pair with radius tips _____ creates a (15 Across)
62. The color of the light in (36 Down) is

Down

41. EF stands for (first word)
43. OTDR stands for (fourth word)
45. (36 Down) stands for (second word)
49. The ____ of the test source matches the same characteristic of the transmitter.
51. If the maximum distance setting on the OTDR is much longer than the length of the cable under test, the installer will experience ____ test time
53. HOML stands for (second word)
54. HOML stands for (fourth word)
58. A (40 Across) is used on ____ fibers
61. When measured in the opposite direction, (16 Across) always shows a ____
63. On an OTDR, power loss is in units of ____
69. The true loss of a splice is the ____ of the losses measured in both directions with an OTDR
70. A multimode mechanical splice ____ creates a (13 Down)
72. In the insertion loss test, the test leads simulate the ____ ____ on the end of a backbone cable (second word)
75. A properly made fusion splice _____ creates a (13 Down)
76. The (4 Across) cable enables testing of the ____ end connector
79. OTDR stands for (second word)
81. On an OTDR trace, an connector with an 8° angle on the end face on the far end of the cable will have ____ (13 Down)
82. On an OTDR trace, an connector pair with 8° end faces will have ____ (13 Down)

Across

64. The multimode test method indicated by input power measurement with one lead and output power measurement with two leads is known as ____ B.
65. (40 Across) stands for (fifth word)
66. HOML stands for (first word)
67. Early multimode transmitters created light with ____
68. (40 Across) stands for (second word)
71. In the insertion loss test, the test leads simulate the ____ ____ on the end of a backbone cable (first word)
73. A (13 Down) is also known as a ____
74. A (44 Across) zone is also known as a ____ zone
77. The ____ calibarates the OTDR to create accurate length measurements
78. The ____ enables idenfication of multiple loss locations in a single teste
80. OTDR stands for (third word)
83. OTDR stands for (first word)
84. On an OTDR trace, a singlemode mechanical splice ____ creates a (13 Down)
85. The slope of the (10 Down) is the ____ ____ (second word)

deliberately blank

26.12 THE AUTHOR

For the last 36 years Mr. Eric R. Pearson has worked in fiber optic communications. This involvement includes a wide variety of business and technical activities, as detailed below.

Mr. Pearson developed and managed two fiber optic cable manufacturing facilities and organizations (Manufacturing Manager, Times Fiber Communications and Business Manager, Whitmor Waveguides). In these positions, Mr. Pearson developed cable designs, manufacturing techniques and qualified designs against performance specifications. As business manager, he was responsible for developing a profitable multi-million dollar business unit.

Mr. Pearson managed fiber manufacturing facilities for Corning Glass Works and Times Fiber Communications.

As of the date of this writing, Mr. Pearson has delivered 532 training presentations and trained more than 8785 personnel in proper installation and design procedures. Between his field installations and training, he has made and supervised more than 47,000 connectors. Mr. Pearson has performed tens of thousands of OTDR, ORL, insertion loss, and dispersion tests on both multimode and singlemode products.

From both field experience and training, Mr. Pearson gained sufficient experience to write three definitive texts on cable and connector installation, The Complete Guide to Fiber Optic Cable Installation (Delmar Publishers, 1997, ISBN #08273-7318-X), Successful Fiber Optic Installation- The Essentials (2005), and this text.

He has written the books: Fiber Optic Network Design, Practical Fiber Optic System Design and Implementation, How to Specify and Choose Fiber Optic Cables, and How to Specify and Choose Fiber Optic Connectors.

Mr. Pearson has been a technical expert in 17 legal proceedings on patent infringement suits, lawsuits between installers and end users, lawsuits regarding technical fraud and defective products.

From 1995-2007, Mr. Pearson was a founding Director and the Director of Certification, of the Fiber Optic Association (FOA). As the latter, he was responsible for developing requirements and examinations for basic and advanced certification of fiber optic installation personnel. These activities require an in-depth knowledge of all aspects of cable, splice, and connector installation.

From the Fiber Optic Association, he has received four advanced certifications: Certified Fiber Optic Specialist certifications (CFOS/T, CFOS/S, CFOS/C, and CFOS/I). In 2013, the FOA designated Mr. Pearson a 'Master Instructor'.

From 1999 to 2014, Mr. Pearson was a Master Instructor for the Building Industry Consultants Services International (BICSI), and the developer of the BICSI fiber optic network design program, FO110.

From 1986-2004, he was been a Member, Editorial Advisory Board, Fiberoptic Product News.

The Academy of Professional Consultants & Advisors (APCA) certified him as a Certified Professional Consultant (CPC).

He has over 100 articles, reports and presentations to his credit. He is frequently quoted in fiber optic and related trade journals.

He has been selected to speak at three Newport Fiber Optic Marketing Conferences.

He is listed in: Who's Who Worldwide, Who's Who of Business Leaders, Who's Who in Technology and Who's Who in California.

Mr. Pearson has provided consulting services to hundreds of companies in the areas of fiber network design and specification, technical and marketing evaluations.

Mr. Pearson received his education at Massachusetts Institute of Technology (BS, 1969) and Case-Western Reserve University (MS, 1970). Both degrees are in Metallurgy and Materials Science.

Mr. Eric R. Pearson, CFOS/C/S/T/I

26.13 CROSSWORD EXERCISES

These crossword puzzles in this text are tools for assessing and improving understanding of the language of fiber optics. For answers, contact:

fiberguru@ptnowire.com

Put 'Crossword answer request' in the subject line.

26.14 PEARSON TECHNOLOGIES' SERVICES

Training Programs
(http://www.ptnowire.com/training-list.htm)
Professional Fiber Optic Installation
Professional Fiber Optic Installation
 With FOA CFOT Certification
Professional Fiber Optic Installation
 With Advanced Connector Installation
Advanced Connector Installation
 With FOA CFOT Certification
Advanced Testing
Advanced Splicing
Advanced Installation And Testing With
 FOA CFOS Certifications
Sales Training
Custom Versions of above programs

Books

Professional Fiber Optic Installation, The Essentials For Success, (2000, 2005, 2010, 2014) ISBN 9780976975434

Successful Fiber Optic Installation, The Essentials For Success, (2005)

Mastering The OTDR- Trace Acquisition And Interpretation, (2011) ISBN 9781466429291

Mastering Fiber Optic Connector Installation, (2011) ISBN 9781466470699

Mastering Fiber Optic Network Design, (2000) ISBN 9781469931296

Successful Fiber Optic Network Installation- A Rapid Start Guide, (2000) ISBN 9781470012304

Mastering Fiber Optic Testing (in development)

The Complete Guide To Fiber Optic Cable System Installation, (1996) ISBN 0-8273-7318-X

Consulting
(http://www.ptnowire.com/services.htm)
 Network Design Review
 Lawsuit Technical Support
 Technical Support For Marketing
 Technical Support

INDEX

A

accuracy .2-2, 2-5, 2-6, 5-6, 8-1, 8-3, 14-1, 16-1, 26-3, 26-7
accurate loss 15-17
adapter 5-8, 5-11, 14-10, 20-2, 20-6, 26-3, 26-5, 26-6, 26-7
air gap 6-4, 12-12, 12-13, 13-7
air polish. 12-14, 22-1, 22-12, 22-14, 23-2, 23-9
analog 1-10, 3-4, 26-3
APC ..5-2, 5-7, 5-8, 5-12, 15-4, **15-7**, **15-8**, **15-9**, 15-21, 15-22, 15-23, 15-24, 15-25, 15-26, 16-2, 16-3, 26-9
aramid ..4-3, 4-4, 4-5, 4-7, 4-8, 21-8, 22-4, 23-5, 26-5
armor 4-4, 4-5, 4-9, 9-3, 13-4, 26-3
Asynchronous transfer mode ATM 26-9
attenuation .. 1-2, 3-7, 3-8, 3-9, 3-10, 3-11, 4-10, 4-11, 4-12, 8-2, 10-2, 11-1, 11-2, 11-5, 13-1, 14-3, 14-6, 14-7, 15-1, 15-2, 15-3, **15-9**, **15-10**, 15-15, 15-17, 15-18, 15-19, 15-21, 15-23, 19-3, 19-4, 26-3, 26-4, 26-5
attenuation rate .. 1-2, 3-7, 3-9, 3-10, 3-11, 4-11, 4-12, 11-1, 11-2, 11-5, 11-6, 14-3, 15-1, **15-9**, **15-10**, 15-17, 15-18, 15-19, 15-20, 15-21, 15-23, 19-3, 19-4, 26-4

B

back light 20-2, 20-6
back reflection 26-3
back shell 4-11, 5-3, 5-4, 5-15, 5-16, 12-2, 12-3, 12-5, 12-6, 12-8, 12-16, 20-1, 22-5, 22-6, 22-9, 22-10, 22-12, 22-13, 23-4, 23-5, 23-7, 23-10, 26-3, 26-4, 26-5, 26-8, 26-9
backscatter 15-3, **15-13**, 15-14, 15-15, 15-16, 15-18, 16-1
bandwidth. 1-1, 2-6, 3-3, 3-4, 3-5, 3-7, 3-9, 6-3, 8-1, 8-2, 15-4, 26-3
barrel 12-2, 22-2, 22-3, 22-6, 26-3
battery 15-1
bend radius4-10, 4-11, 4-13, 5-4, 9-1, 9-3, 9-4, 10-2, 11-1, 11-2, 11-6, 11-7, 11-9, 13-1, 13-5, 15-1, **15-8**, **15-12**, 18-1, 21-2, 25-2, 25-5, 26-3, 26-6
BER 2-7, 5-17, 26-3
Biconic 5-12
binding tape 4-3, 4-4, 21-6, 26-3
blown cable 4-9
blown fiber 4-8
bonding 1-2, 4-10, 25-2, 25-3, 25-14

break out .. 4-8, 4-11, 5-10, 9-1, 22-1, 26-3
buffer coating3-1, 22-11, 23-9, 26-3, 26-4, 26-6, 26-7
buffer tube 4-1, 4-2, 4-3, 4-4, 4-5, 4-6, 4-7, 4-8, 4-11, 4-12, 5-4, 6-1, 6-6, 6-8, 9-1, 9-2, 11-5, 11-8, 12-1, 12-2, 12-3, 12-4, 12-5, 12-6, 12-15, 13-1, 13-2, 13-3, 13-4, 13-5, 13-6, 13-7, 13-12, 13-14, 18-1, 21-1, 21-4, 21-6, 21-7, 21-8, 21-9, 22-1, 22-2, 22-3, 22-4, 22-6, 22-13, 23-1, 23-3, 23-4, 23-5, 23-10, 24-2, 24-4, 24-5, 24-7, 24-8, 25-1, 25-2, 25-4, 25-5, 25-7, 25-11, 25-12, 25-14, 26-3, 26-4, 26-5, 26-6, 26-8
Building Wiring Standard ... 5-11, 6-7, 16-4
bulkhead .... See barrel

C

Category 14-9
CATV 1-2, 2-1, 6-3, 7-2, 26-9
ceramic 5-2, 5-14, 5-15, 12-12, 12-14, 22-1, 23-1, 23-2
Chromatic 2-5, 2-6, 26-4
cladding . 2-6, 3-1, 3-2, 3-3, 3-6, 3-7, 3-11, 4-12, 5-1, 5-6, 5-9, 6-1, 6-2, 12-4, 12-9, 12-12, 12-15, 13-8, 13-10, 14-3, 14-13, **15-8**, 19-4, 20-2, 21-2, 21-3, 22-5, 22-11, 22-12, 22-13, 23-4, 23-9, 25-6, 25-9, 25-13, 26-4
cleaver . 6-1, 12-15, 12-16, 13-6, 13-7, 13-8, 25-1, 25-6, 25-8, 25-9, 25-13, 25-14, 26-4
CO 15-19, 15-20
composite 4-10, 12-12, 24-2
core ...2-5, 2-6, 2-7, 3-1, 3-2, 3-3, 3-4, 3-6, 3-9, 3-10, 3-11, 4-2, 4-3, 4-8, 4-12, 5-1, 5-2, 5-6, 5-9, 5-17, 6-1, 6-2, 6-3, 6-4, 8-2, 12-2, 12-5, 12-9, 12-10, 12-13, 12-14, 13-7, 13-10, 14-3, 14-6, 14-7, 14-10, 14-13, 15-2, 15-3, 15-7, **15-8**, **15-9**, **15-10**, 15-13, 18-1, 18-2, 19-4, 20-2, 20-6, 21-6, 22-11, 22-12, 23-9, 24-2, 26-3, 26-4, 26-5, 26-6, 26-7, 26-8
coupler 15-2, 26-4
crush 4-4, 9-3, 11-2, 11-7, 13-5, **15-8**, 26-4
CWDM 1-1, 2-1, 7-2, 7-6, **15-12**, 26-9

D

D4 .... 5-12, 5-17, 26-4
Diamond 12-12
Dielectric 1-1, 1-2, 4-10
differential modal attenuation 14-3, 26-4
differential modal dispersion 3-5, 3-9

directional differences 14-7
dispersion 1-2, 2-1, 2-5, 2-6, 2-7, 2-9, 3-4, 3-5, 3-6, 3-7, 3-8, 3-9, 8-1, 8-2, 8-3, 14-1, 14-14, 26-4, 26-5, 26-6, 26-7, 26-9, 26-10
Dispersion Shifted 3-7
distribution 4-7, 4-12, 5-1, 15-1, 21-4, 21-8
dry fit 22-13
DWDM . 1-1, 1-2, 2-1, 3-8, 5-7, 5-8, 7-2, 7-3, 7-6, 14-1, **15-12**, 26-9

E

EDFA 7-3, 8-2, 26-9
EMI 1-1, 1-2
enclosure 4-7, 4-8, 5-11, 5-12, 6-4, 6-6, 6-7, 6-8, 9-1, 9-2, 11-8, 11-9, 12-1, 12-3, 12-4, 13-1, 13-2, 13-3, 13-4, 13-5, 13-12, 13-13, 13-14, **15-12**, 18-1, 21-5, 21-6, 21-7, 21-8, 21-9, 22-2, 22-13, 23-2, 23-10, 24-2, 25-1, 25-2, 25-3, 25-4, 25-10, 25-11, 25-12, 25-14, 26-8
epoxy 4-3, 5-13, 5-14, 5-15, 5-16, 10-1, 12-2, 12-5, 12-6, 12-7, 12-8, 12-13, 12-14, 12-16, 22-1, 22-2, 22-3, 22-5, 22-6, 22-7, 22-8, 22-12, 22-13, 22-14, 23-6, 26-5
ESCON 5-8, 5-9, 5-12, 5-13, 26-5
Ethernet 6-4, 10-4
extrinsic 5-6, 7-5, 7-6

F

fast restoration 6-4
fault finder 13-11
FC 5-9, 5-12, 5-13, 5-15, 26-5, 26-9
FDDI 5-8, 5-9, 5-12, 5-13, 26-5, 26-9
feature finder 13-11
ferrule . 5-2, 5-3, 5-4, 5-6, 5-8, 5-9, 5-10, 5-12, 5-14, 5-15, 5-18, 12-2, 12-3, 12-5, 12-6, 12-7, 12-8, 12-9, 12-10, 12-11, 12-12, 12-13, 12-14, 19-4, 20-2, 22-5, 22-6, 22-7, 22-8, 22-9, 22-12, 22-13, 23-1, 23-5, 23-6, 23-7, 23-8, 23-9, 23-10, 24-2, 24-7, 26-4, 26-5
fiber alignment 6-2, 13-8, 25-13
Fiber Jack 5-9, 26-5
FOTP 16-1
Fresnel 2-4, 2-9, 5-6, 26-5
FTTD 1-2, 5-3, 5-9, 5-16, 26-9
FTTH/PON 1-10, 2-1, 3-7, 9-2, 14-7, 14-9, 14-14, **15-12**, 15-19, 15-26, 15-27
FTTP 26-9
furcation 4-2, 4-3, 4-7, 21-7, 25-4
fusion splicing 4-6, 6-1, 6-3, 6-8, 13-6, 13-7, 13-14

G

gainer **15-9**, **15-10**
gel. 4-2, 4-3, 4-5, 4-7, 4-13, 6-4, 9-2, 11-8, 12-8, 12-14, 15-6, **15-7**, **15-9**, 21-1, 21-4, 21-6, 21-7, 25-1, 26-5, 26-8
ghost **15-7**, **15-10**, **15-11**
GI 2-5, 3-4, 3-5, 26-9
graded index 2-4, 3-4, 26-5, 26-9
ground 1-2, 4-3, 4-10, 6-6, 11-1, 11-2, 11-4

H

hybrid 4-10

I

index matching gel .. 6-4, 12-14, 15-6, 25-1
index of refraction . 2-2, 2-8, 2-9, 5-6, 15-3, **15-7**, **15-9**, 25-5, 26-5
insertion loss ... 5-5, 5-6, 10-3, 13-11, 14-4, 14-7, 14-13, 14-14, 15-1, 15-17, 19-1, 19-3, 19-5, 24-1, 24-7, 26-13
installation load . 4-9, 4-10, 4-11, 9-3, 10-1, 11-2, 11-3, 11-4, 11-5, 11-7, 21-1, 21-3, 26-3, 26-6, 26-8
internal reflection 3-4, 15-2, 16-1, 16-2, 26-4, 26-8
intrinsic 3-1, 5-6, 7-5, 7-6, 21-2
IR 2-2, 3-10, 20-1, 26-5

L

laser diode 8-2, 15-2, 16-1, 26-5, 26-8
laser optimized 3-5, 26-3
launch cable 15-13, 15-26, 25-1, 25-5
LC .. 5-7, 5-8, 5-10, 5-15, 5-18, 12-2, 12-6, 12-14, 16-2, 26-6
LCP 12-12, 12-14
LD 8-2, 8-3
LED .. 8-1, 8-2, 8-3, 14-3, 26-6, 26-7, 26-8, 26-9
LID 6-2
Light Emitting Diode .... 8-1, *See* LED
Liquid crystal polymer .... 5-2, *See* LCP
loose tube ... 4-1, 4-2, 4-3, 4-5, 4-6, 4-7, 4-13, 11-5, 13-5, 21-1, 25-1, 26-6, See
LX. 5-7, 5-8, 12-14, 26-6, 26-9
LX.5 .... 5-7, 5-8, 12-14

M

Mandrel 14-9
mass splicing 6-1
mechanical splicing 6-3, 6-4, 6-8, 13-1, 13-2, 13-14, 25-1
MFD 3-6, 3-7, 3-8, 4-12
MFPT 4-4, 4-5, 4-6, 4-13, 26-6, 26-9

microscope... 6-2, 10-4, 13-9, 14-13, 18-2, 20-1, 20-2, 20-6, 22-2, 23-2
mid span. 4-4, 4-5, 4-6, 6-1, 6-3, 9-2, 13-1, 13-5, 14-7, 19-2, 25-1
Mini BNC 5-12
mode field diameter.. 3-6, **15-9**, **15-10**, 15-13, 26-6
moisture resistance 4-1, 4-2, 4-3, 4-4, 4-5, 4-7, 26-8
monomode 3-6, 26-6
MT-RJ 5-8, 5-9, 5-10, 5-17
MU 5-7, 5-8, 26-6, 26-9
multimode.. 2-4, 2-5, 3-3, 3-4, 3-5, 3-11, 4-4, 4-9, 5-6, 5-7, 5-9, 5-11, 6-3, 6-4, 8-2, 12-9, 12-10, 12-12, 12-16, 13-1, 13-10, 14-7, 14-8, 14-9, 15-5, 15-21, 15-22, 15-23, 15-24, 15-25, 15-26, 16-1, 16-4, 19-2, 19-4, 19-5, 20-2, 22-4, 22-9, 22-13, 22-14, 23-1, 23-2, 23-4, 23-7, 23-10, 25-1, 26-3, 26-4, 26-5, 26-6, 26-8, 26-9
multiple reflection 15-5, 15-7, **15-10**

N

NA ... 2-4, 3-2, 3-4, 3-9, 3-10, 4-12, 5-6, 5-17, 8-1, 8-2, 14-6, 14-7, 14-10, 14-13, **15-9**, 15-13, 26-6
National Electric Code 11-8
NEC 4-3, 4-9, 4-10, 4-13, 11-8, 26-9
non-circularity 3-3, 26-4, 26-7
numerical aperture 2-4, 2-8, 26-4, 26-6, 26-9
    NA...2-4, 3-2, 3-4, 3-10, 4-12, 5-6, 5-17, 8-1, 8-2, 14-6, 14-7, 14-10, 14-13, **15-9**, 15-13, 26-6, 26-9

O

OFC 4-9, 11-8
OFCP 4-9, 11-8, 26-9
OFN 4-9, 4-10, 4-13, 11-8, 26-9
OFNP 4-9, 4-10, 11-8, 26-9
OLT 15-19, 15-20
optical power budget... 2-9, 8-3, 8-4, 18-2, 26-6, 26-10
Opti-Jack 5-8, 5-9
OTDR .. 2-2, 3-10, 6-2, 10-3, 13-10, 13-11, 13-13, 15-1, 15-2, 15-3, 15-4, 15-5, 15-6, **15-7**, **15-8**, **15-9**, **15-10**, **15-11**, **15-12**, **15-13**, 15-14, 15-15, 15-17, 15-19, 15-22, 15-23, 15-25, 15-26, 18-1, 19-1, 19-3, 19-4, 19-5, 22-10, 23-8, 25-1, 25-5, 25-10, 25-14, 26-7, 26-10, 26-11, 26-12

P

PAL 6-2, 13-9

passive device 7-5, 7-6, 19-1, 26-8
patch cord . 4-7, 5-9, 9-1, 10-2, 12-1, 16-1, 19-5
patch panel 5-8, 5-10, 5-11, 5-12, 5-15, 6-4, 6-6, 12-2, 12-3, 19-2, 26-7, 26-8
PC 5-7, 26-5, 26-10
photodiode 26-3, 26-7
physical contact 5-7, 15-3, 26-9, 26-10
pigtail 6-1, 13-1, 13-12, 26-7
placement .. 1-3, 9-3, 11-7, 13-5, 13-8, 13-12, 15-14, 15-15, 15-16, 15-17, 15-18, 15-23
POF 3-3, 26-7
PON ... 1-7, 1-10, 2-1, 3-7, 9-2, 14-7, 14-9, 14-14, **15-12**, 15-19, 15-26, 15-27, 26-10
premises 4-7, 4-12, 4-13, 5-9, 12-9, 12-16, 13-3, 21-4, 21-8, 22-1, 22-2, 22-13, 23-3, 23-5, 23-8, 26-9
primary coating .. 3-1, 3-2, 5-9, 10-4, 11-8, 12-1, 12-4, 12-5, 12-15, 13-6, 13-7, 21-1, 21-2, 21-3, 22-1, 22-3, 22-4, 23-1, 23-3, 23-4, 24-2, 24-4, 25-1, 25-6, 25-9, 25-13
profile alignment .. 6-2, See PAL, See PAL
protocol 18-1
pull proof . 5-8, 5-9, 5-11, 5-12, 5-15, 12-2, 12-6

R

radiused 15-5
Raman 7-3
Rayleigh 3-9, 15-2
reflectance .. 5-6, 5-7, 5-12, 5-17, 6-3, 6-4, 7-5, 10-4, 12-12, 12-13, 12-14, 12-15, 13-10, 14-13, 15-3, 15-5, 15-6, **15-7**, **15-8**, **15-10**, 16-1, 16-2, 16-3, 16-4, 18-1, 20-1, 22-1, 22-9, 22-10, 22-13, 23-7, 26-3, 26-7
refractive index .... See Index of Refraction
repeatability 5-4, 5-17, 14-8, 26-7
return loss 26-3, 26-7
RFI 1-2
RI 2-2, 3-10, See Index of Refraction, See Index of Refraction, See Index of Refraction, See Index of Refraction, See Index of Refraction, See Index of Refraction
ribbon 4-6, 6-1, 13-3, 26-7
ribbon splicing 4-6, 6-1
rip cord 21-6
rotary joint 7-4, 26-7

S

safety . 2, 4-10, 5-8, 10-1, 10-4, 10-5, 20-2
SAP 4-2, 4-3, 4-5, 4-7, 26-8, 26-10

SC....5-2, 5-3, 5-7, 5-8, 5-9, 5-10, 5-11, 5-12, 5-15, 5-18, 12-2, 12-6, 12-9, 16-2, 22-1, 22-2, 22-5, 22-8, 23-2, 23-5, 23-7, 23-8, 23-10, 24-1, 24-2, 24-3, 24-7, 26-6, 26-8, 26-10
Scattering 3-9, 15-2
scribing .. 12-10, 12-11, 13-6, 22-12, 23-8, 23-9, 24-7, 25-12, 25-13
SDH 26-10
service loop.... 9-3, 11-8, 11-9, 21-6, 21-7, 21-8
SFF 5-7, 5-8, 5-9, 5-17, 9-1, 26-6, 26-7, 26-8, 26-9
SI 3-3, 3-4, 3-11, 26-10
simplex 4-7, 5-7, 5-8, 5-12, 5-17, 5-18, 22-2, 26-8
singlemode . 1-1, 2-1, 2-5, 2-6, 3-3, 3-6, 3-7, 3-8, 3-9, 3-11, 4-4, 4-10, 5-2, 5-6, 5-7, 5-8, 6-1, 6-4, 7-2, 8-2, 12-10, 12-12, 12-16, 13-1, 13-10, 15-5, **15-7**, **15-8**, **15-9**, 15-21, 15-22, 15-23, 15-24, 15-25, 15-26, 16-1, 19-5, 22-1, 22-4, 22-10, 22-13, 22-14, 23-4, 23-10, 25-1, 26-3, 26-4, 26-6, 26-8, 26-10
SMA 5-4, 5-12, 12-12, 26-3, 26-8
spectral width 2-1, 2-2, 2-7, 2-8, 10-3, 26-8
splitter 7-5, 7-6, 15-19, 15-20, 15-26, 26-8
Storage Area Network
 SAN 26-10
ST™-compatible 5-10
super-absorbent polymer .... *See* SAP
Synchronous Digital Hierarchy 26-10

T

temperature . 4-10, 4-11, 4-12, 4-13, 5-13, 5-14, 5-15, 11-1, 11-2, 11-7, 12-5, 12-7, 12-16, 13-10, **15-8**, 22-1, 22-2, 22-7, 22-12, 26-5, 26-8
TIA/EIA-568 B 6-7
TIA/EIA-568-B 4-12, 19-1, 19-2
tight tube 4-1, 4-2, 4-5, 4-13, 5-9, 11-2, 13-5, 13-6, 13-7, 15-5, 15-6, **15-10**, 18-1, 21-1, 25-8, 26-8

U

undercut... 5-2
UPC 5-2, 5-7, 26-10
Use load 4-10, 11-1

V

vertical rise 11-1, 11-2
Volition 5-7, 5-8, 5-9, 5-18, 26-8

W

Water blocking 4-1, 4-2, 11-8
Waveguide 2-5, 2-6
wavelength . 1-1, 1-2, 2-1, 2-2, 2-5, 2-8, 3-6, 3-7, 3-8, 3-9, 3-10, 3-11, 5-7, 7-1, 7-2, 7-3, 7-6, 8-2, 8-3, 8-4, 10-3, 13-1, 14-9, 14-10, **15-12**, 15-21, 15-26, 16-1, 18-1, 25-5, 26-4, 26-8, 26-9, 26-10
WDM 1-1, 1-10, 2-1, 7-6, 26-10
white light test 22-13, 23-10
wiggle proof 5-8, 5-9, 5-11, 5-12, 5-15, 12-2, 12-6, 26-9

Z

ZDW 3-7, 3-8, 4-12
zero dispersion wavelength 3-7
zip cord 4-11, 22-1, 22-2

Made in the USA
Middletown, DE
05 November 2023